MOLECULAR DYNAMICS SIMULATION

MOLECULAR DYNAMICS SIMULATION

Elementary Methods

J. M. HAILE
Clemson University
Clemson, South Carolina

A Wiley-Interscience Publication
JOHN WILEY & SONS, INC.

New York   •   Chichester   •   Brisbane   •   Toronto   •   Singapore

Library of Congress Cataloging in Publication Data:

Haile, J. M.
  Molecular dynamics simulation: elementary methods/J. M. Haile.
    p. cm.

  "A Wiley-interscience publication."
  Includes bibliographical references and index.
  ISBN 0-471-81966-2
    1. Molecular dynamics—Computer simulation. 2. Causality
(Physics)—Computer simulation. 3. Chemistry. Physical and
theoretical—Computer simulation. I. Title.
QC168.7.H35 1992   91-31963
532'.05—dc20   CIP

Printed and bound in the United States of America by Braun-Brumfield, Inc.

10 9 8 7 6 5 4 3 2 1

...there is a never ending interest
in the definite mathematical problem
of the equilibrium of motion
of a set of points endowed with inertia
and mutually acting upon one another
with any given force.

Sir William Thomson, *Baltimore Lectures
on Molecular Dynamics and the Wave
Theory of Light*, Lecture XI (1884)

And these toys did so firmly possess
his imagination with an infallible opinion
that all that *machina* of dreamed inventions
which he read was true,
as he accounted no history in the world
to be so certain and sincere as they were.

Miguel de Cervantes, *Don Quixote* (1605)

PROLEGOMENA

Molecular dynamics methods are now orthodox means for simulating molecular-scale models of matter. The methods were originally devised in the 1950s, but they only began to receive widespread attention in the mid-1970s, when digital computers became powerful and affordable. Today, these simulation methods continue to attract attention from researchers with new problems and from students new to molecular theory.

The essence of molecular dynamics is simply stated: numerically solve the N-body problem of classical mechanics. Science has its fashions, just as do literature and music and architecture. Since the time of Newton, the N-body problem has been viewed as important, but the *reasons* for its importance have evolved. At the present time, its importance stems from attempts to relate collective dynamics to single-particle dynamics, attempts motivated by the hope that the puzzling behavior of large collections of particles can be explained by examining the motions of individual particles. For example, how does flow of fluid around an object produce a turbulent wake? How do atoms on a protein molecule move together so the protein folds in life-supporting ways? How does swirling fluid produce a long-lived vortex, such as the great red spot on Jupiter? How do chain molecules in solution self-assemble into structures such as micelles, vesicles, and lamellae? How does a local disturbance of a few molecules (say by a laser pulse) propagate throughout a system? How do individual molecules combine to form new molecules?

Such questions suggest that molecular dynamics can enlighten diverse research areas; however, its very diversity of application may have hampered development of an introductory literature that helps train new simulators. Specialists have been too busy exploring new territory to worry much about introducing their colleagues to the basic methodology. In addition to new research, molecular dynamics is now serving as an educational tool in courses in materials science, solid and liquid state physics, statistical mechanics, and physical chemistry. The methods would also be appropriate for study in courses on classical mechanics, advanced engineering thermodynamics, and transport phenomena. It seems that students and instructors, as well as new researchers, need a modest introduction to molecular dynamics methods.

Molecular-scale computer simulation involves a three-step procedure: (a) model individual particles, (b) simulate the movements of a large number of the model particles, and (c) analyze the simulation data for the required collective phenomenon. The emphasis of this book is on step (b). Essentially nothing is discussed concerning step (a), and discussions of step (c) are limited to traditional static and dynamic properties. Steps (a) and (c) depend on the problem to which the simulation is being applied; this book is not about applications but about methodology. Specifically, the book is restricted to molecular dynamics applied to hard spheres and Lennard-Jones atoms because these two substances encompass the two forms of molecular dynamics algorithms. This book contains nothing about molecular substances, quantum effects, multibody interactions, long-range interactions that would require Ewald sums, spectroscopic properties, nonequilibrium simulations, or parallel processing. Instead, the objective is to present a core of fundamentals that should be common to all users of molecular dynamics. From this small core, the reader may build to special applications.

This book is organized around two questions: (a) Why does molecular dynamics work (Chapters 1–5)? (b) How is molecular dynamics used to estimate properties (Chapters 6 and 7)? The emphasis is on *why* because a thorough understanding of fundamentals will enhance the possibility of success when molecular dynamics is applied to situations not explicitly covered here. It is, therefore, perhaps worthwhile to consider what should be included in a thorough response to *why*.

In Book II of his *Physics*, Aristotle proposed a theory of causation in which a complete explanation for a thing necessarily encompasses four causes: (i) material cause, which explains a thing in terms of its component parts; (ii) formal cause, which explains by invoking the design or relation among the components; (iii) efficient cause, which explains a thing as a consequence of an agent that implements the design; and (iv) final cause, which explains by virtue of the use made of the thing. The structure of this book revolves around these modes of explanation:

Chapter 1 considers *final cause*. What are the uses of molecular dynamics? How does molecular dynamics stand in relation to other ways of performing molecular simulation and other ways of doing science? It is commonly held that simulation serves as a competitor for theory and experiment; however, it seems to me that simulation stands in opposition, not necessarily to theory or experiment, but to the broader attitude of reductionism.

Chapters 2 and 4 review *material cause*. Molecular dynamics works because it is a confluence of diverse ideas from nonlinear dynamics, kinetic theory, sampling theory, and numerical methods.

Chapters 3 and 5 present *formal cause*. Molecular dynamics works because of the way it organizes the components discussed in Chapters 2 and 4. One organization of those parts produces an algorithm for simulating hard spheres (Chapter 3), while a different organization produces an algorithm for soft spheres (Chapter 5).

Appendices I, K, and L illustrate *efficient cause*. Molecular dynamics works because computational hardware and software are available for implementing the designs discussed in Chapters 3 and 5. This book contains nothing about hardware and descriptions of software are limited to the codes given in the appendices. This is not to undervalue the role of programming in molecular dynamics; rather, it reflects the attitude that programs themselves can serve as means for communicating among people as well as between people and machines. People can learn to read programs, just as they can learn to read foreign languages or music. The documentation tries to explain what the code is doing or why it is doing it, but *how* the code is accomplishing its tasks is left for the code itself to reveal.

In large measure the objective of the book is for a novice to understand the molecular dynamics codes given in the appendices. *Understand* means that the user will (i) be able to explain what the code does, (ii) be able to have a machine execute the code, and (iii) be able to judge whether the run and its results are meaningful. The programs are not to be used as black boxes; rather, they should be modified, tinkered with, improved—such exploratory play reinforces a learning process. The codes are not written to execute extremely fast: they are intended to be understood. In only a few instances are the codes compromised in favor of execution speed over comprehension. The programs are in Fortran and should execute under any Fortran 77 compilation.

Aristotle enumerated not only four modes of explanation, but he also worried about chance as a causative agent. Although molecular dynamics is often touted as a completely deterministic method, in fact, chance has a healthy and decisive role to play in molecular dynamics. Chance events—random unpredictable events—occur not only as consequences of nonlinearities in the differential equations that prescribe particle trajectories, but, more subversively, as consequences of the hardware and software used in performing the simulations. The role of chance in molecular dynamics is discussed in Chapters 2 and 3.

Molecular dynamics simulation brings to bear a startling range of ideas from several disciplines. This is one reason the topic proves valuable in a classroom. Readers are expected to have some knowledge of classical mechanics, vector analysis, numerical analysis, thermodynamics, and Fortran programming. Previous exposure to kinetic theory, statistics, transport phenomena, and statistical mechanics would also be beneficial.

Computer simulation has become an important and respected research tool, complementing the traditional approaches of theory and experiment. As with any tool, its effectiveness is limited by the skill and dedication of the craftsman who guides its use. But I have come to realize that there are unusual ways in which simulation is something other than simply a tool. Use of a tool suggests one knows at least the goal, if not the path. However, in research one is often diverted from an original goal, and computer simulation can motivate such a diversion: simulation can produce revelation. By this I mean that simulation has the potential to yield the unexpected. Most often the unexpected is readily understood: "Oh yes, of course! Why didn't I think of that before?" But the fact is, we didn't think of it before, and chances are we might never have thought of it, certainly not nearly so soon, without the simulation.[†]

I think this revelatory aspect of simulation is a major reason why it has become so popular and enthralling. People enjoy pleasant surprises, especially if they are also educational. I think there is a latent danger here, though. It is easy to simulate models so abstract that the results have only intrinsic interest: they convey no insight into any aspect of reality. In fact, it is usually easier to simulate *interesting* models rather than realistic ones, and therein lies the danger. Davis and Hersh[‡] have warned of the dangers of abstraction in mathematics and digital computation. This warning applies, as well, to molecular dynamics.

The use of elaborate abstraction to interpret reality was the focal point of a protracted, fourteenth-century battle that pitted both Church and academia against a solitary Franciscan scholar. The principal weapon in that scholar's logical armory has come down to us transmuted into a cornerstone of modern scientific method: *Entities must not be multiplied beyond necessity.* This attitude—Ockham's razor—serves as one theme of this book: it applies not only to interpretations of results, but also to the methods used to obtain results. It should guide us in the models we build, the algorithms we use, and the programs we write. Computer simulation presents exhaustive and exhausting opportunities for making errors. Simple approaches, approaches that work and are readily understood, reduce the chances for error and enhance the chances for finding errors when they are made. Complex, sophisticated approaches are often touted as executing quickly on a computer, but in truth, they rarely prove to be great time savers.

In the fall of 1884 at Johns Hopkins University Sir William Thomson presented a series of 20 lectures under the title *Molecular Dynamics and the Wave Theory of Light*.[§] The technical content of those lectures have little in common with the volume you now hold in your hands: in his lectures Thomson sought to understand light within the framework of classical mechanics. As such, the lectures were washed away in the tide of quantum mechanics, and they failed to contribute to the development of modern

[†]W. Thomson, Lecture X, p. 103: "...I never heard of anomalous dispersion until after I found it lurking in the formulas. I said to myself, 'these formulas would imply that, and I never have heard of it.' And when I looked into the matter I found to my shame that a thing which had been known by others for eight or ten years I had not known until I found it in the dynamics."

[‡]P. J. Davis and R. Hersh, *Descartes' Dream*, Houghton Mifflin Co., Boston, 1986.

[§]W. Thomson, reprinted in *Kelvin's Baltimore Lectures and Modern Theoretical Physics*, R. Kargon and P. Achinstein, Eds., MIT Press, Cambridge, MA, 1987.

science. Nevertheless, if you read the lectures, not for technical content but to gain insight into the workings of a nineteenth-century scientific mind, you find that Thomson's mind-set—the way he tackled problems—was modern indeed. Or, to twist this observation around, the way we tackle problems is not at all modern—it's merely the power of the computer that's new and that allows us to push an established methodology farther than ever before. I think there is value in placing what we do, including computer simulation, within a historical context, and so I have embellished this text with quotes from Thomson's Baltimore lectures. In 1892 Thomson was named Baron Kelvin of Largs.

Buried within this book is the attitude that simulation only makes manifest possible realities that are latent within the human mind. There are, after all, no little spheres doing their dance in the computer; to the machine, a simulation is merely an elaboration of electronic impulses responding to the directions of a rudimentary arithmetic. Computer graphics do not reveal what is happening in the machine, but rather what humans *interpret* as happening. To contrive a beneficial understanding of these interpretations, these latent realities, we must make contact with the world of our senses: when isolated from theoretical and experimental forms of science, simulation is sterile. It is my hope that this book will help at least one novice keep clear the distinction between the simulated and the real; that it will stimulate one enthusiastic novice to pause and think, then simulate, then think again; that it will provoke one novice to abandon simulation for something more worthy of his talents; that it will challenge one novice to improve any one of the methods described herein. By such small steps does science advance.

Acknowledgments

The following individuals unselfishly took time from their productive schedules to read and criticize portions of the manuscript: J. J. Erpenbeck, C. G. Gray, P. A. Monson, F. H. Ree, and W. A. Steele. I am grateful to them all. Special thanks to Sohail Murad and Ariel Chialvo who read the entire manuscript and corrected scores of errors.

But in spite of the constructive comments provided by these experts, I have not always followed their advice. This is an unconventional book to the extent that objective discussions of methods and results are mingled with subjective views that illuminate those objective perceptions. It presumes a skeptical reader for, as Umberto Eco reminds us in *The Name of the Rose*,

"Books are not made to be believed, but to be subjected to inquiry."

HAILE@CLEMSON

Aboard the Sloop Passepartout
Lake Keowee, South Carolina
June 1991

CONTENTS

A NOTE TO THE USER

At the end of each chapter I have posed a variety of problems: some exercise the material in the chapter, some go beyond the material explicitly presented, some try to connect to ideas from previous chapters, some try to connect to ideas from supporting disciplines. To meet the needs of a diverse readership, the problems vary in complexity and topic. Few of the problems are especially difficult, but properly executed, many are time consuming. Rather than work many problems cursorily, it seems to me more profitable to work a few problems thoroughly. The test for thoroughness is that students begin to use a problem as a springboard for devising their own problems.

To warn the reader (and instructor) that certain problems require special consideration, I have marked them with a flag from the set of international maritime signals. The flags are explained in the following table.

Flag	Alphabetic Character	Maritime Meaning	Usage in This Book
⊠	V (victor)	*Require assistance*	A computer is needed for these problems. Aside from the effort needed in attacking the problem, sufficient time should be allowed by the reader (and instructor) for coping with the vagaries of machine and programming language.
◪	U (uniform)	*Standing into danger*	These problems will challenge the average reader, either because of difficulty or because of the dedicated effort required.

MOLECULAR DYNAMICS SIMULATION

1

INTRODUCTION

Molecular dynamics simulations compute the motions of individual molecules in models of solids, liquids, and gases. The key idea here is *motion*, which describes how positions, velocities, and orientations change with time. In effect, molecular dynamics constitutes a motion picture that follows molecules as they dart to and fro, twisting, turning, colliding with one another, and, perhaps, colliding with their container.

This usage is not unique: *molecular dynamics* may also refer to the motions of real molecules when studied primarily by molecular beam [1] or spectroscopic [2] techniques. This terminological confusion is compounded by *lattice dynamics* [3], which refers to the study of vibratory motions of atoms in solids, and by *molecular mechanics* [4], also called force field calculations, which refers to quantum mechanical calculations of the structure of individual molecules. This book is concerned with molecular dynamics solely in the sense of simulation.

Molecular dynamics simulation is the modern realization of an old, essentially old-fashioned, idea in science; namely, the behavior of a system can be computed if we have, for the system's parts, a set of initial conditions plus forces of interaction. From the time of Newton to the present day, this deterministic mechanical interpretation of Nature has dominated science [5]. In 1814, roughly a century after Newton, Laplace wrote [6]:

Given for one instant an intelligence which could comprehend all the forces by which nature is animated and the respective situation of the beings who compose it—an intelligence sufficiently vast to submit these data to analysis—it would embrace in the same formula the movements of the greatest bodies of

1

the universe and those of the lightest atoms; for it, nothing would be uncertain and the future, as the past, would be present to its eyes.

If this approach is thwarted by the complexities of reality, then we replace reality with a model. In one of his Baltimore lectures (Lecture XI), roughly a century after Laplace, Thomson observed [7]:

> It seems to me that the test of "Do we or not understand a particular subject in physics?" is, "Can we make a mechanical model of it?"

Today, roughly a century after Thomson, we remain undeterred from Laplace's dream: the requisite "intelligence" is provided by the digital computer, the "respective situation" is a set of initial positions and velocities, "the same formula" though not literally true could be interpreted as the same algorithmic program, and Laplace's universe has given way to model universes. Now, deterministic mathematical models pervade not only the physical sciences and engineering, but the life and social sciences [8] as well.†

This attitude is old-fashioned in the sense that, while often successful, it is nevertheless simplistic. In spite of Laplace's claim, we can still identify systems that are unpredictable—stock markets and the weather, for example. Why should this be? If deterministic mathematical models can help us successfully land *Apollo XI* on the moon, why can't they help us predict next month's weather on earth?

The resolution of this dilemma is based on the kind of forces acting among system components: when a system contains objects that interact *nonlinearly*, the system's behavior may be unpredictable. In the past few years studies in nonlinear dynamics have decoupled deterministic from predictable [9]. *Deterministic* situations have system outputs causally connected to system inputs. *Calculable* situations are those deterministic situations in which an algorithm allows us to compute system outputs if the inputs were known. *Predictable* situations are those calculable situations in which the algorithm can be numerically implemented to actually compute the outputs. Calculable situations may be unpredictable because of the large number of inputs needed, because of an unrealistically high precision with which the inputs must be known, and/or because the algorithm's stability is sensitive to intermediate calculations. In pool, *Eight ball in the side pocket* is deterministic, calculable, and predictable; however, whether it will rain in two weeks is deterministic but unpredictable.

The overriding theme of this book is predicated on the decoupling of predictability from determinism. Be warned—that you use a machine to

---

†In fairness, paleontologists, at least, have discovered deterministic unpredictability. Thus, Stephen Jay Gould [10] posits that if the tape of life were rewound to some previous, sufficiently removed condition and then replayed, the result would be life unlike life as we know it. For more technical conjectures on connections between life and deterministic unpredictability, see Fox [11].

compute the behavior of a many-body model does not guarantee that the computed behavior is representative of that model, much less that the model mimics reality. To my mind it is this deterministic unpredictability that makes molecular dynamics fascinating and challenging. Is the fun (aka intellectual stimulation) merely in making a model, writing some differential equations, and loading it all into a computer? No. The fun, it seems to me, begins when we have completed a simulation, when we have a number. Now there arise all the old familiar questions characteristic of science: How good is this number? How could it be wrong? Is it representative, that is, reproducible? What does it mean? Do I believe it? How do I test it? If it is right, what must follow? This book should not only help you learn how to simulate but also make you aware that questioning the results is part of the procedure.

Computer simulations are performed on models, not on real things, and so the science of simulation, while distinct from, is necessarily bound to the art of model building. The purposes of this chapter are to clarify the distinctions between models and simulations and to discuss how together they contribute to new understanding.

1.1 SYSTEMS AND ALL *THAT*

The portion of the physical world on which we focus our attention is called the *system*; it is a subset of the universe. The system may be composed of any number of similar or dissimilar parts and the condition of those parts identifies the *state* of the system. For example, the door to my office constitutes a system to which I can ascribe two states: *open* or *shut*. To analyze and describe the behavior of the system, we need ways for assigning numerical values either to the state or to functions of the state; such assignments are called *observables*. Thus, to my door we can ascribe an observable called openness, to which I assign the value 1 if the door is open and 0 if the door is shut. As another example, let the system be 10^{24} molecules of a gas. Its state is specified by the position and momentum of each molecule, and the state gives rise to such observables as temperature and pressure.

The state of a system can be manipulated and controlled from the environment via *interactions*. For example, I may change the state of my office door by an interaction, specifically, by exerting a force. Allowed interactions are constrained by the nature of the boundary that separates the system from its surroundings. Various kinds of boundaries and interactions are possible, but in this book we will limit our attention to *isolated systems*: systems that can exchange neither matter nor energy with their surroundings.

We cannot usually study a system by directly observing the state; instead, we probe states indirectly by manipulating, controlling, and measuring observables. Thus, studies of a gas may involve controlling the system volume, manipulating its temperature, and measuring how the pressure responds. The isolated system is special in the sense that we do not interact with it, and therefore we can manipulate its observables only before the system is isolated.

To organize, describe, and perhaps even predict observables, we create theories. Theories may operate at one of several levels. At the simplest level, theories merely provide relations among observables. For example, the ideal-gas law

$$PV = NkT \qquad (1.1)$$

was originally obtained by organizing results from measurements of the pressure, volume, and temperature of low-density gases.

At the next level of complexity are theories that relate observables to the underlying state. For example, at this level we have kinetic theory, which teaches that the observable temperature is related to the state through the molecular velocities. Theories at this level provide interpretations or explanations for observables, but if the state itself is unobservable, these theories cannot be used to compute numerical values for observables.

To overcome this computational dead end, two strategies have been devised: (1) concoct theories at still higher levels or (2) perform computer simulations. Higher level theories try to resolve the computational difficulty by reorganizing and reducing the detailed information about the state needed to compute values for observables. Such is the objective of statistical mechanics, in which observables are related not to the underlying state itself, but rather to the probability of the system being in particular states.

The alternative strategy includes molecular dynamics. Molecular dynamics assigns numerical values to states, thereby making states observable, at least for model substances. With numerical values assigned to states, theoretical relations from kinetic theory can be used to compute values for experimentally accessible observables. Thus molecular dynamics is closely tied to kinetic theory and not as closely related to statistical mechanics.[†] In particular, molecular dynamics is less sophisticated, less elegant, but more direct than statistical mechanics.

1.2 MODELING VERSUS SIMULATION

Whether we study systems theoretically or experimentally, the general procedure is the same: we manipulate and control certain observables (inputs), the

---

[†]W. Thomson, Lecture I, p. 1: "...the kinetic theory of gases is a part of molecular dynamics, is founded upon molecular dynamics, works wholly within molecular dynamics, to it molecular dynamics is everything, and it must be advanced by molecular dynamics..."

system responds, and then we measure or compute other observables (outputs):

Manipulate and control   measure or compute
certain observables   → | SYSTEM | →   other observables
(inputs)   (outputs)

Since theoretical analyses are now largely done via modeling or simulation or both, it is instructive to clarify how modeling differs from simulation. For a discussion more complete than what follows, see Casti [8].

The goal of theoretical work is to establish connections between measurable outputs and controlled inputs. In Section 1.1 we discussed how this may be done at different levels of complexity; in particular, sophisticated theories use an underlying state to connect outputs to inputs. Part of the theoretical problem is to define the state in such a way that complicated interactions among state variables are decoupled, or at least weakened, so that observable outputs can in fact be computed.

A model is an attempt to decouple and remove interactions that have little or no influence on the observables being studied. Thus, a model is simpler than the system it mimics: it has access to fewer states. Decoupling interactions means relaxing constraints; hence, a model has access to some states not available to the original system and vice versa. In other words, a model is a subset or subsystem of the original system: outputs from a model will be consistent with those of the original system, but only for a restricted set of inputs. For those restricted inputs, since the model is a subsystem of the original, states visited by the model correspond to those visited by the original system.

In contrast, a simulation is more complicated than the system it simulates: a simulation generally can reach many more states than can the original system. A simulation imposes constraints so that the simulated output is consistent with the output of the original system, at least for a restricted set of inputs. A simulation will typically bear no structural relation to the original system; for example, the way constraints are imposed in the simulation may differ from the mechanism that confines the original system to certain states. Hence, states in the simulation may bear no correspondence to states of the original system. Although a simulation is more complex than the original system, it does *not* follow that the original system is a model of the simulation.

An example should clarify these ideas. As the real system, consider a simple ball-and-spring arrangement. One end of the spring is attached to the ball, the other end is fixed to a wall. In response to a displacement from its equilibrium position, the ball slides on a floor. The problems are to, in turn, model and simulate the motion of the ball that results from a displacement.

Let R represent this real system, that is, the spring as mover plus the ball as the thing moved.

To study the motion of R, we might construct a device M that is a one-dimensional harmonic oscillator (ODHO) having spring constant γ. The motion of M is described by the differential equation

$$\frac{d^2x}{dt^2} = -\gamma x(t) \tag{1.2}$$

where x is the distance of the ball from its equilibrium position and t is time. Writing this equation presumes several simplifying assumptions: (a) the motion of the ball is restricted to a line, (b) the spring is perfectly harmonic, (c) the ball experiences no sliding friction on the floor, and (d) the ball has no internal states that exchange energy with the spring. These assumptions imply that M is a simple subsystem of R. For some initial displacements M will mimic the motion in R; however, M cannot mimic all the behavior available to R. For example, in R we might initially raise the ball from the floor, allowing the ball to move in the xz-plane, motion not allowed to M. The one-dimensional harmonic oscillator M is a *model* of the real system R.

An alternative scheme S for studying the motion of R would be to remove the spring and use a person as the mover. This person might be a well-trained graduate student who has the uncanny ability to move the ball, for many sets of initial conditions, so as to reproduce the motion of the ball in R. This situation S is more complex than R because a person is more sophisticated than a spring. Moreover, S involves imposing constraints on the student's arm to make the motion of the ball mimic the motion in R. Otherwise, more states are possible in S than in R; for example, the student might absent-mindedly drop the ball in his pocket when he stops for coffee. The situation S is a *simulation* of the real system R.

Note that as well as simulating the real system R, we could also simulate the model M. For example, we might have another mover (a professor or, equivalently, a robot) that can move the ball through states visited by the perfect ODHO model M. This simulation of M is more complex than M itself, but the added complexity does not make the model more realistic, that is, more like the real system R.

How do these distinctions relate to molecular simulations? Well, what do we typically do? We identify a substance and its observables that we want to study, say, thermodynamic properties of argon. Then we construct a model of the substance, say, the spherically symmetric, pairwise additive Lennard-Jones potential. This is a true model. The Lennard-Jones potential is simpler than the argon potential because argon atoms are not perfect spheres and their interactions are certainly not only pairwise additive. With the model chosen, we then perform a simulation—but a simulation of what? It can only be a simulation of the model, of the Lennard-Jones substance. We do not simulate argon. The simulation is more complex than the model, but the added complexity does not add to the realism of the resulting observable outputs. In error are those who claim that molecular dynamics simulates argon, or water, or proteins, or whatever. We simulate molecular models of such substances.

1.3   THEORY VERSUS EXPERIMENT

At a scientific conference in the 1970s there broke out a heated debate as to whether computer simulations like molecular dynamics are theories or are experiments. The theory side argued that simulation is clearly not experiment because no measurements are done on real systems; molecular simulations are pure calculation. The experiment side countered that simulation results are *used* like experiments, namely, to test theories; it isn't sensible to test one theory with another theory is it? Moreover, this side noted that simulation results, like experiments, are prone to problems of reproducibility and statistical error. Hence, pervading the literature is the interpretation of molecular simulations as computer experiments.

What's the resolution of this dilemma? And does it really matter how we think of simulation? Consider the following example. To perform an experiment, to take a measurement, the observer must interact with the system: some type of probe necessarily has to cross the system boundary. Thus, truly isolated systems cannot be studied experimentally: once our probe crosses the boundary, the system is no longer isolated. However, we can perform theoretical calculations, such as simulations, on truly isolated systems and obtain meaningful results. The resolution of our dilemma has to be that molecular simulations are forms of theory; they do not involve measurements on real systems.

How we think of simulation is important, indeed crucial, because of the consequences: if we accept that simulations are experiments, then it follows that the models simulated are real. Armed with this attitude plus the ease of actually doing simulations, we may be tempted to abandon laboratory experiment altogether. The danger lies in severing simulation from reality.

Is this only an academic issue? Recently a public presentation was made by the chief executive officer of a major computer manufacturer. In the talk *Voyager* photos of Jupiter were compared with images produced from a computer simulation of Jovian weather. The speaker's punch line was that the simulated images were actually "more real than real life." But if this were true, then there would be no need for further explorations by spacecraft —we would simply perform simulations. Note that, in fact, reality doesn't enter this picture at all. The CEO can be faulted on two counts: not only did he erroneously claim that a simulation can supplant reality, but he also confused a photographic image with reality.[†]

To illustrate this point in a more mundane situation, consider: Police have been summoned to the scene of a domestic quarrel. In the kitchen the patrolmen find a bewildered wife standing over the prostrate form of her husband.

"Okay, lady, what happened?"

"I don't rightly know. Herb riled me, so I hit'em in the head with a tomato."

"Sure, lady. There's no tomato on the body or the floor."

"Isn't there? O'course not. T'was a decorative ceramic."

And lest you feel that this contrived example[‡] is only impractical philosophical quibbling, consider the intense ethical debates that were prompted by American television broadcasts of selected news events in the form of interpolative reenactments (aka *simulations*).

1.4 REDUCTIONISM VERSUS SIMULATION

Since the time of Newton, scientific theories have nearly all been developed in a reductionistic mode: a complex system is reduced to one or more simple subsytems and the subsystems are analyzed. Subsystems ultimately take the form of models, and today models are almost exclusively mathematical. Before about 1960 mathematical models had to be simple enough to be tractable analytically, but now this constraint is relaxed by the availability of digital computers.

As an example of reductionism, consider the study of matter in simpler and simpler forms:

$$\text{Matter} \rightarrow \text{compounds} \rightarrow \text{elements} \rightarrow \text{molecules} \rightarrow \text{atoms}$$

$$\rightarrow \text{elementary particles} \rightarrow \text{quarks}$$

Modeling may occur at any stage of such a reduction. Successful modeling requires a construction that forces the behavior of interest to remain *invariant* when the subsystem is replaced by the model. The goal of reductionism is

---

[†]A principal feature of science is the apparently endless disentangling of images from reality. Well over 800 years ago in his *Questiones Naturales*, the scholar Adelard of Bath was moved to write, "Wherefore, if you want to hear anything more from me, give and take reason. For I'm not the sort of man that can be fed on a picture of a beefsteak" [12].

[‡]This example is a slight modification of a vaudeville routine used by James Thurber [13] to illustrate *confusing the container with the thing contained*.

to explain system behavior by combining explanations for the behaviors of its subsystem models. Simulation provides an alternative to reductionism because simulation allows us to study the behaviors of classes of systems or subsystems. Thus, while reductionism emphasizes structural analysis, simulation emphasizes behavioral classification. As an example, consider study of the fluid–solid phase transition. One simulation approach to this problem would be to load marbles into a drum. To force the marbles to move, we rotate the drum (imagine a cement mixer). As the drum rotates, we add more and more marbles, until finally enough are added to freeze the motion. From the number of marbles N and the drum volume V, we obtain the density N/V needed for solidification. We then repeat the simulation with balls of other diameters: golf balls, baseballs, soccer balls, basketballs. By inspecting the solidification densities for balls of various diameters, we conclude that one mode of solidification is a geometric packing effect, controlled by the packing fraction V_{balls}/V_{drum}. Note the features of this simulation: the controlled input observables are the ball diameter, the number of balls, and the drum volume; the measured output observable is the density at which motion ceases. During the experiments, the states of the balls—positions and velocities—bear no relation to states of molecules in any real substance. Further, the entire study is not just of one substance, but rather a systematic progression through a class of substances: spheres of increasing diameters.

In contrast, a reductionist would study this problem by combining a model with a theory to predict the solidification of a particular substance. The theory might involve only relations among observables, such as PVT equations of state, or the theory might include underlying system states, such as is done in statistical mechanics. In any case, the validity of the model and the theory would be tested by comparing predictions with experimental measurements on a real substance. However, the connection between input and output observables would remain *implicit* in the mathematical apparatus used to make the prediction.

But the goal of the simulations is not so much to predict solidification as to make *explicit* how input and output observables are connected. In other words, rather than predictions, the goal is more in the nature of providing explanations: idealized models "*explain* nature even while they do not describe it" [14]. Both reductionism and simulation contribute to science; however, for a specific problem one or the other may be more appropriate. In particular, simulation is not always the best method.

As shown in Figure 1.1, we identify two distinct roles that simulation can play in scientific investigation. At the higher level, simulation, including computer simulation, serves as an alternative to reductionism. At this level, as popularly claimed, simulation is a new way of doing science. In addition, computer–simulation can be used at a lower level, as a tool in reductionism. It is this second use of simulation that is implied by the more familiar triangular diagram shown in Figure 1.2 [15]. That diagram suggests two reductionistic uses for simulation: (a) simulation data on models can be used to test

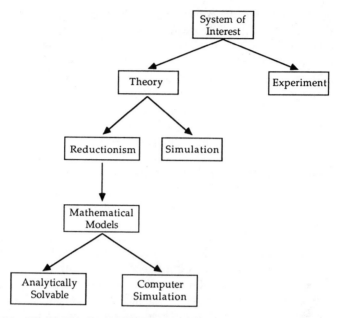

FIGURE 1.1 Hierarchy of scientific modes of investigation. Note that the system of interest may be real, or it may itself be a model.

theories and (b) simulation data can be compared with experimental data to test the realism of simulated models.

In Figure 1.1, the dual use—as an alternative to reductionism or as a reductionistic tool—explains, perhaps, early debates over whether computer simulation is theory or experiment. If you identify all theory as reductionism and sense that simulation is something different, then you may interpret simulation as experiment. Conversely, if you see computer simulation as a tool for studying reductionistic models, then you may interpret simulation as theory.

The theme of this section is that computer simulation offers possibilities more instructive and more far-reaching if it is used as an alternative to reductionism rather than as merely a servant to reductionism. To make this

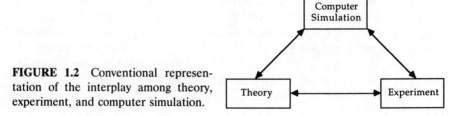

FIGURE 1.2 Conventional representation of the interplay among theory, experiment, and computer simulation.

statement concrete: using computer simulation to map out a phase diagram for a model of methane (or whatever), merely to test how realistic the model may be, is to misunderstand and underutilize the power of simulation.

1.5 MODELS FOR MOLECULAR SIMULATIONS

A computer simulation is valuable because it is applied to a precisely defined model for the material of interest. The model is actually a composite of two: one for interactions among the molecules making up the system and another for interactions between the molecules and their environment:

Simulated model	=	model for molecular interactions

	+	model for system–environment interactions

Note the decoupling implied by this schematic—intermolecular interactions are presumed to be independent of interactions with the environment.

The model for molecular interactions is contained in an intermolecular force law or, equivalently, an intermolecular potential energy function. This potential function implicitly describes the geometric shapes of individual molecules or, more precisely, their electron clouds. Thus when we specify the potential function, we establish the symmetry of the molecules, whether they are rigid or flexible, how many interaction sites occupy each molecule, and so on. A detailed characterization of intermolecular potential functions may be given analytically or numerically; in any case, a quantitative form for the potential function defines a molecular model and hence the form must be chosen before a simulation can be performed.

In this book we consider only spherically symmetric molecules (atoms). For N such atoms the intermolecular potential function is represented by $\mathscr{U}(\mathbf{r}^N)$. The notation $\mathbf{r}^N$ represents the set of vectors that locate the atomic centers of mass, $\mathbf{r}^N = \{\mathbf{r}_1, \mathbf{r}_2, \mathbf{r}_3, \ldots, \mathbf{r}_N\}$. When we establish values for the set $\mathbf{r}^N$, we define the *configuration* of a system. Macroscopic properties that are averages over only the set $\mathbf{r}^N$ are called *configurational* properties.

In most simulations the intermolecular potential energy is taken to be a sum of isolated pair interactions; this assumption is called *pairwise additivity*. Hence,

$$\mathscr{U} = \sum_{i<j}\sum u(r_{ij}) \tag{1.3}$$

where $u(r_{ij})$ is a pair potential energy function whose form is known and r_{ij} is the scalar distance between molecules i and j. Since no dissipative forces

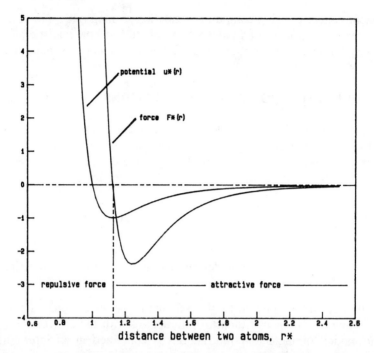

FIGURE 1.3 Typical potential energy $u^*(r)$ and force $F^*(r)$ functions used to model interactions between two atoms separated by distances r^*. Positive forces are repulsive; negative forces attractive. Here $r^* = r/\sigma$, $u^* = u/\varepsilon$, and $F^* = F\sigma/\varepsilon$, where σ is a unit of length and ε is a unit of energy.

act among molecules, intermolecular forces are conservative, and therefore the force on molecule i is related to the potential by

$$\mathbf{F}_i = -\frac{\partial \mathscr{U}(\mathbf{r}^N)}{\partial \mathbf{r}_i} \tag{1.4}$$

where $\partial/\partial\mathbf{r}$ represents the gradient operator. Typical forms for the pair potential and its resultant force are shown in Figure 1.3. By convention, repulsive forces are positive, attractive forces are negative.

The second part of the simulated model encompasses boundary conditions, which describe how the molecules interact with their surroundings. Characteristics of boundary conditions are largely dictated by the physical situation to be simulated; however, some freedom usually exists in the way in which boundary conditions are realized. For example, if a bulk fluid is to be studied, we want to avoid hard boundaries, which obscure bulk–fluid behavior in small systems. If nonuniform regions such as fluid–fluid or fluid–solid interfaces are to be simulated, then boundary conditions that mimic these situations are needed. If the shear or bulk viscosity is to be determined, we may introduce moving boundaries that shear or compress the system. In this book we consider only boundaries that isolate the system, so no interactions occur between system and surroundings. In any event, setting boundary conditions completes the definition of the model system to be simulated.

1.6 STOCHASTIC VERSUS DETERMINISTIC

A molecular-scale simulation consists of three principal steps: (1) construction of a model, (2) calculation of molecular trajectories, and (3) analysis of those trajectories to obtain property values. The second step constitutes the simulation proper. We use the way in which molecular positions $\mathbf{r}^N$ are computed in step 2 to discriminate among simulation methods.

In molecular dynamics the positions are obtained by numerically solving differential equations of motion and, hence, the positions are connected in time—the positions reveal dynamics of individual molecules as in a motion picture. In other simulation methods the molecular positions are not temporally related. For instance, in *Monte Carlo* simulations the positions are generated stochastically such that a molecular configuration $\mathbf{r}^N$ depends only on the previous configuration. When the outcome of a random event in a sequence depends only on the outcome of the immediately previous event, the sequence is called a *Markov chain*. In still other simulation methods, the positions are computed from hybrid schemes that involve some stochastic features, as in Monte Carlo, and some deterministic features, as in molecular dynamics. These various methods can be arranged, as in Figure 1.4, according to the degree of determinism used in generating molecular positions. Here some of these methods are briefly described to place molecular dynamics within the wider context of molecular-scale simulation methods.

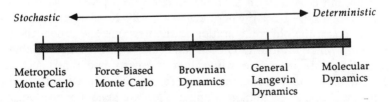

FIGURE 1.4 Relative degree of determinism in various molecular simulation methods. (Adapted from Ceperley and Tully [16] and used with permission.)

1.6.1 Monte Carlo

The purely stochastic method, Monte Carlo, is typically performed on a fixed number of molecules N placed in a fixed volume V and maintained at a constant absolute temperature T. The simulation procedure is adapted from general Monte Carlo methods for evaluating multidimensional integrals. Here the integrals of interest are statistical–mechanical ensemble averages, that is, configurational properties $\langle A \rangle$ of the N-body system. For atomic substances these integrals are of the form

$$\langle A \rangle = \frac{1}{\mathscr{Z}} \int \cdots \int \exp\left[-\beta \mathscr{U}(\mathbf{r}^N)\right] A(\mathbf{r}^N)\, d\mathbf{r}_1 \cdots d\mathbf{r}_N \qquad (1.5)$$

where $\beta = 1/kT$, k is Boltzmann's constant, and $\mathscr{Z}$ is the configurational integral

$$\mathscr{Z} = \int \cdots \int \exp\left[-\beta \mathscr{U}(\mathbf{r}^N)\right] d\mathbf{r}_1 \cdots d\mathbf{r}_N \qquad (1.6)$$

The integrals in (1.5) and (1.6) are $3N$-fold because each differential volume element contains three components; thus, $d\mathbf{r}_1 = dx_1\, dy_1\, dz_1$.

In Monte Carlo simulations, ensemble averages such as (1.5) are evaluated by accumulating the integrand at randomly generated values of the independent variables—the atomic positions $\mathbf{r}^N$. Note that because of the "Boltzmann" factor $\exp(-\beta \mathscr{U})$, some configurations make large contributions to the integral while others contribute nothing. Thus we seek to bias the sampling in favor of those configurations most likely to occur. The importance sampling scheme devised by Metropolis et al. [17] does just that.

The Metropolis method involves the following principal steps. First, we assign initial positions $\mathbf{r}_i$ to the N molecules and compute the total potential energy $\mathscr{U}$ using (1.3). Then we hypothesize a new configuration by arbitrarily choosing one molecule and proposing that it be moved from $\mathbf{r}$ through a randomly chosen distance and direction to a new position $\mathbf{r}'$. We compute the total potential energy $\mathscr{U}'$ for this new configuration and accept the proposed move (allow it to occur) based on the following criteria. If $\mathscr{U}' < \mathscr{U}$, then we accept the move. If $\mathscr{U}' > \mathscr{U}$, then we accept it with probability proportional to the factor $\exp[-\beta \Delta \mathscr{U}]$, where $\Delta \mathscr{U} = \mathscr{U}' - \mathscr{U}$. If we reject the proposed move, we count the old configuration as the new one and repeat the process using some other arbitrarily chosen molecule. For each new configuration generated by this procedure, we evaluate property integrands such as (1.5) and accumulate them in running sums. To obtain adequate statistical precision for averages, a few million configurations are generally needed.

Several variations on the Metropolis Monte Carlo method have been proposed. One of these is the force-biased algorithm [18] in which the proposed move of a molecule is no longer completely random; rather, its direction is taken to be that of the force exerted on the molecule by all other molecules. Such a procedure reduces the number of configurations needed for adequate statistical precision, but it increases the amount of computation per configuration [19].

A general presentation of Monte Carlo methods, including the Metropolis scheme, can be found in the book by Kalos and Whitlock [20]. The classic article on Monte Carlo applied to statistical mechanical problems is that by Wood [21]. More recent developments are contained in articles by Valleau and co-workers [22]. The books edited by Binder [23, 24] are good sources for applications of the method.

1.6.2   Molecular Dynamics

The molecular dynamics method encompasses two general forms: one for systems at equilibrium, another for systems away from equilibrium. As devised by Alder and Wainwright in the late 1950s [25], equilibrium molecular dynamics is typically applied to an isolated system containing a fixed number of molecules N in a fixed volume V. Because the system is isolated, the total energy E is also constant; here E is the sum of the molecular kinetic and potential energies. Thus the variables N, V, and E determine the thermodynamic state.

In NVE–molecular dynamics molecular positions $\mathbf{r}^N$ are obtained by solving Newton's equations of motion:

$$\mathbf{F}_i(t) = m\ddot{\mathbf{r}}_i(t) = -\frac{\partial \mathscr{U}(\mathbf{r}^N)}{\partial \mathbf{r}_i} \qquad (1.7)$$

Here $\mathbf{F}_i$ is the force on i caused by the $N-1$ other molecules, the dots indicate total time derivatives, and m is the molecular mass. In writing (1.7) we have used (1.4), which relates the force to the intermolecular potential energy. Integrating (1.7) once yields the atomic momenta; integrating a second time produces the atomic positions. Repeatedly integrating for several thousand times produces individual atomic trajectories from which time averages $\langle A \rangle$ can be computed for macroscopic properties

$$\langle A \rangle = \lim_{t \to \infty} \frac{1}{t} \int_{t_0}^{t_0 + t} A(\tau) \, d\tau \qquad (1.8)$$

At equilibrium this average cannot depend on the initial time t_0. Since positions and momenta are obtained, the time average (1.8) represents both static properties, such as thermodynamics, and dynamic properties, such as transport coefficients.

According to the ergodic hypothesis, the time average (1.8) provided by molecular dynamics should be the same as the ensemble average (1.5) provided by Monte Carlo. Although a rigorous proof of the ergodic hypothesis exists only for the hard-sphere gas [26], it can be tested by comparing Monte Carlo results with molecular dynamics results. One such comparison is presented at the end of Chapter 3.

In Figure 1.5 the dynamic modeling problem is shown divided into its two great tasks: developing a suitable model for the problem at hand and applying molecular dynamics to that model. In this book we are not concerned with model development; instead, we concentrate on the second problem. That is, having developed a model, how do we perform the simulation? The simulation problem itself divides into two tasks: solving the equations of motion to generate molecular trajectories and then analyzing those trajectories for the properties of interest. In the figure we show how this book is organized to address those major tasks.

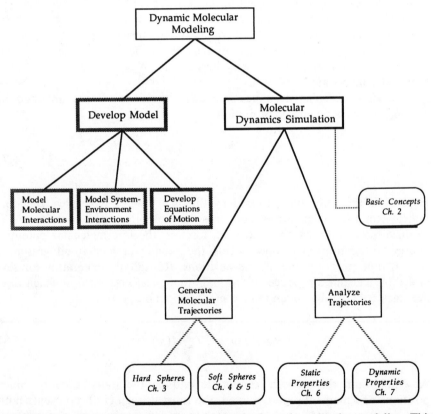

FIGURE 1.5 Hierarchy of the principal steps in dynamic molecular modeling. This book is not concerned with developing molecular models, but only with the application of molecular dynamics to existing models.

Molecular dynamics simulations are limited largely by the speed and storage constraints of available computers. Hence, simulations are usually done on systems containing 100–1000 particles, although calculations involving as many as 10^6 particles have been performed [27]. Because of this size limitation, simulations are confined to systems of particles that interact with relatively short-range forces; that is, intermolecular forces should be small when molecules are separated by a distance equal to half of the smallest overall dimension of the system. Because of the speed limitation, simulations are confined to studies of relatively short-lived phenomena, roughly, those occurring in less than 100–1000 psec. The characteristic relaxation time for the phenomenon under investigation must be small enough so that one simulation generates several relaxation times.

In addition to equilibrium molecular dynamics, nonequilibrium methods have been developed. These methods first appeared in the early 1970s [28–30], initially as alternatives to equilibrium simulations for computing transport coefficients. In these methods an external force is applied to the system to establish the nonequilibrium situation of interest, and the system's response to the force is then determined from the simulation. Nonequilibrium molecular dynamics has been used to obtain the shear viscosity, bulk viscosity, thermal conductivity, and diffusion coefficients [31].

Books devoted to molecular dynamics include that on simple fluids by Vesely [32] and those on nonequilibrium simulations by Hoover [33, 34] and by Evans and Morriss [35]. Simulations of plasmas, galactic evolution, and electron flow are emphasized in the book by Hockney and Eastwood [36]. The most comprehensive expositions of methods and applications are contained in the collection edited by Ciccotti and Hoover [37] and in the book by Allen and Tildesley [38].

1.6.3 Monte Carlo versus Molecular Dynamics

Although the physical and mathematical basis of Monte Carlo may be less transparent to a novice than that for molecular dynamics, Monte Carlo is usually easier than molecular dynamics to code in a high-level language such as Fortran. Monte Carlo is also easier to implement for systems in which it is difficult to extract the intermolecular force law from the potential function. Systems having this difficulty include those composed of molecules that interact through discontinuous forces; examples are the hard-sphere and hard convex-body models. Similar difficulties arise in systems for which the potential function is a complicated multidimensional surface, such as might be generated by ab initio calculations.

For determination of simple equilibrium properties such as the pressure in atomic fluids, Monte Carlo and molecular dynamics are equally effective: both require about the same amount of computer time to attain similar levels of statistical precision. However, molecular dynamics more efficiently evaluates such properties as the heat capacities, compressibilities, and interfacial properties. Besides configurational properties, molecular dynamics also provides access to dynamic quantities such as transport coefficients and time correlation functions. Such dynamic quantities cannot generally be obtained by Monte Carlo, although certain kinds of dynamic behavior may be deduced from Monte Carlo simulations [23].

Molecular dynamics also offers certain computational advantages because of the deterministic way in which it generates trajectories. The presence of an explicit time variable allows us to estimate the length needed for a run: the duration must be at least several multiples of the relaxation time for the slowest phenomenon being studied. No such convenient guide is available for estimating the length required for a Monte Carlo calculation. Finally, many kinds of small errors in a molecular dynamics program tend to accumulate with time and so become apparent as violations of conservation principles; in contrast, subtle errors in a Monte Carlo program may not blatantly advertise their presence.

1.6.4 Stochastic Dynamics

Molecular dynamics simulations of flexible molecules such as alkanes and proteins are more complex and time consuming than simulations of atomic and rigid-molecule systems. Not only are the equations of motion more complicated because of internal degrees of freedom, but also the internal modes, such as bond vibration and rotation, tend to relax on time scales very different from those of molecular collisions, which dominate external translational and rotational modes. Consequently, lengthy molecular dynamics simulations must be performed to capture many relaxation processes typical of flexible molecules.

If the objective is to study the dynamics of individual molecules, rather than determine properties of an aggregate, then Langevin and Brownian dynamics methods offer inexpensive alternatives to molecular dynamics. These methods reduce the amount of computation by modeling the system as one flexible molecule immersed in a continuum. For an interaction site i on a flexible molecule, the Langevin equation of motion is

$$m\ddot{\mathbf{r}}_i = -\zeta\dot{\mathbf{r}}_i + \mathbf{F}_{intra} + \mathbf{F}_{inter} \tag{1.9}$$

Here m is the mass of site i, ζ is a macroscopic friction coefficient, and $\mathbf{F}_{intra}$ is the force on i caused by interactions with other sites on the molecule. The intermolecular force $\mathbf{F}_{inter}$ is approximated as a stochastic interaction with the continuum.

In Brownian dynamics the momentum degrees of freedom are removed from (1.9) by arguing that over long times the positions of the interaction sites change very little. Thus setting the left-hand side of (1.9) to zero and solving explicitly for the positions $\mathbf{r}_i$, we obtain the Brownian equation of motion

$$\mathbf{r}_i(t) - \mathbf{r}_i(0) = \frac{1}{\zeta} \int_0^t \left[\mathbf{F}_{\text{intra}}(\tau) + \mathbf{F}_{\text{inter}}(\tau) \right] d\tau \tag{1.10}$$

Although (1.10) is approximate, it offers the advantage of allowing simulations of flexible molecules to be performed in a reasonable amount of computer time. Of course, if we need values for collective quantities such as thermodynamic properties or spatial distribution functions, then the full molecular dynamics method must be used. General discussions of these methods and sample results are provided in the review by Evans [39].

1.7 RIGHT VERSUS WRONG

In previous sections we have discussed how simulation compares with theory and experiment, how computer simulation compares with modeling, and how molecular dynamics compares with other forms of molecular-scale simulation. At this point it would be natural to give sample applications that show the kinds of problems to which molecular dynamics has been applied. However, this last section is not going to give such examples; if you need sample applications, the Bibliography should help you access the literature. Instead of trying to convince you that simulation can produce important results, this section tries to convince you that there is importance in the *way* in which simulation produces results. Those ways—those methods—are in fact worthy of our attention.

1.7.1 Realism, Accuracy, and Validity

To have a context for discussion, we review the reductionistic approach to problems. In reductionism, to predict the behavior of a physical system, we first create a mathematical model. The model mimics the real system only to the extent that it incorporates those features that determine the behavior under investigation. Thus a difficult problem, the real system, is replaced by a simpler problem, the model. We use the term *realism* to mean the extent to which the model reproduces behavior of the real system.

Now the behavior of the model might be calculated via analytic theory. Traditionally, models were kept simple so that their theoretical descriptions could be posed in a closed mathematical form and solved exactly. Models without this simplicity were avoided because, in the absence of computing machines, numerical solutions were too tedious and prone to error. Simple models can usually be accurately described by theory, so comparing a theoretical result with a measurement on the real system provides an unambiguous test of the realism of the model. As suggested in Figure 1.2, such tests advance science through the classic interplay of theory and experiment.

But with the availability of reliable high-speed computers, mathematical models need not be as simple as before and their theoretical descriptions need not be restricted to the analytically solvable. Theories may themselves involve unknown quantities or functions that must be estimated before the theoretical calculation can proceed. Such theories are not necessarily faithful representations of the model; we use the term *accuracy* to refer to the degree to which a theory reproduces behavior of a model. To fix these ideas, the relation $PV = NkT$ is an accurate theoretical result, derivable in statistical mechanics, for the ideal-gas model of matter; however, the ideal gas is an unrealistic model for liquid water.

This new complexity in models and theories means that realism of a model is now an issue separate from accuracy of a theory. Both issues cannot be simultaneously resolved by simply comparing theory with experiment. If the comparison is unfavorable, it may be because the model is unrealistic, or because the theory is inaccurate, or both. If the comparison is favorable, it may be because unrealistic features of the model fortuitously compensate for inaccuracies in the theory.

For determining the behavior of a proposed model, an alternative to analytic theory is provided by computer simulation. However, simulations themselves are not necessarily reliable because they can incur many kinds of errors: uncertainties in reaching an equilibrium or steady state, statistical errors in drawing a finite number of partially correlated samples, numerical errors in the algorithms used to perform the simulation, and round-off errors due to the finite word length of the computer. If, in spite of these errors, the simulated behavior of the model is within some allowable tolerance of its actual behavior, then we say the simulation is *valid*. Therefore, the reductionistic program in Figure 1.1 for studying a real system must confront either the combined effects of realism and accuracy or those of realism and validity; see Figure 1.6.

1.7.2 On the Importance of Being Valid

For the remainder of this section and in the rest of the book, we focus on establishing the validity of a simulation. In some instances, validity can be determined by comparing simulated behavior with that calculated from an

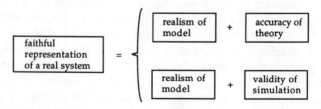

FIGURE 1.6 In a reductionistic study of a real system, we must confront either the combined effects of realism of a model plus accuracy of a theory or of realism of a model plus validity of a simulation.

analytic theory of known accuracy. Such a comparison constitutes the *strong test* for validity. In exceptional cases we may be able to simulate a model whose behavior is calculable from a completely accurate theory; such cases, in which we demonstrate perfect validity, are called *ideal simulations* [40]. An example is the one-dimensional harmonic oscillator (ODHO). A numerical simulation of the ODHO can be shown to be valid by comparing the simulated motion with that provided by exact analytic theory. However, note that a simulation of the ODHO may not be valid; for example, we might use a crude finite-difference method that erroneously solves the equations of motion. As a model the ODHO may be a realistic representation of the motion of a mass on a spring but be less realistic as a model for the vibratory motion of a carbon bond on a polymer chain.

In practice the strong test for validity is of limited use because we rarely have an analytic theory of known accuracy with which to compare. In fact we often want to use simulation results to test a theory's accuracy rather than use a theory to confirm a simulation's validity. A weaker test for validity is to estimate each of the possible simulation errors and argue that their accumulated effect is within the allowed tolerance for a valid simulation. It is this weak test for validity that must usually be employed and that requires so much attention from the simulator.

Although the validity of a simulation is discussed above in the context of a reductionistic approach to science, note that in fact validity is an issue no matter how the simulation is being used. We must always worry about the reliability of any simulation. And although it is argued in Section 1.3 that simulations are not experiments, nevertheless, the issue of validity is analogous to the issue of accuracy in laboratory measurements. We can probably do no better than take the same attitude to simulation results that the careful experimentalist takes to measurements: What are the possible sources of error or uncertainty? Are sampled populations representative? Are successive samples correlated? Are the results reproducible? Can I do a different simulation that should produce the same results? Are the results internally consistent? Do the results confirm results from previous simulations done elsewhere? How accurate and how precise are the results? In short, simulations done in a deterministic manner on reliable machines are not necessarily correct.

To emphasize this last statement, consider the data in Figure 1.7. Figure 1.7(*a*) shows simulation results for an observable *y* obtained at particular values of input observable *x*. These *y*-values happen to be from 1983 Monte Carlo runs [41], but neither the kind of simulation nor the identities of *x* and *y* affect this lesson. It is sufficient to say that the quantity *y* is notoriously difficult to extract reliably from molecular simulation and that *y* should pass through zero when *x* does. For our purpose the question is, what is the nature of the curve that we would draw through the data in Figure 1.7(*a*)?

Before proceeding, perform this little experiment: use a sheet of paper (or your hand) to cover panels 1.7(*b*) and (*c*) and also cover the rightmost two points in Figure 1.7(*a*). What kind of curve would you draw through the

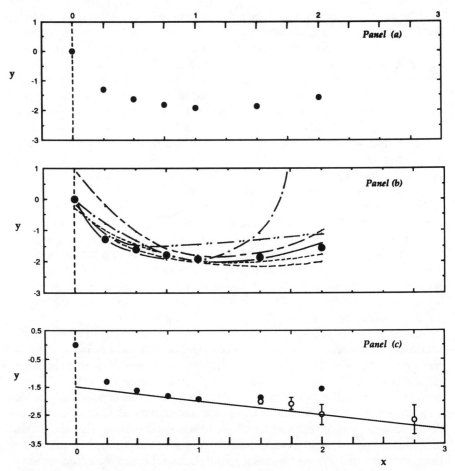

FIGURE 1.7   A single spurious value from simulation can have serious ramifications. *Panel (a)*: Original Monte Carlo results [41] suggesting a minimum in the quantity $y(x)$. *Panel (b)*: Results from six efforts to model the simulation data; all show a minimum: —— [42]; — — — [43]; — — — — [44]; —·· —·· [45]; ---- [46]; —·— · [47]. *Panel (c)*: Original and additional Monte Carlo results [48] finding that, in fact, $y(x)$ has no minimum. The solid line is from a simple classical theory. The error bars are those provided by the original authors.

remaining five points, a monotone or one with a minimum? I think your answer must be a monotone. Now, shift your paper to the right to expose the next point, the one at $x = 1.5$. Does this additional point change the shape of your imagined curve? Remember, there are uncertainties in these y-values, although the authors haven't given us direct guidance as to the size of the uncertainties. I think your response should be that the point at $x = 1.5$ changes your curve a bit, but its shape is still monotonic. Now uncover the point at $x = 2$. Does this point affect the shape of your imagined curve? If we accept that the y-value at $x = 2$ is as reliable as any of the others (and the authors give us no reason not to), then you conclude that the curve is now changed: the points apparently define a minimum. Note, however, that this conclusion is weakly based because the shape of the curve has been determined primarily by one point, that at $x = 2$. End of experiment. Are you comfortable with a conclusion based on essentially one sample?

The minimum in the data in Figure 1.7(a) posed a challenge to modelers. Figure 1.7(b) shows various results for y obtained from analytic calculations using several very different models: local composition [42], empirical quartic PVT equation of state [43], a kind of mean density approximation [44], variational theory [45], corresponding states [46], and another empirical equation of state [47]. All of these, as well as other models, manage, at least, to generate a minimum.

Now comes the twist in the plot: in 1988 there appeared additional Monte Carlo results for y [48], published by the authors of the 1983 data of Figure 1.7(a). These new data are shown, along with the old data, in Figure 1.7(c). The new data indicate that the original y-value at $x = 2$ was in error and that y is apparently monotone in x: there is no minimum.

So, what are we to make of the models represented in Figure 1.7(b)? Did they really establish a connection between the output y and the input x? Could the authors of those models now adjust parameters so their models fit the data in Figure 1.7(c) as well as (or better than) they fit the original data? And if they could make those adjustments, what would they teach us? Would we believe that the revised models now establish connections between y and x? Or are the models merely elaborate curve fits, flexible enough to accommodate many different functional forms?

The lesson here is not that you can necessarily avoid mistakes—mistakes are, after all, a powerful way to learn—rather, the lesson is that you should tread warily. Combining models with simulation can be a tricky and subtle business. To accept simulation results as valid demands careful consideration of the available evidence and often a search for more evidence. Simulation involves more than obtaining a program, performing some runs, and publishing the results. In short, simulations done in a deterministic manner on reliable machines can be wrong.

REFERENCES

[1] W. H. Miller, Ed., *Dynamics of Molecular Collisions*, Parts A and B, Plenum, New York, 1976.

[2] M. Evans, G. J. Evans, W. T. Coffey, and P. Grigolini, *Molecular Dynamics and Theory of Broad Band Spectroscopy*, Wiley, New York, 1982.

[3] G. K. Horton and A. A. Maradudin, Eds., *Dynamical Properties of Solids*, 3 vols., North-Holland, Amsterdam, 1974.

[4] U. Burkert and N. L. Allinger, *Molecular Mechanics*, ACS Monograph 177, American Chemical Society, Washington, DC, 1982.

[5] I. B. Cohen, *Revolution in Science*, Harvard University Press, Cambridge, MA, 1985, Chapter 10.

[6] P. S. Laplace, *Philosophical Essay on Probabilities* (1814); English translation by F. W. Prescott and F. L. Emory (1902), reprinted in *The Beginnings of Modern Science*, H. Boynton, Ed., Walter J. Black, Roslyn, NY, 1948.

[7] W. Thomson (Lord Kelvin), *Baltimore Lectures on Molecular Dynamics and the Wave Theory of Light*, C. J. Clay and Sons (Cambridge University Press), London, 1904. Reprinted as *Kelvin's Baltimore Lectures and Modern Theoretical Physics*, R. Kargon and P. Achinstein, Eds., MIT Press, Cambridge, MA, 1987.

[8] J. L. Casti, *Alternate Realities*, Wiley-Interscience, New York, 1989.

[9] R. Penrose, *The Emperor's New Mind*, Oxford University Press, New York, 1989, Chapter 5.

[10] S. J. Gould, *Wonderful Life: The Burgess Shale and the Nature of History*, Norton, New York, 1989, Chapter 1 and pp. 282–283.

[11] R. F. Fox, *Energy and the Evolution of Life*, Freeman, New York, 1988.

[12] L. Thorndike, *A History of Magic and Experimental Science*, Vol. II, Macmillan, New York, 1929, p. 29.

[13] J. Thurber, "Here Lies Miss Groby," *The Thurber Carnival*, Harper & Brothers, New York, 1945.

[14] A. Funkenstein, *Theology and the Scientific Imagination from the Middle Ages to the Seventeenth Century*, Princeton University Press, Princeton, NJ, 1986, p. 156.

[15] C. G. Gray and K. E. Gubbins, *Theory of Molecular Fluids*, Vol. 1, Oxford University Press, Oxford, 1984.

[16] D. Ceperley and J. Tully, Eds., "Stochastic Molecular Dynamics," *Proc. Nat. Resource Comp. Chem.*, **6**, xiii (1979).

[17] N. A. Metropolis, A. W. Rosenbluth, M. N. Rosenbluth, A. H. Teller, and E. Teller, "Equation of State Calculations by Fast Computing Machines," *J. Chem. Phys.*, **21**, 1087–1092 (1953).

[18] C. Pangali, M. Rao, and B. J. Berne, "On a Novel Monte Carlo Scheme for Simulating Water and Aqueous Solutions," *Chem. Phys. Lett.* **55**, 413–417 (1978).

[19] D. L. Beveridge, M. Mezei, P. K. Mehrotra, F. T. Marchese, G. Ravi-Shanker, T. Vasu, and S. Swaminathan, "Monte Carlo Computer Simulation Studies of the Equilibrium Properties and Structure of Liquid Water," *ACS Adv. Chem. Ser.*, **204**, 297–352 (1983).

[20] M. H. Kalos and P. A. Whitlock, *Monte Carlo Methods*, Vol. 1: *Basics*, Wiley, New York, 1986.

[21] W. W. Wood, "Monte Carlo Studies of Simple Liquid Models" in *Physics of Simple Liquids*, H. N. V. Temperley, J. S. Rowlinson, and G. S. Rushbrooke, Eds., North-Holland, Amsterdam, 1968, Chapter 5.

[22] J. P. Valleau and S. G. Whittington, "A Guide to Monte Carlo for Statistical Mechanics: 1. Highways" in *Statistical Mechanics*, Part A, B. J. Berne, Ed.,

Plenum, New York, 1977, Chapter 4; J. P. Valleau and G. M. Torrie, "A Guide to Monte Carlo Methods for Statistical Mechanics: 2. Byways," ibid., Chapter 5.

[23] K. Binder, Ed., *Monte Carlo Methods in Statistical Physics*, 2nd ed., Springer-Verlag, Berlin, 1986.

[24] K. Binder, Ed., *Applications of the Monte Carlo Method in Statistical Physics*, Springer-Verlag, Berlin, 1984.

[25] B. J. Alder and T. E. Wainwright, "Studies in Molecular Dynamics. I. General Method," *J. Chem. Phys.* **31**, 459–466 (1959).

[26] J. Ford, "The Transition from Analytic Dynamics to Statistical Mechanics," *Adv. Chem. Phys.*, **24**, 155–185 (1973).

[27] W. C. Swope and H. C. Andersen, "10^6-Particle Molecular-Dynamics Study of Homogeneous Nucleation of Crystals in a Supercooled Atomic Liquid," *Phys. Rev. B*, **41**, 7042 (1990).

[28] A. W. Lees and S. F. Edwards, "The Computer Study of Transport Processes under Extreme Conditions," *J. Phys. C*, **5**, 1921–1929 (1972).

[29] E. M. Gosling, I. R. McDonald, and K. Singer, "On the Calculation by Molecular Dynamics of the Shear Viscosity of a Simple Fluid," *Mol. Phys.*, **26**, 1475–1484 (1973).

[30] W. T. Ashurst and W. G. Hoover, "Argon Shear Viscosity via a Lennard-Jones Potential with Equilibrium and Nonequilibrium Molecular Dynamics," *Phys. Rev. Lett.*, **31**, 206–208 (1973).

[31] W. G. Hoover, "Nonequilibrium Molecular Dynamics," *Ann. Rev. Phys. Chem.*, **34**, 103–127 (1983).

[32] F. Vesely, *Computerexperimente an Flüssigkeitsmodellen*, Physik Verlag, Weinheim, 1978.

[33] W. G. Hoover, *Molecular Dynamics*, Lecture Notes in Physics, **258**, Springer-Verlag, Berlin, 1986.

[34] W. G. Hoover, *Computational Statistical Mechanics*, Elsevier, Amsterdam, 1991.

[35] D. J. Evans and G. Morriss, *Statistical Mechanics of Nonequilibrium Liquids*, Academic, New York, 1990.

[36] R. M. Hockney and J. W. Eastwood, *Computer Simulation Using Particles*, McGraw-Hill, New York, 1981.

[37] G. Ciccotti and W. G. Hoover, Eds., *Molecular-Dynamics Simulation of Statistical–Mechanical Systems*, North-Holland, Amsterdam, 1986.

[38] M. P. Allen and D. J. Tildesley, *Computer Simulation of Liquids*, Clarendon, Oxford, 1987.

[39] G. T. Evans, "Liquid State Dynamics of Alkane Chains," *ACS Adv. Chem. Ser.*, **204**, 423 (1983).

[40] G. A. Bekey, "Simulability of Dynamical Systems," *Trans. Soc. Comp. Simul.*, **2**, 57–71 (1985).

[41] K. S. Shing and K. E. Gubbins, "The Chemical Potential in Non-Ideal Mixtures. Computer Simulation and Theory," *Mol. Phys.*, **49**, 1121 (1983).

[42] Y. Hu, D. Lüdecke, and J. Prausnitz, "Molecular Thermodynamics of Fluid

Mixtures Containing Molecules Differing in Size and Potential Energy," *Fluid Phase Equilibria*, **17**, 217 (1984).

[43] W. L. Kubic, Jr., "A Quartic Hard Chain Equation of State for Normal Fluids," *Fluid Phase Equilibria*, **31**, 35 (1986).

[44] M. L. Huber and J. F. Ely, "Improved Conformal Solution Theory for Mixtures with Large Size Ratios," *Fluid Phase Equilibria*, **37**, 105 (1987).

[45] E. Z. Hamad and G. A. Mansoori, "Variational Theory of Mixtures," *Fluid Phase Equilibria*, **37**, 255 (1987).

[46] E. C. Meyer, "A One-Fluid Mixing Rule for Hard Sphere Mixtures," *Fluid Phase Equilibria*, **41**, 19 (1988).

[47] B. Smit and K. R. Cox, "A New Approach for Calculating the Accessible Volume in Equations of State for Mixtures. II. Application to Lennard-Jones Mixtures," *Fluid Phase Equilibria*, **43**, 181 (1988).

[48] K. S. Shing, K. E. Gubbins, and K. Lucas, "Henry Constants in Non-Ideal Fluid Mixtures. Computer Simulation and Theory," *Mol. Phys.*, **65**, 1235 (1988).

EXERCISES

1.1 In later chapters we will have intercourse with one-dimensional peri-
☒ odic systems; an example is the board game Monopoly. The Monopoly universe is a closed line divided into 40 spaces. Play starts from the space labeled Go and moves are clockwise, with the length of each move determined by the throw of two dice. Moves also result from instructions provided by Community Chest and Chance cards.

(a) The strategy used in a game might be enhanced if we knew the spaces most likely to be visited. Write a program that simulates the game and identifies the 10 spaces most likely visited. One player should be sufficient. The program can be developed in stages: (i) first generate moves only by throwing dice, then (ii) add moves prompted by the Community Chest cards, then (iii) add moves prompted by the Chance cards. The random-number generator in Appendix H can be used to simulate throwing dice and shuffling cards.

(b) Does your program simulate Monopoly or does it simulate a model of Monopoly? Justify your answer.

(c) Repeat the calculation in (a), but use only one die; then repeat using three dice.

(d) To determine whether your answer in (a) is affected by initial conditions, repeat the calculation using various starting points other than the Go space.

(e) To determine whether your answer in (a) is invariant under time reversal, repeat the calculation performing moves counterclockwise rather than clockwise.

Of the 40 spaces, the following are either destinations or points of departure for special moves:

Number	Label	Number	Label
2	Community Chest	24	Illinois Ave.
5	Reading RR	25	B&O RR
7	Chance	28	Water Works
10	Jail	30	Go to Jail
11	St. Charles Place	33	Community Chest
12	Electric Co.	35	Short Line RR
15	Pennsylvania RR	36	Chance
17	Community Chest	39	Boardwalk
22	Chance	40	Go

There are 16 Community Chest cards and 16 Chance cards.
Community Chest: Two of these 16 cards direct moves

One Go to Jail (space 10)
One Advance to Go (40)

Chance: Ten of these 16 cards direct moves

One Go to Reading RR (space 5)
One Go to Jail (10)
One Go to St. Charles Place (11)
One Go to Illinois Ave. (24)
One Go to Boardwalk (39)
One Advance to Go (40)
One Go back three spaces
One Advance to nearest utility (12 or 28)
Two Advance to nearest RR (5 or 10 or 15 or 20)

In the last two, note that the instruction is *Advance to nearest* (i.e., move ahead to), not merely Go to.

1.2 The objective of modeling is to decouple interactions; therefore, part of the success of a model hinges on the choice of variables for representing state space, so the desired decoupling is realized. As an example, consider the gravitational two-body problem: masses m_1 and m_2 attract one another with the force model

$$F = -\frac{gm_1m_2}{r^2}$$

where $r = |\mathbf{r}_1 - \mathbf{r}_2|$ is the distance between the centers of mass and g is Newton's gravitational constant. The motion of the masses is restricted to a plane that is defined by the initial positions $\{\mathbf{r}_1(0)$ and $\mathbf{r}_2(0)\}$ and initial velocities $\{\mathbf{v}_1(0)$ and $\mathbf{v}_2(0)\}$. For this problem, write the two components of Newton's second law in Cartesian coordinates and in polar coordinates. Which representation achieves the necessary decoupling that makes possible the analytic determination of the motion? Why is the gravitational three-body problem unsolved analytically?

1.3 An ideal gas is composed of molecules that are mathematical points and have no intermolecular forces; thus, the potential energy of such a gas is zero, $\mathscr{U}(\mathbf{r}^N) = 0$. The ideal-gas model is the ultimate model for matter because it decouples each molecule from every other molecule in the system. Explain this statement. Evaluate the NVT configurational integral (1.6) for the ideal gas. Then use the statistical mechanical equation of state

$$P = kT\left(\frac{\partial \ln \mathscr{Z}}{\partial V}\right)_{NT}$$

to derive the ideal-gas law. Here T is the absolute temperature, k is Boltzmann's constant, V is the system volume, and N is the number of atoms present.

1.4 Some reductionists (and some simulators) claim that the choice between two models is based on which model best represents experimental data. Let us test this claim. Here are experimental data for an output observable y measured for certain values of input observable x:

x	1	1.75	2	4	5	6
y	10.1	9.37	10.6	9.34	9.67	9.8

These data have been carefully, but not perfectly, measured—each y-value contains some error. Two models have been proposed for these data:

Model A: $y = 59.819 - 107.14x + 80.297x^2 - 26.683x^3$

$$+ 4.0391x^4 - 0.2275x^5$$

Model B: $y = 10.08 - 0.08x$

For model A, the sum of squares of deviations of the model from the measured y-values is $\sum \Delta^2 = 0$, while for model B it is $\sum \Delta^2 = 1$. That is, model A essentially passes through the measured y-values.

(a) Which of these models would you choose as more representative of how observable y changes with x? Defend your choice with arguments as strong as you can devise.

(b) Plot the given data and on the same graph also plot the curves given by models A and B. Plot the model curves from $x = 0.5$ to $x = 6.5$ at intervals of $\Delta x = 0.25$. Now which model would you choose as more representative of how observable y changes with x?

1.5 We noted in Section 1.2 that in creating model subsystems of a system, we decouple selected interactions and therefore we lose access to certain states of the original system. Rosen points out that this decoupling means that we relax selected constraints and therefore the model may have access to states that are not accessible to the original system (R. Rosen, *Fundamentals of Measurement and Representation of Natural Systems*, North-Holland, New York, 1978, p. 118). Thus, in replacing a system with a model, not only do we *lose* access to some states of the original system, but we also *gain* access to other states not available to the original system.

At the normal boiling point of water, 100°C and 1.0133 bars, steam tables give $v = 1.044$ and $v = 1673$ cm^3/g for the saturated liquid and vapor volumes, respectively. At this pressure and temperature, compute the volume given by the ideal-gas law. Schematically sketch the 100°C isotherm for water on a $P-V$ diagram from 0.2 to 5 bars. Include on your sketch the same isotherm predicted by the ideal-gas law.

Discuss how your sketch illustrates that the ideal-gas model has access to some thermodynamic states not available to water and conversely water has access to some states not available to the ideal-gas model.

1.6 Some simulators (and some reductionists) test the validity of models by curve fitting the model's functional from to available experimental data. If the fit is acceptable, then so is the model; that is, it is presumed that the fitted model establishes a connection between the output and input observables. Let us test this presumption.

In studying PVT behavior of methane, a modeler notices that at 100 bars, 500 K, the volume of methane is reliably estimated by the ideal-gas law. He therefore presumes that methane is an ideal gas at 100 bars and uses the ideal-gas law to estimate methane volumes on this isobar from 100 to 900 K. Repeat his calculations and also compute the percentage of errors in the calculated volumes relative to the following experimental data for methane

T(K)	100	300	500	700	900
v(cm^3/g)	2.24	13.3	26.1	37.4	53.4

(These data are from N. B. Vargaftik, *Tables on the Thermophysical Properties of Liquids and Gases*, 2nd ed., Halsted Press (Wiley), New York, 1975.) Why is $T = 500$ K a special state in this situation? What mistakes has our modeler made in extending the behavior at this special state to other, more general, states? In testing models by comparing with experiment, what procedures could you employ to guard against drawing faulty conclusions based on special cases?

1.7   Consider a system S that responds to input observable f by producing output observable g:

$$f \rightarrow \boxed{\text{system } S} \rightarrow g$$

The reductionist program for studying S, simply stated, is divide and conquer. For example, S might first be replaced by two subsystem models, A and B:

$$f \rightarrow \boxed{\text{model } A} \rightarrow k \qquad \text{and} \qquad l \rightarrow \boxed{\text{model } B} \rightarrow g$$

The next step would be to combine these two models and use the composite as a model of the original system S. This reductionist program will only be successful if observables k and l are equivalent; $k \Leftrightarrow l$. Since k depends on states of A and on the input f, while l depends on states of B and output g, the required equivalence implicitly involves both models as well as the input and output of system S,

$$k(f, A) \Leftrightarrow l(g, B)$$

Thus, if the composite is to model the original system S, we cannot in general create models A and B independently of one another. Rosen formally proves that composite models do not necessarily model composite systems (R. Rosen, *Fundamentals of Measurement and Representation of Natural Systems*, North-Holland, New York, 1978, p. 190).

In engineering thermodynamics, one strategy for reducing the amount of experimentation needed to study substances is to use one substance as a model for another. For example, can we use known vapor pressures of argon to estimate vapor pressures of methane? The real system here is methane, and since vapor pressure depends only on temperature, we have the symbolic transformation

$$T \rightarrow \boxed{\text{methane}} \rightarrow P$$

The question is whether we can replace this system with models that involve known information about argon,

$$T \to \boxed{\begin{array}{c}\text{model } A \\ \text{for argon}\end{array}} \to P^{\text{argon}} \quad \text{and} \quad ? \to \boxed{\begin{array}{c}\text{model } B \\ \text{for methane}\end{array}} \to P^{\text{methane}}$$

For some pairs of substances, this can in fact be done: The procedure involves an appropriate scaling of the state space, aka *corresponding states* (CS). The scaling is usually done using critical properties, T_c and P_c, though other choices are possible. The modeling is then of the form

$$T \to \boxed{\begin{array}{c}\text{CS model} \\ \text{argon}\end{array}} \to P_R^{\text{argon}} \Leftrightarrow P_R^{\text{methane}} \to \boxed{\begin{array}{c}\text{CS model} \\ \text{methane}\end{array}} \to P^{\text{methane}}$$

where $T_R = T/T_c$ and $P_R = P/P_c$. This method succeeds when the equivalence

$$P_R^{\text{substance 1}}(T_R) \Leftrightarrow P_R^{\text{substance 2}}(T_R)$$

is obeyed; otherwise, the approach fails. The vapor pressure of argon (for 84 K $< T <$ 140 K) obeys

$$\ln P \text{ (bar)} = 9.379 - \frac{849.6}{T \text{ (K)} + 3.437}$$

Critical properties of three substances are

	T_c (K)	P_c (bar)
Argon	150.8	48.7
Methane	190.6	46
Water	647.3	220.5

(a) Use corresponding states with the above expression for the vapor pressure of argon to estimate the vapor pressure of methane at 150 K. Compare your estimate with the experimental value of 10.33 bars.

(b) Use corresponding states with the above expression for the vapor pressure of argon to estimate the vapor pressure of water at 100°C. Compare your estimate with the known experimental value.

Background to Exercises 1.8 and 1.9

The modeling strategy in Exercise 1.7 can be varied by identifying aspects of the system behavior that are independent of one another. These aspects can be decoupled by using separate models A and B, as in Figure E1.8. The composite behavior is then taken to be some algebraic combination of the model outputs, g_A and g_B; often the combining form is simply a superposition. To be successful, there can be no interaction between the real behaviors that are decoupled by the modeling. The fundamental problem in this strategy is to find a representation of the state space of S that allows the decoupling.

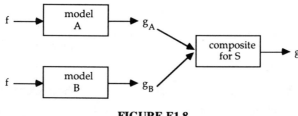

FIGURE E1.8

1.8 A signal $g(\theta)$ from an electronic device is found to have the following values:

θ	0	$\frac{1}{2}\pi$	$\frac{3}{4}\pi$	π	$\frac{3}{2}\pi$
g	1.754	1.461	-0.207	-1.754	-1.461

It is proposed to model this signal as the superposition of two sine waves,

$$g(\theta) = A_1 \sin(\theta + \omega_1) + A_2 \sin(\theta + \omega_2)$$

(a) How would you test whether the proposed composite for g models the measured data?

(b) How would you test whether the proposed composite models the full signal g, not just the measured values?

(c) Devise at least one consistency test that the measured data must satisfy if the proposed composite model applies. Your test should be applicable independent of the values for the parameters A_1, A_2, ω_1, and ω_2. Do the data pass your test? Discuss why your test is only a necessary condition for the validity of the proposed model and is not a sufficient condition.

1.9 (*See background statement immediately prior to Exercise 1.8*) The holy grail of chemical engineering thermodynamics is the ability to calculate thermodynamic properties of fluid mixtures from knowledge of just pure component properties. The ultimate model in this quest is the *ideal solution*, in which each component in a mixture contributes to a property independently of all other components.

For example, at low pressures, the pressure exerted by a vapor mixture that is in equilibrium with a binary ideal (liquid) solution is obtained by a scheme represented schematically in Figure E1.9.1, in which

$$P_{is} = x_A P_A + x_B P_B$$

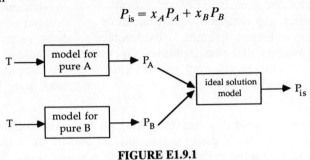

FIGURE E1.9.1

and P_i is the vapor pressure of pure component i at the temperature of the mixture, while x_i is the mole fraction of i in the liquid (not the vapor) mixture.

(a) For some mixtures the ideal-solution model reliably approximates reality; examples are mixtures of benzene–toluene and of ethane–propane. However, most mixtures are not ideal solutions. Explain.

(b) Although the ideal solution fails to model most fluid mixtures, the modeling strategy is appealing, so the chemical engineering community preserves the basic modeling process by trying to decouple the ideal-solution behavior from the rest of the behavior of a mixture. Schematically, the process now looks as shown in Figure E1.9.2.

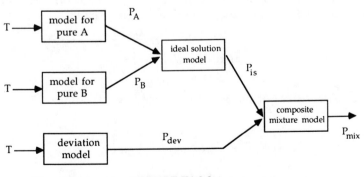

FIGURE E1.9.2

Now let T and P each represent arbitrary thermodynamic observables, not necessarily temperature and pressure. The new quantity here is P_{dev}, which measures deviation from ideal-solution behavior, and the composite model is simply a superposition

$$P_{mix} = P_{is} + P_{dev}$$

The deviation quantity will depend on thermodynamic state variables, like temperature and pressure and also, perhaps implicitly, on molecular quantities, like molecular shape factors, multiple moments, and polarizabilities. Discuss why, in spite of this elaborate strategy, the grail remains elusive, that is, why the deviation function P_{dev} cannot be obtained solely from pure component data.

1.10 (a) In a system of N atoms interacting via a pairwise additive potential (1.3), how many different pair interactions are there if no interaction is counted twice (i.e., $u_{12} = u_{21}$ is counted as one interaction)?

(b) Consider four identical balls lying in a plane. At one instant their position vectors have components $r_1 = (0,0)$, $r_2 = (1.5, 0)$, $r_3 = (0, 1.5)$, and $r_4 = (1.5, 1.5)$. Each ball is connected by springs to the others. The potential energy between any two balls, i and j, obeys

$$u_{ij} = \tfrac{1}{2}\gamma(r_{ij} - 1.)^2$$

where γ is a constant and $r_{ij} = |r_i - r_j|$ is the distance between i and j (in units of the ball diameter). Compute the total potential energy U/γ.

1.11 Devise and implement on a computer a Monte Carlo–like scheme for ⊠ obtaining the numerical value of π. Do runs of increasing length to study how run length affects the number of significant digits in your answer.

1.12 Determine the average depth $\langle z \rangle$ of a pond if the local depth $z(x, y)$ ⊠ at point (x, y) on the water's surface is given by

$$z(x, y) = 20\sqrt{1 - \left(\frac{x}{300}\right)^2 - \left(\frac{y}{50}\right)^2} \qquad \text{(E1.12.1)}$$

The surface of the water is bounded by

$$\left(\frac{x}{300}\right)^2 + \left(\frac{y}{50}\right)^2 = 1 \qquad \text{(E1.12.2)}$$

Distances here are all measured in feet. Use a Monte Carlo–like procedure: imagine standing on a bank and randomly throwing a lead line into the pond. After M throws, an estimate of the average depth would computed by

$$\langle z \rangle = \frac{1}{M} \sum_i^M z(x_i, y_i) \qquad \text{(E1.12.3)}$$

If the lead line falls onto a bank rather than into the water, then, of course, that throw is not used in computing the average. Does the sequence of samples obtained from this procedure form a Markov chain? Simulate this procedure by using a random-number generator to throw random values of the position (x, y) on $\{-300 < x < 300\}$ and $\{-50 < y < 50\}$, checking whether the throws fall on the surface defined by (E1.12.2) and, if they do, accumulating the average (E1.12.3). Use the random-number generator in Appendix H and perform the simulation. Study how the average $\langle z \rangle$ and its standard deviation vary with the number of samples used. Try, say, $10 < M < 10{,}000$. (Exercises 1.12 and 1.13 are inspired by a figure of D. Frenkel's in *Molecular-Dynamics Simulation of Statistical-Mechanical Systems*, G. Ciccotti and W. G. Hoover, Eds., North-Holland, Amsterdam, 1986.)

1.13 Use a molecular dynamics–like procedure for estimating the average depth of the pond in Exercise (1.12). Imagine rowing a boat over the surface in a systematic way and at regular time intervals Δt using a lead to sample the local depth $z(x, y)$. The average depth over our path would be computed as an integral over time.

$$\langle z \rangle = \frac{1}{t} \int_0^t z[x(\tau), y(\tau)] \, d\tau \approx \frac{1}{M} \sum_k^M z[x(k\,\Delta t), y(k\,\Delta t)]$$

where $t = M\Delta t$ is the total sampling duration and M is the number of samples.

(a) To simulate this process, consider a set of N concentric rings, creating annuli of thickness Δr on the pond's surface. Image rowing your boat around each annulus, sampling the depth at the same interval of arc length in each annulus. The time-average depth can be written as

$$\langle z \rangle = \frac{1}{t} \sum_{i=1}^N \int_0^{t_i} z[x(\tau), y(\tau)] \, d\tau$$

where t_i is the time spent in annulus i and $t = \sum t_i$ is the total sampling time. Starting with four annuli, a time step of $\Delta t = 1$ min, and a constant speed along an arc $ds/dt = 20$ ft of arc per minute, evaluate the average depth. Study how the average is affected by changing the number of annuli, the time step Δt, and the speed ds/dt. Compare your estimate for $\langle z \rangle$ with that obtained in Exercise 1.12.

(b) Repeat the calculation in (a) using the sinusoidal path

$$y(t) = 50\sqrt{1 - \left(\frac{x}{300}\right)^2} \sin\left(\frac{\pi t}{10}\right)$$

with $x(t) = x(0) + v_x t$ and v_x constant. Use $x(0) = -300$, $y(0) = 0$, $v_x = 10$ ft/min, and $\Delta t = 1$ min to compute the average depth along this path. Study how the average is affected by the number of samples M and the frequency of the sinusoidal path. Compare your estimate for $\langle z \rangle$ with those obtained in (a) and in Exercise 1.12 and explain any disagreement.

1.14 The steps at the heart of the Metropolis Monte Carlo method are described in Section 1.6.1; in particular, a proposed move is either accepted or rejected according to the value of $\Delta \mathcal{U} = \mathcal{U}' - \mathcal{U}$. Using the pairwise additive potential (1.3), relate the new energy $\mathcal{U}'$ to the old energy $\mathcal{U}$ and show that $\Delta \mathcal{U}$ can be computed without resampling all possible pair interactions.

1.15 In a typical Monte Carlo simulation a parameter d represents the maximum relocation of any atom k due to a proposed move from position $\mathbf{r}_k$ to $\mathbf{r}'_k$; that is

$$\mathbf{r}'_k = \mathbf{r}_k + \boldsymbol{\xi} d$$

where $\boldsymbol{\xi}$ is a unit vector of random components. Typically the value of d is adjusted throughout a Monte Carlo run, so that a prechosen fraction of proposed moves are accepted (say, 50% of them). How is the quality of the run degraded if the value selected for d is too large? Too small?

1.16 A system is composed of N identical atoms that interact through pairwise additive forces. At any instant the center of mass is located by

$$\mathbf{r}_c(t) = \left(\frac{1}{N}\sum_i^N x_i(t)\right)\hat{\mathbf{i}} + \left(\frac{1}{N}\sum_i^N y_i(t)\right)\hat{\mathbf{j}} + \left(\frac{1}{N}\sum_i^N z_i(t)\right)\hat{\mathbf{k}}$$

No external forces act on the atoms and initially the center-of-mass velocity was zero. Determine the time-average position, velocity, and acceleration of the center of mass.

1.17 Consider a special case of the Langevin equation (1.9) in which only the macroscopic friction force is present: the intra- and intermolecular forces are zero.

  (a) Analytically integrate the resulting simplified Langevin equation once and obtain the time-dependent particle velocity in terms of its initial velocity. Then use that result to obtain the time-average particle velocity.

  (b) Analytically integrate a second time to obtain the time-dependent particle position in terms of its initial position. Then determine the particle's time-average position.

2

FUNDAMENTALS

For as few as 100 atoms in three dimensions, a molecular dynamics simulation produces 600 values of positions and momenta every time the equations of motion are integrated forward one step. The integration typically proceeds for thousands of steps. How can we mentally grapple with such a huge amount of data? One way is provided by the concept of a phase-space trajectory.

The phase-space trajectory serves as the common thread that binds diverse tools into what we call molecular dynamics. We compute the trajectory using classical mechanics (Sections 2.1 and 2.2) and we describe features of the trajectory in terms of classical nonlinear dynamics (Sections 2.3–2.5). To obtain properties, we analyze the trajectory by appealing to kinetic theory (Section 2.6), statistical mechanics (Section 2.7), and sampling theory (Section 2.8). We test the trajectory by probing constraints imposed by periodic boundary conditions (Section 2.9) and by conservation principles (Section 2.10). Taken together, these tools form the foundation of molecular dynamics simulation.

2.1 NEWTONIAN DYNAMICS

In the Newtonian interpretation of dynamics the translational motion of a spherical molecule i is caused by a force $\mathbf{F}_i$ exerted by some external agent. The motion and the applied force are explicitly related through Newton's second law,

$$\mathbf{F}_i = m\ddot{\mathbf{r}}_i \tag{2.1}$$

Here m is the mass of the molecule; it is assumed to be independent of position, velocity, and time. The acceleration is given by

$$\ddot{\mathbf{r}}_i = \frac{d^2\mathbf{r}_i}{dt^2} \tag{2.2}$$

where $\mathbf{r}_i$ is a vector that locates the molecule with respect to a laboratory-fixed set of coordinate axes, as in Figure 2.1. For N spherical molecules, Newton's second law (2.1) represents $3N$ second-order, ordinary differential equations of motion.

If no external force acts on molecule i, then the second law (2.1) reduces to

$$\dot{\mathbf{r}}_i = \text{const} \tag{2.3}$$

That is, a molecule initially at rest will remain at rest and a molecule moving with a specified velocity will continue to move with that velocity until a force acts on it. This is Newton's first law.

The second law (2.1) can also be used to obtain Newton's third law. Consider an isolated system that contains two spherical molecules, 1 and 2. By definition, an isolated system has no external forces; hence, the total force

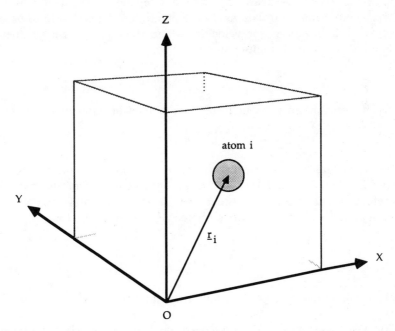

FIGURE 2.1 Cartesian, laboratory-fixed reference frame $\{X, Y, Z\}$ used to define a position vector $\mathbf{r}_i$ that locates atom i in a system.

is zero,

$$\mathbf{F}_{total} = 0 \tag{2.4}$$

Therefore, any force exerted by molecule 1 on molecule 2 must be balanced by a force exerted by 2 on 1,

$$\mathbf{F}_{total} = \mathbf{F}_1 + \mathbf{F}_2 = 0 \tag{2.5}$$

Hence,

$$\mathbf{F}_1 = -\mathbf{F}_2 \tag{2.6}$$

This is Newton's third law. With these ideas, we can define kinetic energy[†] as the work required to move a spherical molecule from rest to velocity $\dot{\mathbf{r}}$ (see Exercise 2.1)

$$E_k = \tfrac{1}{2}m\dot{\mathbf{r}}^2 \tag{2.7}$$

2.2 HAMILTONIAN DYNAMICS

Although molecular forces and positions change with time, the functional form of Newton's second law (2.1) is time independent: the form $\mathbf{F}_i = m\ddot{\mathbf{r}}_i$ is invariant under time translations. Consequently, we expect there to be some function of the positions and velocities whose value is constant in time; this function is called the Hamiltonian $\mathscr{H}$,

$$\mathscr{H}(\mathbf{r}^N, \mathbf{p}^N) = \text{const} \tag{2.8}$$

Here, the momentum $\mathbf{p}_i$ of molecule i is defined in terms of its velocity by

$$\mathbf{p}_i = m\dot{\mathbf{r}}_i \tag{2.9}$$

For an isolated system, we know that one conserved quantity is the total energy E: the combined kinetic and potential energies of the molecules. Therefore, for an isolated system, we identify the total energy as the Hamiltonian; then, for N spherical molecules, $\mathscr{H}$ takes the form

$$\mathscr{H}(\mathbf{r}^N, \mathbf{p}^N) = \frac{1}{2m} \sum_i \mathbf{p}_i^2 + \mathscr{U}(\mathbf{r}^N) = E \tag{2.10}$$

where the potential energy $\mathscr{U}$ results from intermolecular interactions.[‡] To

---

[†]W. Thomson, Lecture XVI, p. 160: "*Actual energy* was Rankine's word. The expression, kinetic energy, I am answerable for."

[‡]W. Thomson, Lecture XIII, p. 131: "The words of 'strain and stress' are due [to] Rankine; 'potential' energy also."

obtain equations of motion, first consider the total time derivative of the general Hamiltonian (2.8)

$$\frac{d\mathcal{H}}{dt} = \sum_i \frac{\partial\mathcal{H}}{\partial\mathbf{p}_i} \cdot \dot{\mathbf{p}}_i + \sum_i \frac{\partial\mathcal{H}}{\partial\mathbf{r}_i} \cdot \dot{\mathbf{r}}_i + \frac{\partial\mathcal{H}}{\partial t} \tag{2.11}$$

If, as is (2.10), $\mathcal{H}$ has no explicit time dependence, then the last term on the right-hand side (rhs) of (2.11) vanishes and we are left with

$$\frac{d\mathcal{H}}{dt} = \sum_i \frac{\partial\mathcal{H}}{\partial\mathbf{p}_i} \cdot \dot{\mathbf{p}}_i + \sum_i \frac{\partial\mathcal{H}}{\partial\mathbf{r}_i} \cdot \dot{\mathbf{r}}_i = 0 \tag{2.12}$$

Equation (2.12) is a general result. Now consider the total time derivative of the isolated-system Hamiltonian given in (2.10),

$$\frac{d\mathcal{H}}{dt} = \frac{1}{m} \sum_i \mathbf{p}_i \cdot \dot{\mathbf{p}}_i + \sum_i \frac{\partial\mathcal{U}}{\partial\mathbf{r}_i} \cdot \dot{\mathbf{r}}_i = 0 \tag{2.13}$$

On comparing (2.12) and (2.13), we find for each molecule i,

$$\boxed{\frac{\partial\mathcal{H}}{\partial\mathbf{p}_i} = \frac{\mathbf{p}_i}{m} = \dot{\mathbf{r}}_i} \tag{2.14}$$

and

$$\frac{\partial\mathcal{H}}{\partial\mathbf{r}_i} = \frac{\partial\mathcal{U}}{\partial\mathbf{r}_i} \tag{2.15}$$

Substituting (2.14) into (2.12) gives

$$\sum_i \dot{\mathbf{r}}_i \cdot \dot{\mathbf{p}}_i + \sum_i \frac{\partial\mathcal{H}}{\partial\mathbf{r}_i} \cdot \dot{\mathbf{r}}_i = 0 \tag{2.16}$$

$$\sum_i \left(\dot{\mathbf{p}}_i + \frac{\partial\mathcal{H}}{\partial\mathbf{r}_i} \right) \cdot \dot{\mathbf{r}}_i = 0 \tag{2.17}$$

Since the velocities are all independent of one another, (2.17) can be satisfied only if, for each molecule i, we have

$$\boxed{\frac{\partial\mathcal{H}}{\partial\mathbf{r}_i} = -\dot{\mathbf{p}}_i} \tag{2.18}$$

Equations (2.14) and (2.18) are Hamilton's equations of motion. For a system of N spherical molecules, (2.14) and (2.18) represent $6N$ first-order differential equations that are equivalent to Newton's $3N$ second-order equations (2.1). To demonstrate this, take the first time derivative of the definition of the momentum (2.9) and use it to eliminate $\dot{\mathbf{p}}_i$ from (2.18). The result is

$$\frac{\partial \mathscr{H}}{\partial \mathbf{r}_i} = -m\ddot{\mathbf{r}}_i \qquad (2.19)$$

Using (2.15) in (2.19) and then comparing with Newton's law (2.1) yields

$$\mathbf{F}_i = -\frac{\partial \mathscr{H}}{\partial \mathbf{r}_i} = -\frac{\partial \mathscr{U}}{\partial \mathbf{r}_i} \qquad (2.20)$$

But this is merely the usual expression for a conservative force (1.4); that is, any conservative (nondissipative) force can be written as the negative gradient of some potential function $\mathscr{U}$.

The above development is straightforward and emphasizes the difference between Newtonian and Hamiltonian dynamics. In the Newtonian view motion is a response to an applied force. However, in the Hamiltonian view forces do not appear explicitly; instead, motion occurs in such a way as to preserve the Hamiltonian function. Although the development given above is brief, it unnecessarily mixes these two distinct viewpoints. Such mixing can be avoided in the traditional development that starts from Lagrangian mechanics [1].

Here is a summary of the assumptions used above in obtaining Hamilton's equations of motion:

(a) An isolated system was used. If, instead, the system could exchange energy with its surroundings, then the Hamiltonian would contain additional terms to account for those interactions. In such cases $\mathscr{H}$ would no longer be the system's total energy; $\mathscr{H}$ would still be conserved, but E would not.

(b) The momentum and velocity were taken to be related by (2.9).

(c) The Hamiltonian was not allowed to contain any explicit time dependence. If it did, then $\mathscr{H}$ would not be a conserved quantity.

Hamilton's equations of motion (2.14) and (2.18) are valid regardless of whether or not these assumptions are satisfied; however, this is not apparent from the above development. Nevertheless, these assumptions are satisfied by the situations considered in this book, and so the above development is sufficient for our immediate needs.

2.3 PHASE-SPACE TRAJECTORIES

In Chapter 1 we stated that the first objective of a molecular dynamics simulation is to generate molecular trajectories over a finite time. We now extend the idea of a trajectory to include not only positions but also molecular momenta. Consider a collection of N spherical atoms that interact through a prescribed potential energy function. The center of each atom is located by a position vector $\mathbf{r}_i$, as in Figure 2.1. For an isolated system (one that exchanges no mass or energy with its surroundings), the atoms move in accordance with Newton's law, tracing out trajectories that can be represented by time-dependent position vectors $\mathbf{r}_i(t)$. As an atom follows its trajectory, its momentum also changes in response to interactions with other atoms; thus, we also have time-dependent momentum vectors $\mathbf{p}_i(t)$.

At one instant, imagine plotting the positions and momenta of the N atoms in a $6N$-dimensional hyperspace. Such a space, called *phase space*, is composed of two parts: a $3N$-dimensional *configuration space*, in which the coordinate axes are the components of the position vectors $\mathbf{r}_i(t)$, and a $3N$-dimensional *momentum space*, in which the coordinate axes are the components of the momentum vectors $\mathbf{p}_i(t)$. At one instant, the positions and momenta of the entire N-atom system are represented by one point in this space. As the positions and momenta change with time, the point moves, describing a trajectory in phase space. We may now restate the first objective of a molecular dynamics simulation: compute the phase-space trajectory. The trajectory is obtained by numerically solving Newton's second law (2.1) or, equivalently, Hamilton's equations (2.14) and (2.18).

A simple example of a phase-space trajectory is provided by the ODHO. Let the ODHO represent a mass m attached to a harmonic spring, as shown in Figure 2.2(a). The mass–spring system is isolated from its surroundings (an obvious idealization). When the spring is expanded or compressed, it resists the displacement with a stiffness measured by the constant γ. Moreover, when the spring is displaced to a position r, the mass is a distance $x = r - r_0$ from its equilibrium position r_0, and the system has potential energy $u(x)$ given by

$$u(x) = \tfrac{1}{2}\gamma x^2 \qquad (2.21)$$

In an isolated system Newton's laws apply, so we have the equations of motion

$$F(x) = \dot{p} = m\ddot{x} = -\frac{du(x)}{dx} \qquad (2.22)$$

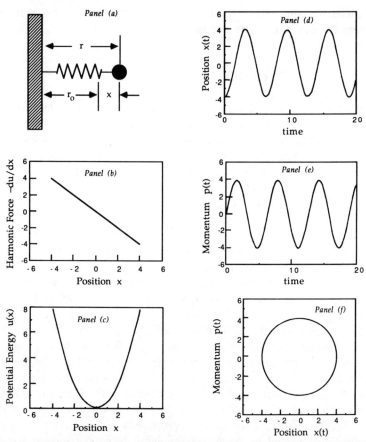

FIGURE 2.2   Motion of the one-dimensional harmonic oscillator. (*a*) Ball has mass $m = 1$ and the spring has stiffness $\gamma = 1$. The natural position of the spring places the ball at r_0 and a displacement from r_0 is measured by x. (*b*) Force exerted by the spring is linear in x; the slope of this line is γ. (*c*) Potential energy is quadratic in x and symmetric about $x = 0$. (*d*, *e*) Time-dependent position $x(t)$ and momentum $p(t)$ resulting from initial conditions $x(0) = -4$ and $p(0) = 0$. Both $x(t)$ and $p(t)$ are sinusoidal, but their oscillations are out of phase, making the total energy constant. (*f*) Phase-plane trajectory obtained from (*d*) and (*e*). In general the trajectory is an ellipse with semiaxes $(2mE)^{1/2}$ and $(2E/\gamma)^{1/2}$, but here it degenerates to a circle because $m = 1$ and $\gamma = 1$. (All units are arbitrary.)

Hence

$$m\ddot{x} = -\gamma x \tag{2.23}$$

Thus the force $F(x)$ acting on the mass is linear in its displacement; see Figure 2.2(*b*). From initial conditions $\{x(0), p(0)\}$, (2.23) can be solved by inspection to obtain the time-dependent position $x(t)$ and momentum $p(t)$.

However, to determine the phase-space trajectory for this simple case, we need not solve (2.23) explicitly. Since the system is isolated, its total energy E must be a constant of the motion,

$$E = E_k + \mathcal{U} = \text{const} \tag{2.24}$$

where E_k and $\mathcal{U}$ are, respectively, the kinetic and potential energies of the mass,[†]

$$E = \frac{1}{2m}p^2 + \tfrac{1}{2}\gamma x^2 \tag{2.25}$$

The phase space for this situation is two dimensional: one position coordinate x and one momentum coordinate p. In this plane the phase point, according to (2.25), traces out an ellipse that has semiaxes $(2mE)^{1/2}$ and $(2E/\gamma)^{1/2}$; see Figure 2.2(f). Note that the initial conditions $x(0)$ and $p(0)$ determine the constant value for the energy E; thus, for a given set of initial conditions, only a restricted region of phase space is accessible to an isolated system.

We can develop (2.25) more formally from the definition of momentum (2.9) and a combination of the two forms of the second law, (2.22) and (2.23). Thus, we have

$$\dot{x} = \frac{p}{m} \quad \text{and} \quad \dot{p} = -\gamma x \tag{2.26}$$

To obtain the phase-plane trajectory, we combine the two equations in (2.26) to eliminate time,

$$\frac{dx}{dp} = \frac{p}{m}\left(\frac{-1}{\gamma x}\right) \tag{2.27}$$

This equation can now be integrated by separating variables, yielding (2.25).

As a second example of phase space, consider N atoms in an isolated system forming an ideal gas. The phase space for this system is $6N$ dimensional. Since the system is isolated, the total energy E is again conserved, but since the system is an ideal gas, it has no intermolecular potential energy—the total energy is just the kinetic energy of the atoms,

$$E = \frac{1}{2m}\sum_{i=1}^{N} \mathbf{p}_i \cdot \mathbf{p}_i = \text{const} \tag{2.28}$$

[†]W. Thomson, Lecture VII, p. 65: "For the moment, take the expression for the simple harmonic motion, and you see at once that that comes out in terms of the energy. ... The energy, which at any time is partly kinetic and partly potential, will be all kinetic at the moment of passing through the middle position."

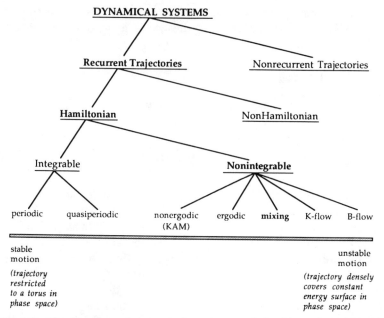

FIGURE 2.3   Classification of dynamical systems and the hierarchy of motion in phase space.

Thus, the phase point may wander anywhere in configuration space, but in momentum space it is restricted to the surface of a $3N$-dimensional hypersphere of radius $(2mE)^{1/2}$.

In both examples we see that external conditions imposed on a system (e.g., isolation) restrict the volume of phase space available to a phase point. We have already observed that such restrictions determine values for equilibrium macroscopic properties; therefore, it appears that restricted phase space is the mechanism by which externally imposed constraints are connected to particular values for macroscopic properties. A more general and nicely illustrated discussion of phase space is given by Penrose [2].

2.4   CLASSIFICATION OF DYNAMICAL SYSTEMS

Features of phase-space trajectories can be used to classify the motion of objects in dynamical systems [3, 4]. The classification takes a hierarchical form, as suggested by Figure 2.3. At the first level of the hierarchy, systems having *recurrent trajectories* are separated from those having *nonrecurrent trajectories*. Examples of nonrecurrent trajectories are provided by the motion of most comets: they pass only once through the solar system. Halley's comet is an exception, since it follows a recurrent trajectory, albeit a special one.

More generally, recurrent systems obey Poincaré's recurrence theorem [5]: for a phase space of finite volume, the phase point will pass, arbitrarily closely and indefinitely often, almost every accessible configuration on the trajectory. However, for systems of more than a few objects, the time required for a Poincaré recurrence is huge—it can exceed the age of the universe [4, 6] (see Exercise 2.18).

Depending on whether an explicit Hamiltonian governs motion, recurrent trajectories are either Hamiltonian or non-Hamiltonian. Examples of non-Hamiltonian situations include dissipative systems, which are those that involve friction in one of its various forms. Non-Hamiltonian systems can be studied by nonequilibrium molecular dynamics methods. In such cases, particle trajectories are generated from equations of motion; however, those equations are not consistent with any Hamiltonian function.

In Hamiltonian systems the trajectory is constrained in phase space to the hypersurface of constant $\mathcal{H}$. Hamiltonian systems can be further divided into *integrable* and *nonintegrable* systems. Integrable systems, such as the ODHO, are those for which the number of constants of the motion equals the number of degrees of freedom. In most situations the number of conserved quantities is small, the equations of motion contain nonlinear terms, and so systems containing even a very few objects are usually nonintegrable. For example, the classic three-body problem of astronomy cannot be solved analytically.

2.4.1 Integrable Systems

Integrable systems can be subdivided into those that have *periodic* motion and those that can have either periodic or *quasiperiodic* motion. Periodic motion means that within a finite time the phase trajectory will intersect itself. Because the motion of the phase point is completely determined by differential equations, once the phase point encounters its own trajectory, that trajectory is repeated indefinitely. All Hamiltonian systems having one degree of freedom are integrable and periodic.

Quasiperiodic motion results from integrable Hamiltonian systems that are not periodic. In such systems the phase-space trajectory is confined to the surface of a torus (a doughnut-shaped object in the case of three dimensions). This torus is a subspace of the constant-Hamiltonian surface.

We illustrate quasiperiodic motion using a ball connected to two harmonic springs that are at right angles to one another, as in Figure 2.4(a). The spring constants are γ_x and γ_y, but the motion of one spring does not affect the motion of the other: the springs are not coupled. Such a system has two degrees of freedom[†] (x and y), and its phase space is four dimensional (x, y, p_x, p_y). Because the springs are uncoupled, the energy of each is

---

[†]W. Thomson, Lecture XVIII, p. 178: "...we have motion in two dimensions alone, and our formulae belong therefore to the general formulae with that limitation to two dimensions..."

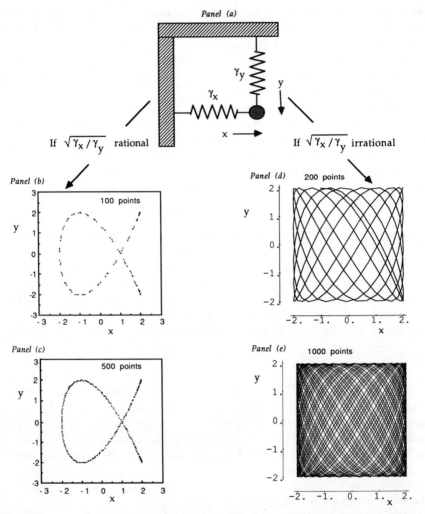

FIGURE 2.4 (*a*) Two kinds of motion may be generated by a two-dimensional harmonic oscillator. The trajectories here are for $m = 1$, $x(0) = y(0) = 2$, and $p_x(0) = p_y(0) = 0$. (*b*, *c*) $\gamma_x / \gamma_y = \frac{4}{9}$, and the motion is periodic so the mass continually retraces a restricted path. (*d*, *e*) $\gamma_x / \gamma_y = \frac{4}{10}$, the motion is quasiperiodic, and the mass covers an area. Units are arbitrary.

constant with time,

$$E_x = \frac{1}{2m}p_x^2 + \tfrac{1}{2}\gamma_x x^2 = c_1 \tag{2.29}$$

$$E_y = \frac{1}{2m}p_y^2 + \tfrac{1}{2}\gamma_y y^2 = c_2 \tag{2.30}$$

These two equations define the two-dimensional surface of a torus in the four-dimensional phase space. Since the system is isolated, the Hamiltonian is the total energy

$$E = E_x + E_y = \text{const} \tag{2.31}$$

which defines a three-dimensional surface.

Because the number of constants of the motion (E_x and E_y) equals the number of degrees of freedom, the system is integrable. Further, if the spring constants make the frequency ratio $\omega_x / \omega_y = \sqrt{\gamma_x / \gamma_y}$ a rational number, then the motion is periodic. In this case the ball repeatedly moves back and forth on a path in the real xy-plane; see Figures 2.4(b) and (c). In phase space the trajectory lies on the surface of the torus defined by (2.29) and (2.30).

However, if the frequency ratio $\sqrt{\gamma_x / \gamma_y}$ is irrational, then the motion is quasiperiodic. In this case the phase-space trajectory winds around the two-dimensional surface of the torus; the trajectory densely covers that surface but it never intersects itself. In the xy-plane, the ball is not confined to a line; rather, as time evolves, the ball covers an area, as shown in Figures 2.4(d) and (e).[†]

We emphasize that although the trajectory of a quasiperiodic system densely covers the surface of a torus, that torus does not constitute the full constant-energy surface. In our example, the phase space is four dimensional, the constant-energy surface is three dimensional, and the toroidal surface is two dimensional. For a given set of initial conditions, many sets of positions (x, y) and momenta (p_x, p_y) lie on the constant-energy surface but do not lie on the surface of the torus. Hence, trajectories from quasiperiodic systems cannot be made compatible with calculations from statistical mechanics because statistical mechanics assumes, for an isolated system, an equal a priori probability for sampling *any* point on the constant-energy surface.

2.4.2 Nonintegrable Systems

We identify five subdivisions of nonintegrable systems: nonergodic, ergodic, mixing, K-flows, and B-flows. The distinction among these five is based primarily on the idea of stability, on how a trajectory responds to a small perturbation, and is illustrated in Figure 2.5. *Nonergodic* motion (aka KAM motion) is exhibited by systems that are "slightly" nonintegrable: they obey Hamiltonians of the form

$$\mathscr{H} = \mathscr{H}_o + \mathscr{H}_p \tag{2.32}$$

[†]The motions of two-dimensional harmonic oscillators were extensively studied by J. A. Lissajous (1822–1880) and figures, such as Figure 2.4(e), generated by their quasiperiodic trajectories bear his name.

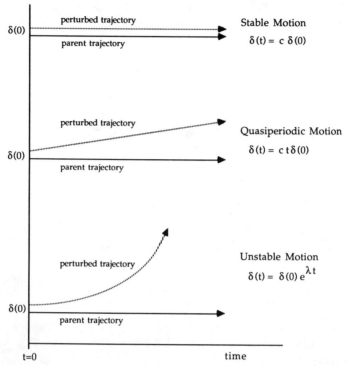

FIGURE 2.5 Sensitivity to initial conditions can be used to identify the stability of a trajectory [13]. Here, the solid lines, parallel to the time axis, represent three parent (unperturbed) trajectories. To each of these a displacement (perturbation) $d(0)$ is applied at time zero; the degree of instability is determined by the rate at which the perturbed trajectory diverges from its parent. Thus, along a stable trajectory the displacement is constant (*top*), along a quasiperiodic trajectory the displacement grows slowly (*middle*), but along an unstable trajectory the displacement grows exponentially (*bottom*). In quasiperiodic motion the growth of the displacement is not necessarily linear, as shown here, but whatever the form of its growth, the rate is slow compared to the exponential growth of unstable motion.

where $\mathscr{H}_o$ is an integrable Hamiltonian and $\mathscr{H}_p$ is a small perturbation. In our example of the two-dimensional harmonic oscillator (Figure 2.4), a perturbation could be added in the guise of a small nonlinear coupling between the motion in the x-direction and that in y. For such systems, Kolmogorov [7] stated and Arnol'd [8] and Moser [9] proved that so long as the nonintegrable perturbation is small, the phase-space trajectory is still confined to a torus, except for a small set of special trajectories. These tori may be slightly distorted from the integrable ($\mathscr{H}_p = 0$) shapes, but nevertheless the motion is essentially periodic or quasiperiodic. A consequence of KAM motion is that in a system of multiple harmonic oscillators, the introduction of small anharmonic perturbations does not make the motion ergodic. This nonergodicity of coupled oscillators was discovered independently of KAM in the first molecular dynamics simulation, performed by Fermi, Pasta, and Ulam in 1955 [10]. For a nontechnical description of that early simulation, see Ulam [11].

From nonergodic motion, we move to *ergodic* motion. Ergodicity has had numerous meanings ascribed to it; the original meaning, usually attributed to Boltzmann (but see Brush [12], pp. 364 and 369), is that over a sufficiently long time the phase point passes through all configurations on the constant-Hamiltonian surface. A related concept is *quasi-ergodic* motion, which means that over a sufficiently long time the phase point passes arbitrarily close to all configurations. Under these original definitions, real dynamical systems in three dimensions are neither ergodic nor quasi-ergodic. The proof of nonergodicity involves mathematical tools from set theory and measure theory; a readable discussion is given by Brush [12]. Suffice it here to make the hand-waving argument that for the trajectory to actually pass through every available point (or close to nearly every point) on a constant-Hamiltonian surface, the trajectory must eventually cross itself [3, 12]. At the point of intersection the motion necessarily becomes periodic.

In light of this, modern writers in statistical mechanics often retrench and take ergodicity to mean that the time average (1.8) of a function over the trajectory is equal to an ensemble average (1.5), such as is computed in a Monte Carlo simulation. This is sometimes called the *ergodic hypothesis* and is the meaning of ergodic that pertains in Figure 2.3. However, this definition of ergodicity does not teach us anything about the phase-space trajectory; in particular, it says nothing about the stability of trajectories. An ergodic trajectory could be stable: a perturbed ergodic trajectory could remain close to and correlated with its unperturbed parent trajectory. Thus, ergodicity does not guarantee that an isolated system started from an arbitrary nonequilibrium state will irreversibly evolve to equilibrium states.

For a system to evolve from nonequilibrium states to equilibrium states, the phase-space trajectory must be unstable to small perturbations: a perturbed trajectory must drift away from its unperturbed parent trajectory, see Figure 2.5. The rate of divergence determines the degree of the instability; the lowest level of unstable motion is called *mixing* motion. In mixing the rate of divergence may be fast or slow. At the next level is the *K-flow* (after Kolmogorov): in response to a small perturbation, a K-flow trajectory diverges exponentially from its unperturbed parent. Some authors refer to such trajectories as Lyapunov unstable [14]. The highest level of unstable trajectories is the *B-flow* (Bernoulli).[†] In a B-flow the state observed at an instant on

[†]Inspired by reports from his father, Johann (1667–1748), Daniel Bernoulli (1700–1782) conducted extensive studies of the dynamics of one-dimensional systems composed of N connected masses. Theirs was the first systematic study of the dynamics of N particles rather than systems of only two or three.

a perturbed trajectory is completely uncorrelated with the state on the parent trajectory observed at the same time. A B-flow trajectory appears to have been generated at random, despite the underlying determinism of the equations of motion.

Now, for a system to evolve from nonequilibrium to equilibrium states, two conditions must be satisfied. First, the full constant-Hamiltonian surface must be accessible. But this is not sufficient: just because the full surface is accessible does not guarantee that our trajectory will fully cover that surface. Therefore, the second condition is that the motion must be at least mixing so that the trajectory actually samples, eventually, all portions of the constant-Hamiltonian surface. Systems that do not satisfy both conditions are grouped under the label *nonmixing*. Thus, quasiperiodic motion is nonmixing because, although a perturbed trajectory diverges (slowly) from its parent, quasiperiodic motion is always restricted to a subspace of the constant-$\mathcal{H}$ surface.

Even if trajectories are neither periodic nor quasiperiodic, we are not assured that a computed trajectory will sample a representative portion of the constant-$\mathcal{H}$ surface during the finite time of a simulation: the trajectory might become trapped in a region of the surface. Such situations are said to be metastable. To illustrate, we consider a one-dimensional oscillator that obeys the potential function

$$u(x) = \begin{cases} \frac{1}{2}\gamma_1 x^2 & x < x_1 \\ \frac{3}{4}\left(\frac{1}{2}\gamma_1 x_1^2\right) + \frac{1}{2}\gamma_2(x - x_2)^2 & x \geq x_1 \end{cases} \tag{2.33}$$

The form of this function is shown in Figure 2.6(a). The possibly troublesome feature of this potential is the secondary minimum at x_2, which is separated from the primary minimum by a barrier at x_1. For most choices of initial conditions $[x(0), p(0)]$, the total energy exceeds that at the barrier, the phase point samples both potential wells, and the trajectory appears as in Figure 2.6(e).

However, for certain sets of initial conditions, the trajectory is confined to the region around one minimum. Consider, for example, the situation having parameter values $m = 1$, $\gamma_1 = 1$, $\gamma_2 = 5$, $x_1 = 2$, and $x_2 = 2.5$ (all in arbitrary units). For this case, any set of initial positions and momenta from the region in Figure 2.6(b) confines the trajectory to the region of the secondary minimum; the resulting phase-space trajectories are like that in Figure 2.6(c). In contrast, other sets of initial conditions will confine the trajectory to the region around the primary minimum, with a trajectory as in Figure 2.6(d). Note that the trajectories in 2.6(c) and (d) both have the same total energy; they are both portions of the same constant-energy surface. However, the trajectory in 2.6(c) provides averages for properties that differ from those computed over the trajectory in 2.6(d). This simple example suggests that, for some systems, there may be critical energies at which the trajectory undergoes a transition from one level of unstable motion to another level.

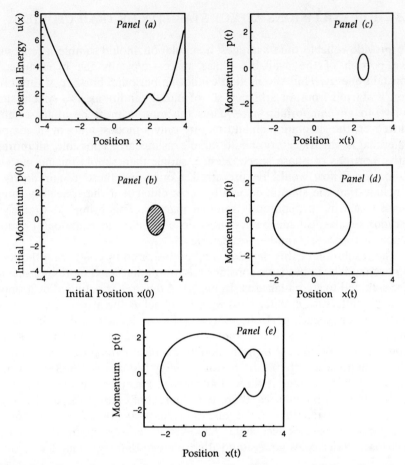

FIGURE 2.6 Example of a phase point trapped in phase space. (*a*) One-dimensional potential having a secondary minimum at $x = 2.5$. (*b*) Range of initial positions and momenta that causes the motion to be trapped in the secondary minimum: the corresponding phase plane trajectory is confined to a portion of the available space, as in panel (*c*). (*c*) Obtained from $p(0) = 0$, $x(0) = 2.8$ and has $E = 1.725$. (*d*) Trajectory trapped in the primary minimum. It was obtained from $p(0) = 0$, $x(0) = -1.875$ and it also has $E = 1.725$. (*e*) Trajectory that samples all available phase space; it was obtained from $p(0) = 0$, $x(0) = 3.1$. All units are arbitrary.

In Figure 2.3, the seven classes (from periodic to B-flow) of Hamiltonian dynamics are not mutually exclusive; for example, the motion of the ODHO is ergodic even though periodic [4]. More importantly, higher levels of instability encompass the desirable properties of lower levels; thus, a K-flow is necessarily ergodic and mixing. In molecular dynamics simulations, the equations of motion are nonintegrable and the computed trajectories are recurrent, Hamiltonian, at least mixing, and preferably K-flows.

2.5   HOW COLLISIONS AFFECT STABILITY OF TRAJECTORIES

To provide reliable time averages, a simulation should sample a representative portion of the available phase space—typically the hypersurface of constant energy. Thus the motion cannot be periodic. Further, if the simulation is started from an arbitrary set of initial conditions, the system should evolve to an equilibrium state. Thus the motion must be at least mixing. Moreover, the simulation should require only a modest amount of computer time; that is, the trajectory should sample, as quickly as possible, all representative portions of phase space. Hence, rather than simple mixing motion, a K-flow or B-flow would be preferred. Combining these requirements, we conclude that the simulation will be most efficient if the trajectory samples phase space in an apparently random manner. This brings us to a central paradox of classical mechanics: how do deterministic equations of motion generate apparently random trajectories?

The resolution of this paradox can be illustrated by comparing the following two models. In both, we consider the two-dimensional motion of one hard disc confined to a circular area. In model A the area is bounded by a smooth, perfectly reflecting, circular hard wall. In model B the area is bounded by 32 fixed hard discs, each perfectly reflecting. Since, in each model, only one disc moves, the phase space is four dimensional: two position coordinates and two momentum coordinates. In both models the disc undergoes elastic collisions with its boundaries; therefore, its total energy is conserved and the available constant-energy surface is three dimensional. Except when the movable disc collides with its boundaries, there is no potential energy: the total energy is just the kinetic energy of the disc and consequently all of the bounded configuration space is accessible. The exercise is to compare the configuration-space trajectories generated within the two different boundaries.

2.5.1   Behavior of Model A

In model A the disc is given an initial position and velocity and its subsequent motion is followed. In configuration space the motion will be straight lines interrupted by collisions. When the disc collides with a perfectly reflecting wall, the collision reverses the component of the velocity normal to the circular surface and leaves the tangential component unchanged. Let v_r and v_t be the radial and tangential components, respectively, of the disc velocity *before* a collision, and let v_r and v_t be the same components *after* the collision. Then a collision causes

$$v_r = - v_r \qquad v_t = v_t \qquad (2.34)$$

For the tests discussed here, the inner face of the circular wall was given a radius $R = 5\sigma$, where σ is the diameter of the movable disc. The motion was started from initial positions

$$x(0) = y(0) = -\frac{R}{\sqrt{2}} \tag{2.35}$$

and initial (dimensionless) velocities

$$v_x(0) = 2 \qquad v_y(0) = 1.4 \tag{2.36}$$

Therefore, the total kinetic energy, which is the constant of the motion, was 2.98 (taking the mass of the disc as unity). The resulting trajectory over 100 collisions is shown in Figure 2.7(a). For these initial conditions, the motion is almost periodic; in particular, the trajectory and points of collision form a nine-pointed star that slowly rotates around the circular area. This is the parent trajectory for model A.

We then repeated the trajectory calculation using initial conditions slightly perturbed from those used in Figure 2.7(a). In fact, the initial position was the same as (2.35) and the perturbation was applied only to the direction of the initial velocity. Specifically, the initial velocity was changed to

$$v_x(0) = 2.0010 \qquad v_y(0) = 1.3986 \tag{2.37}$$

This maintained the same magnitude of velocity as in Figure 2.7(a) and hence the same kinetic energy $E_k = 2.98$. Thus both the parent [2.7(a)] and the perturbed [2.7(b)] trajectories fell on the same constant-energy surface. If we let $\phi = \tan^{-1}(v_y/v_x)$ measure the direction of the velocity vector, then the change in direction of the initially assigned velocity was from $\phi_a = 34.992°$ to $\phi_b = 34.952°$, a change of only 0.12%. The trajectory shown in Figure 2.7(b) is also over 100 collisions: the trajectory is still a nine-pointed star, rotating around the circular area at a somewhat faster rate than the parent in 2.7(a).

Stability of trajectories can be determined by the response to an imposed perturbation. Thus we calculated, at the time of each collision, the difference in positions between the two trajectories,

$$d = \sqrt{[x_a(t) - x_b(t)]^2 + [y_a(t) - y_b(t)]^2} \tag{2.38}$$

This difference is measured in units of the disc diameter σ and is plotted, for the first 100 collisions, in Figure 2.7(c). The deviation grows slowly and linearly with time,

$$d(t) \propto t$$

That is, each collision increases the deviation by a constant amount that is independent of previous values of the deviation. After 100 collisions, the

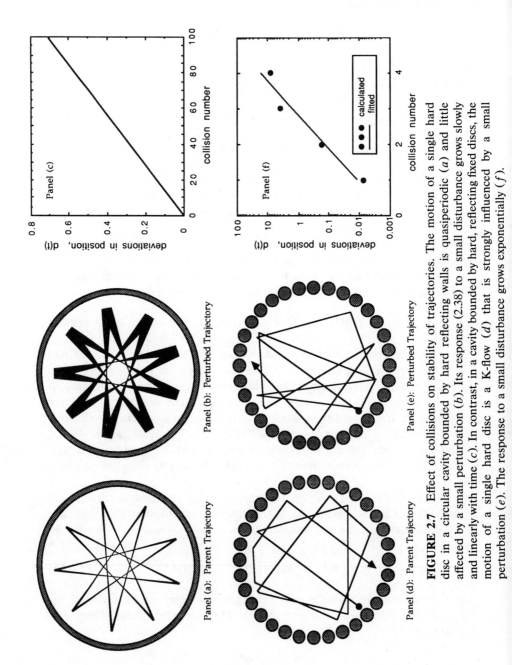

FIGURE 2.7 Effect of collisions on stability of trajectories. The motion of a single hard disc in a circular cavity bounded by hard reflecting walls is quasiperiodic (*a*) and little affected by a small perturbation (*b*). Its response (2.38) to a small disturbance grows slowly and linearly with time (*c*). In contrast, in a cavity bounded by hard, reflecting fixed discs, the motion of a single hard disc is a K-flow (*d*) that is strongly influenced by a small perturbation (*e*). The response to a small disturbance grows exponentially (*f*).

deviation in position is less than one disc diameter. The motion is quasiperiodic in configuration space, with the trajectory slowly and regularly evolving over a portion of the available space. For these initial conditions, the trajectory never samples the central circular area.

Giving other initial conditions to model A will generate configuration-space trajectories of shapes other than those in Figure 2.7. Some of those will sample more of the available space, others less; however, the motion in model A is always either periodic or quasiperiodic. It is periodic if the initial velocity has components such that $\phi/180$ is rational, where $\phi = \tan^{-1}(v_y/v_x)$ is the direction of the initial velocity vector. Otherwise, it is quasiperiodic.

2.5.2 Behavior of Model B

In model B the disc also moves in straight lines interrupted by collisions. The collisions are again with perfectly reflecting circular surfaces, so a collision still reverses the radial velocity component and leaves the tangential component unchanged, as in (2.34). The new feature in model B is that the reflecting surfaces are fixed discs, each of the same diameter σ as the moving disc. These fixed discs are arranged in a circle, as shown in Figure 2.7(d). The inner faces of the discs define a (nearly) circular area of radius $R = 5\sigma$, as was used in model A.

The parent trajectory for model B was generated from the same initial conditions, (2.35) and (2.36), used in the parent of model A. Thus the motion is confined to the same constant-energy surface used in model A. The results after 12 collisions are shown in Figure 2.7(d). The small black dot identifies the initial position of the moving disc, while the arrow identifies its position at the twelfth collision. The trajectory is obviously not a regular geometric shape.

The perturbed trajectory was then computed for model B in exactly the same way as for model A; that is, the initial conditions were (2.35) and (2.37). The results after 12 collisions are shown in Figure 2.7(e). This perturbed trajectory bears little resemblance to its parent. The deviations between disc positions on collision were computed via (2.38), and the results are shown, for the first four collisions, on a semilogarithmic plot in Figure 2.7(f). While in model A these deviations grow linearly with time, now in model B the response to the perturbation grows exponentially with time,

$$d(t) \propto e^t$$

That is, each collision increases the deviation by an amount that is proportional to the previous value of the deviation. In four collisions the deviation has reached the diameter of the container (10σ), while in model A the deviation had not reached even a tenth the container diameter in 100 collisions!

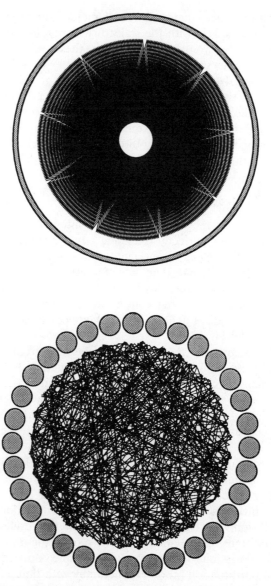

FIGURE 2.8 Comparison of quasiperiodic and chaotic trajectories. The paths are configuration-space trajectories from models *A* (*top*) and *B* (*bottom*), each for 400 collisions. The trajectory for model *A* is a continuation of the calculation shown in Figure 2.7(*b*); that for model *B* is a continuation of Figure 2.7(*e*).

The trajectory of model B is strongly unstable and the motion is a K-flow. The trajectory covers all available configuration space in an apparently random manner; moreover, the trajectory is sensitive to small perturbations. Very small disturbances in the initial conditions will be quickly magnified to large modifications in the computed trajectory. Of course, the deviations in position cannot increase beyond the size of the container. Formally, we have no guidance about the response to large perturbations; the concept of stability addresses the response to *small* perturbations. In model A the correlation time, over which the position of the disc is related to previous positions, is hundreds of collisions, but in model B the correlation time is no more than 10 collisions.

Models A and B differ only in how the disc interacts with its boundary, and therefore, the two models illustrate that molecular interactions are responsible for unstable trajectories. Model A has two constants of the motion. Because the radial and tangential components of the velocity are independent, the two corresponding components of the total energy are each separately constant: the system has two degrees of freedom and two constants of the motion, and hence the motion is periodic or quasiperiodic. In model B the boundary of fixed discs couples the radial and tangential components of the velocity, so there is now only one constant of the motion —the total energy—and the motion is chaotic. To emphasize the difference, Figure 2.8 compares trajectories from models A and B, each after 400 collisions. Again, model A samples most (but not the central portion) of the available configuration space in a regular and orderly way, while model B samples the full space in a haphazard manner.

This comparison is instructive, but not conclusive; it is a demonstration, not a proof. Multibody dynamics and multibody interactions, which are the focus of molecular dynamics, are necessarily more complicated. For example, on the one hand we expect that instabilities in multibody dynamics will be enhanced by multiple interactions. Furthermore, when a body collides with another moving body, changes of velocity occur that are more elaborate than those in model B, in which the collisions are with fixed objects. On the other hand, we also expect competing effects; for example, at high densities a molecule may become trapped in a cage of neighboring molecules, resulting in a prolonged correlation time for the motion of that individual molecule. Nonlinearities in the equations of motion are not sufficient to guarantee that chaotic motion will develop. Nevertheless, the lesson here is that in the presence of molecular interactions, classical equations of motion can produce random trajectories.

2.6   DETERMINATION OF PROPERTIES

From a description of phase-space trajectories, we now turn to a discussion of how those trajectories are used. Ultimately, our objective is to determine macroscopic properties and the phase-space trajectory constitutes the raw data from which those properties are calculated.

Consider N spherical atoms confined to an isolated system of volume V at thermodynamic equilibrium with total energy E. According to the molecular theory of matter, macroscopic properties result from the behavior of collections of individual molecules; in particular, any measurable property A can be interpreted in terms of some function $A(\mathbf{r}^N, \mathbf{p}^N)$ that depends on the position of the phase point in phase space $\{\mathbf{r}^N, \mathbf{p}^N\}$. Now a measured value of A, call it A_m, is not obtained from an experiment performed at an instant; rather the experiment requires a finite duration. During that measuring period individual atoms evolve through many values of positions and momenta; that is, the phase point moves along its trajectory in phase space. Therefore the measured value A_m is the phase function $A(\mathbf{r}^N, \mathbf{p}^N)$ averaged over a time interval t [15]:

$$A_m = \frac{1}{t} \int_{t_0}^{t_0 + t} A(\mathbf{r}^N(\tau), \mathbf{p}^N(\tau)) \, d\tau \qquad (2.39)$$

For systems in thermodynamic equilibrium this average must be independent of the location of the starting time t_0. Moreover, at equilibrium we assume that this interval average reliably approximates the time average $\langle A \rangle$, which would be obtained from a measurement performed over an essentially infinite duration,

$$\langle A \rangle = A_m \qquad (2.40)$$

where

$$\langle A \rangle = \lim_{t \to \infty} \frac{1}{t} \int_{t_0}^{t_0 + t} A(\mathbf{r}^N(\tau), \mathbf{p}^N(\tau)) \, d\tau \qquad (2.41)$$

One situation in which the assumption (2.40) is obviously valid is when the integrand $A(\mathbf{r}^N, \mathbf{p}^N)$ is a constant of the motion. In that case A does not vary along the phase-space trajectory and any interval average (2.39) is equal to the time average (2.41). However, as the phase point journeys along the hypersurface of constant energy E, most quantities are not constant; instead, their values fluctuate. That is, because the molecules are in continual motion and are colliding with one another, the positions and momenta of individual molecules are continually changing, and therefore most functions that depend on the positions and momenta are fluctuating.

An example of a fluctuating quantity is provided in Figure 2.9. The quantity plotted is the translational kinetic energy, computed during a simulation of a fluid composed of 108 atoms in an isolated system. Although

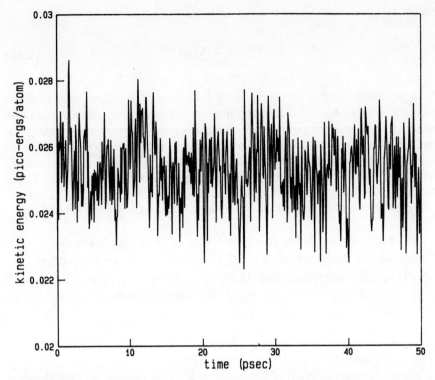

FIGURE 2.9   An example of fluctuations occurring in a molecular dynamics simulation. The quantity plotted is the instantaneous kinetic energy from a simulation of 108 Lennard-Jones atoms. The energy and time have been made dimensional using values of the mass and potential parameters (ε and σ) characteristic of argon.

the total energy E is constant, the kinetic energy E_k and potential energy $\mathscr{U}$ fluctuate; but those fluctuations must preserve the value of E,

$$E = E_k(\mathbf{p}^N) + \mathscr{U}(\mathbf{r}^N) = \text{const} \qquad (2.42)$$

These fluctuations occur on a time scale that is significantly less than 1 psec (10^{-12} sec). Nevertheless, the fluctuations in the figure appear to be stable about an average value. In the case of the kinetic energy, this average is proportional to the kinetic temperature,

$$\langle E_k \rangle = \lim_{t \to \infty} \frac{1}{t} \int_{t_0}^{t_0 + t} E_k(\mathbf{p}^N) \, d\tau = \tfrac{3}{2} NkT \qquad (2.43)$$

where

$$E_k = \frac{1}{2m} \sum_i^N \mathbf{p}_i^2 \qquad (2.44)$$

N is the total number of atoms, k is Boltzmann's constant, and T is the absolute temperature. In the absence of degrees of freedom within the atoms, this kinetic temperature is identical to the thermodynamic temperature [16]. To relate the average kinetic energy to the absolute temperature, as in (2.43), the molecular momenta $\mathbf{p}_i$ must be based on the "peculiar" velocities, that is, on the molecular velocities relative to a reference frame moving with the system's center of mass [16].

Here are the basic steps for obtaining property values from simulation:

(a) Determine the phase space trajectory $\{\mathbf{r}^N(t), \mathbf{p}^N(t)\}$.
(b) Evaluate the function $A(\mathbf{r}^N(t), \mathbf{p}^N(t))$ for the property of interest.
(c) Form the interval average (2.39).
(d) Assume (2.40); that is, assume the computed interval average equals the time average.

Therefore, the subscript on A_m denotes not only "measured" but also "molecular dynamics."

When is the assumption (2.40) valid? It appears reasonable provided that the limiting value $\langle A \rangle$ exists for all trajectories of interest and that, for any trajectory, the average $\langle A \rangle$ is independent of the initial time t_0. According to Birkhoff [17] and von Neumann [18], these stipulations are met by many trajectories on the constant-energy surface. They are always assumed to apply to trajectories computed by molecular dynamics.

In spite of the Birkhoff–von Neumann theorems, certain operational difficulties arise in equating molecular dynamics results (2.39) to the more generally desired time average (2.41). Prominent difficulties include the following:

1. How long must the sampling time be? We obviously cannot compute over an infinitely long time and actually evaluate (2.41). A standard answer is to sample over a duration equal to several multiples of the relaxation time for the property. Unfortunately, different properties have different relaxation times.

2. How do we ensure that the phase point samples regions of phase space that are "representative" of the equilibrium state? That is, how do we determine that the phase-space trajectory is not metastable—that the phase point is not confined to subregions of accessible phase space. Along a metastable trajectory normal equilibrium fluctuations of the quantities $A(t)$

are usually suppressed, leading to values for A_m that are not necessarily equal to the time average $\langle A \rangle$.

These metastable situations are troublesome because for multibody systems we do not know whether local minima in the energy surface exist. We presume they do exist, but still we do not know where they are in phase space or how deep they are. Therefore, we try to test whether a computed trajectory is localized near such minima. Two features of the metastable situation of Figure 2.6(c) suggest ways for doing this: (i) its existence is sensitive to the initial conditions used to start the motion and (ii) its existence is sensitive to small perturbations. Thus we can test for such metastabilities by repeating the trajectory calculation using different initial conditions and/or by stimulating the system with increases in energy that should drive the phase point over any energy barriers. For example, we can often test for metastability by introducing an external perturbation, such as a temperature excursion. This is not fool-proof; in some cases, the natural time scale for a metastability may simply exceed the typical duration of a molecular dynamics run.

3. Finally, how do we identify equilibrium in molecular dynamics simulations? Being on the constant-energy surface is not a sufficient condition; the system could easily be in a nonequilibrium state and yet be isolated so that its total energy is conserved. For isolated systems, the macroscopically sufficient condition for equilibrium is that the system's entropy be a maximum. However, entropy is not readily calculated in a simulation, so we are forced to seek other criteria for identifying equilibrium. These issues pose vexing problems in the analysis of molecular dynamics trajectories.

Item 3 raises a final issue: some properties are not measurable. That is, not all properties are defined as time averages over some function of the phase-space trajectory. Properties that are not measurable include the entropy, the free energies, and the chemical potential. These quantities are formally related not to the phase-space trajectory but rather to the accessible phase-space volume. To evaluate these properties, we must transform their definitions into some function that can be averaged over the trajectory. The resulting functions are not wholly satisfactory because they typically involve integrals or derivatives that must be computed numerically. These problems will be discussed in Chapter 6.

Equating an interval average to a time average, (2.40), is central to molecular dynamics. In the first place, (2.40) distinguishes molecular dynamics from statistical mechanics. In statistical mechanics we avoid evaluating the time average by replacing it with an ensemble average; that is, we invoke the ergodic hypothesis. This replacement is motivated by our inability to actually compute the phase-space trajectory for a real system containing Avogadro's number of molecules. In contrast, molecular dynamics confronts this problem directly by calculating an approximation to the time average for a manageable number of molecules. More importantly, the assumption (2.40) poses a significant problem in testing the validity of a simulation: How do we convince ourselves that the finite-time averages calculated from a molecular dynamics run are in fact reliable approximations to infinite-time averages?

2.7   FUNDAMENTAL DISTRIBUTIONS

Consider a system composed of N atoms occupying volume V and having absolute temperature T. The atoms have translational kinetic energy E_k and exert forces on one another according to the intermolecular potential energy function $\mathscr{U}$; thus, the total energy is

$$E(\mathbf{p}^N, \mathbf{r}^N) = E_k(\mathbf{p}^N) + \mathscr{U}(\mathbf{r}^N) \tag{2.45}$$

At equilibrium, the atoms will be distributed among available momenta and positions, and the distribution will be consistent with the values of N, V, and T. Let that distribution be represented by

$$N(\mathbf{r},\mathbf{p})\,d\mathbf{r}\,d\mathbf{p} = \text{Average number of atoms having position} \qquad (2.46)$$
$$\text{vectors in the range } \mathbf{r} \text{ to } \mathbf{r}+d\mathbf{r} \text{ and momentum}$$
$$\text{vectors in } \mathbf{p} \text{ to } \mathbf{p}+d\mathbf{p}$$

If we specify the momentum at a value $\mathbf{p}$ to $\mathbf{p}+d\mathbf{p}$ and integrate over all possible positions, then

$$N(\mathbf{p})\,d\mathbf{p} = \text{Average number of atoms having momenta in} \qquad (2.47)$$
$$\text{the range } \mathbf{p} \text{ to } \mathbf{p}+d\mathbf{p} \text{ irrespective of their positions}$$

Alternatively, if we specify the atomic position at $\mathbf{r}$ to $\mathbf{r}+d\mathbf{r}$ and integrate over all possible momenta, then

$$N(\mathbf{r})\,d\mathbf{r} = \text{Average number of atoms having positions in} \qquad (2.48)$$
$$\text{volume element } \mathbf{r} \text{ to } \mathbf{r}+d\mathbf{r} \text{ irrespective of their}$$
$$\text{momenta}$$

If we integrate over the complete range of accessible positions and momenta, then we obtain the total number of atoms in the system,

$$\int N(\mathbf{r},\mathbf{p})\,d\mathbf{r}\,d\mathbf{p} = N \tag{2.49}$$

2.7.1   Distribution of Velocities

Typically, as in (2.45), the total energy is separable; that is, the kinetic energy is independent of atomic positions and the potential energy is independent of atomic momenta. Then the equilibrium distribution of velocities (2.47) obeys the Maxwell–Boltzmann law [16]

$$N(\mathbf{p}) = \int N(\mathbf{r}, \mathbf{p}) \, d\mathbf{r} = \frac{N}{C} \exp\left(-\frac{p^2}{2mkT}\right) \qquad (2.50)$$

where k is Boltzmann's constant and C is a constant that ensures the normalization (2.49) is satisfied. The Maxwell–Boltzmann distribution is a fundamental result from the kinetic theory of dilute gases and, more generally, from the statistical mechanics of the NVT (aka canonical) ensemble. Using $\mathbf{v} = \mathbf{p}/m$, the fraction of atoms having velocities between $\mathbf{v}$ and $\mathbf{v} + d\mathbf{v}$ is

$$f(\mathbf{v}) \, d\mathbf{v} = \frac{N(\mathbf{v}) \, d\mathbf{v}}{N} = \frac{1}{C} \exp\left(-\frac{mv^2}{2kT}\right) d\mathbf{v} \qquad (2.51)$$

Because the velocity components are mutually independent and obey $v^2 = v_x^2 + v_y^2 + v_z^2$, the components of the velocity are also separable, and for any one component of the velocity, we can write

$$f(v_x) \, dv_x = \frac{N(v_x) \, dv_x}{N} = \frac{1}{C_x} \exp\left(-\frac{mv_x^2}{2kT}\right) dv_x \qquad (2.52)$$

According to (2.49), the fraction $f(v_x) \, dv_x$ must accumulate to unity; therefore,

$$C_x = \int_{-\infty}^{\infty} \exp\left(-\frac{mv_x^2}{2kT}\right) dv_x \qquad (2.53)$$

This definite integral can be evaluated analytically, giving

$$f(v_x) \, dv_x = \frac{N(v_x) \, dv_x}{N} = \sqrt{\frac{m}{2\pi kT}} \exp\left(-\frac{mv_x^2}{2kT}\right) dv_x \qquad (2.54)$$

Analogous expressions are obtained for the y- and z-components. The distribution (2.54) is Maxwell's velocity distribution. It is a Gaussian with standard deviation $\sigma = (kT/m)^{1/2}$ about the mean value $\langle v_x \rangle = 0$, and as shown in Figure 2.10, it becomes more nearly uniform as temperature increases. Although the Maxwell distribution is formally derived in a fixed NVT system, the velocity distribution should be Maxwellian in any equilibrium situation because, in the thermodynamic limit, properties in different ensembles become equivalent. (The thermodynamic limit means that both N and V are made large but the ratio N/V is held fixed.) In an isolated system, the Maxwell distribution (2.54) applies at the equilibrium temperature $\langle T \rangle$.

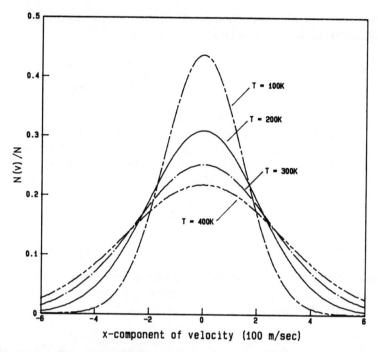

FIGURE 2.10 Effect of temperature on the distribution of one component of the molecular velocities, as computed from the Maxwell distribution (2.54).

The separable velocity components leading to (2.52) are mutually independent. Therefore, since each has the same distribution (2.54) and the same standard deviation σ, they each also have the same mean-square values,

$$\langle v_x^2 \rangle = \langle v_y^2 \rangle = \langle v_z^2 \rangle = \sigma^2 = \frac{kT}{m} \qquad (2.55)$$

Thus, each component of the average kinetic energy makes the *same* contribution to the temperature. This is the *equipartition* of kinetic energy, a special case of a more general equipartition theorem of statistical mechanics.

A Gaussian distribution usually results from a large number of independent events. Does this explain why the velocity distribution is Gaussian? It is true that the velocity of a molecule, at one instant, results from a large number of previous collisions (or, for soft bodies, interactions) with other molecules. But collisions in a sequence, as experienced by one molecule, are not independent. And indeed, at an instant the velocity distribution is not necessarily Gaussian.

Maxwell's distribution is obeyed only over a finite duration, during which velocities, on the average, assume the Gaussian form. This suggests that beyond a short time, an atomic velocity is nearly independent of previous collisions. ("Nearly" independent because evidence provided by long-time tails of velocity autocorrelation functions suggests that small dependencies remain after long times.) Formally, these ideas are contained in the concept of a *correlation time* [19], during which the position of the system on its phase-space trajectory is related to previous positions. Because of the finite duration of simulations, it is only when the correlation time is short that molecular dynamics reliably simulates model systems.

With Maxwell's distribution, we can compute the average value for any function A that depends only on the velocities,

$$\langle A(v_x) \rangle = \int_{-\infty}^{\infty} A(v_x) f(v_x)\, dv_x \tag{2.56}$$

For example, we may compute the average speed, the average square of the velocity, and the average kinetic energy.

During a simulation it is convenient to follow the development of the Maxwell distribution (and, hence, the approach to equilibrium) by monitoring the time behavior of the single number provided by

$$H_x(t) = \int_{-\infty}^{\infty} f(v_x) \ln f(v_x)\, dv_x \tag{2.57}$$

This is the kinetic portion of Boltzmann's celebrated H-function. Its time average, according to (2.56), is $\langle \ln f \rangle$. Since $f(v_x)$ is a positive fraction, its logarithm is negative, and therefore H_x is negative. Moreover, for an isolated system, equilibrium is the most probable state, and therefore the equilibrium distribution of velocities is the most probable distribution. Thus, for a simulation started from an arbitrarily assigned nonequilibrium situation and undergoing mixing dynamics, we expect $H_x(t)$ to decrease to its value given by the Maxwell distribution (2.54). In the kinetic theory of low-density gases the previous sentence is Boltzmann's H-theorem [20].

2.7.2 Distribution of Instantaneous Property Values [21, 22]

In an isolated system at equilibrium, the entropy S is a maximum and is connected to the phase-space volume Ω (microcanonical partition function) by

$$S = k \ln \Omega(NVE) \tag{2.58}$$

Other dynamic quantities, such as $A(t)$, fluctuate about averages $\langle A \rangle$, as illustrated in Figure 2.9; the average $\langle A \rangle$ can be obtained by accumulating, from all possible values of $A(t)$, only those that occur on the constant-energy surface. This interpretation of $\langle A \rangle$ is represented by Equation (C.4) in Appendix C.

Here we want to determine how the instantaneous A-values are distributed about their average. To do so, we use an alternative to (C.4) for the average; specifically, we compute $\langle A \rangle$ by summing over all values of A, with each weighted by its probability of occurrence,

$$\langle A \rangle = \int A \mathscr{P}(A)\, dA \tag{2.59}$$

Here $\mathscr{P}(A)\, dA$ is the probability of finding a particular A-value on the surface of constant energy E. It is a true probability and therefore is properly normalized,

$$\int \mathscr{P}(A)\, dA = 1 \tag{2.60}$$

Our objective is to find the functional form for $\mathscr{P}(A)$. Now, for any particular value of $A(t)$, call it $\hat{A}$, the probability density $\mathscr{P}(\hat{A})$ can be expressed as the fraction of the constant-energy surface that has $A(t) = \hat{A}$ (this is the principle of equal a priori probabilities),

$$\mathscr{P}(\hat{A}) \equiv \frac{\Omega(E, \hat{A})}{\Omega(E)} = \frac{\text{portion of constant-}E \text{ surface having } A(t) = \hat{A}}{\text{total constant-}E \text{ surface}}$$

$$\tag{2.61}$$

where

$$\Omega(E, \hat{A}) = \frac{1}{h^{3N}N!} \int d\mathbf{p}^N\, d\mathbf{r}^N\, \theta[E - \mathscr{H}]\, \delta[\hat{A} - A(\mathbf{p}^N, \mathbf{r}^N)] \tag{2.62}$$

We do not show on the left-hand side (lhs) the dependence on N and V because they are constant and are not involved in this discussion. For states on the constant-E surface that have a particular A-value, we write the entropy, analogous to (2.58), as

$$S(E, A) = k \ln \Omega(E, A) \tag{2.63}$$

Then from (2.61),

$$\mathscr{P}(A) = \frac{\exp[S(E, A)/k]}{\Omega} \tag{2.64}$$

where Ω is effectively a normalization constant that ensures (2.60) is satisfied.

We now expand $S(E, A)$ in a Taylor series in A about the equilibrium value $\langle A \rangle$ (E is fixed in the expansion, so we do not show it explicitly),

$$S(A) = S(\langle A \rangle) + (A - \langle A \rangle)\left(\frac{\partial S}{\partial A}\right)_{\langle A \rangle} + \tfrac{1}{2}(A - \langle A \rangle)^2\left(\frac{\partial^2 S}{\partial A^2}\right)_{\langle A \rangle} + \cdots$$

(2.65)

Since S is a maximum when $A = \langle A \rangle$, we must have

$$\left(\frac{\partial S}{\partial A}\right)_{\langle A \rangle} = 0$$

(2.66)

and

$$\frac{1}{k}\left(\frac{\partial^2 S}{\partial A^2}\right)_{\langle A \rangle} \equiv -c < 0$$

(2.67)

where c is a positive constant. Thus, neglecting higher order terms, the expansion (2.65) becomes

$$\frac{S(A)}{k} = \frac{S(\langle A \rangle)}{k} - \tfrac{1}{2}c(A - \langle A \rangle)^2$$

(2.68)

and putting (2.68) into (2.64) yields

$$\mathscr{P}(A)\, dA = C\exp\left[-\left(\tfrac{1}{2}c\right)(A - \langle A \rangle)^2\right] dA$$

(2.69)

where C accumulates all the constant terms from (2.68) and (2.64). Applying the normalization condition (2.60), we obtain

$$C = \sqrt{c/2\pi}$$

(2.70)

Finally, combining (2.69) and (2.70) gives the probability distribution

$$\boxed{\mathscr{P}(A)\, dA = \sqrt{\frac{c}{2\pi}}\ \exp\left[-\tfrac{1}{2}c(A - \langle A \rangle)^2\right] dA}$$

(2.71)

This is a Gaussian (aka the normal) distribution centered at $\langle A \rangle$ and having variance σ^2 given by its second moment,

$$\sigma^2 = \langle (A - \langle A \rangle)^2 \rangle = \frac{1}{c}$$

(2.72)

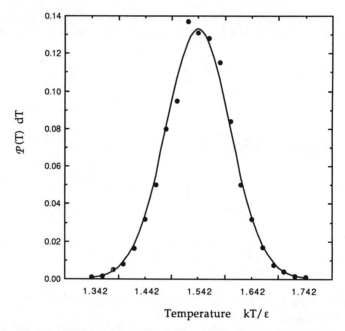

FIGURE 2.11 Distribution of instantaneous temperatures about their average value $\langle kT/\varepsilon \rangle = 1.542$ with standard deviation $\sigma = 0.060$ for 108 Lennard-Jones atoms at density $\rho\sigma^3 = 0.6$. Points are simulation results averaged over 10,000 integration time-steps; line is the Gaussian distribution (2.71) with $\sigma = 0.06$.

An example of (2.71) is given in Figure 2.11. When A is an extensive property, that is of $O(N)$, then $\langle (A - \langle A \rangle)^2 \rangle$ is also of $O(N)$, and consequently, the standard deviation of any value of $A(t)$ relative to the average $\langle A \rangle$ is

$$\frac{\sigma}{\langle A \rangle} = O(N^{-1/2}) \tag{2.73}$$

As a result, in the thermodynamic limit ($N \rightarrow$ large, $V \rightarrow$ large, with N/V fixed), the distribution becomes more sharply peaked about the average: the Gaussian approaches a delta symbol (Appendix A).

2.8   ELEMENTS OF SAMPLING THEORY

A primary use of molecular dynamics is to estimate time averages (Section 2.6). During a run we evaluate, at regular intervals Δt, a time-dependent quantity $x(t)$, and after M such intervals we approximate the time average

(2.41) by

$$\langle x \rangle = \frac{1}{M} \sum_{k}^{M} x(k\,\Delta t) \tag{2.74}$$

This computed average is an approximation because it is a finite accumulation over a fluctuating quantity. Thus, the average computed over $M\,\Delta t = 10$ psec may differ somewhat from that computed over 20 psec.

The principal purpose for generating numbers is to facilitate comparisons. Say we are told $A = 100 \pm \sigma_A$ and $B = 95 \pm \sigma_B$ and then asked whether $A = B$. The question cannot be answered with confidence unless we know something about the uncertainties σ_A and σ_B. Thus, our answer will depend on whether A and B are known to within 1% or to within 10%. Those who generate averages from simulations have the responsibility of assessing the uncertainties in their calculations. If uncertainties are not assessed and reported with averages, then the averages are essentially worthless for making comparisons.

In this section we summarize the statistical foundations that allow us to claim that molecular dynamics computes averages reliably. In particular, how do we resolve the apparent disparity between the averages we want, which are defined for systems of macroscopic size and macroscopic time scales, and the averages we compute, which are obtained from simulations involving only a few hundred atoms and extending over only a few picoseconds? The problem naturally embraces means for assessing how well the finite sum (2.74) approximates the infinite-time average (2.41). To clarify the problem and its resolution, we introduce in Section 2.8.1 three kinds of statistical averages. These averages will differ, in general, and in Section 2.8.2 we show that these differences depend on the sampling method. In particular, statistically unbiased sampling provides certain desirable relations among the three statistical averages. In Section 2.8.3 we show that simple random sampling is unbiased and provides a clear resolution of the statistical uncertainties in computed averages. Then in Section 2.8.4 we argue that the sampling actually done in molecular dynamics is effectively random.

2.8.1   Sampling Statistics

In Section 2.6 we showed that in a molecular dynamics simulation time-dependent quantities fluctuate about average values. The fluctuations are natural processes, consequences of the constraints placed on the system and of the continuously varying strength of interactions among the atoms. For example, in an isolated system the total kinetic and potential energies continually redistribute themselves so as to keep their sum constant. Because of the chaotic nature of the system's path through phase space, those fluctuations are essentially random, even on a picosecond time scale (see Figure 2.9). As direct experimental evidence for random fluctuations in molecular trajectories, we cite Brownian motion: when immersed in a fluid of microscopically invisible molecules, a small but microscopically visible particle displays random motion.

Although an instantaneous quantity $x(t)$ fluctuates randomiy, values of $x(t)$ describe a definite distribution about their average. Thus, kinetic theory teaches that atomic velocities are distributed in the Maxwell distribution (Section 2.7.1). Likewise, statistical mechanics teaches that instantaneous values of thermodynamic properties also describe Gaussian distributions about their averages (Section 2.7.2); an example is shown in Figure 2.11.

The central question, then, is this: to reliably sample the distribution of $x(t)$, how long must a simulation be? To broach this question, we distinguish among three types of statistical averages: the trajectory average, the sample average, and the expected value of the sample average. A nontechnical discussion of these ideas can be found in the book by Williams [23].

Trajectory Statistics. Consider the complete equilibrium phase-space trajectory for an isolated system and imagine sampling that trajectory at M_∞ discrete points. The sampling points are uniformly distributed along the trajectory with adjacent points separated by a time interval Δt. Since the trajectory is the full equilibrium one and Δt is small, M_∞ is exceedingly large. For some time-dependent quantity $x(t)$, the time average (2.41) can be approximated by

$$\langle X \rangle = \frac{1}{M_\infty} \sum_k^{M_\infty} x(k\,\Delta t) \qquad (2.75)$$

The trajectory average $\langle X \rangle$ is the quantity we hope to estimate. For any one value of $x(t)$ relative to the average $\langle X \rangle$, the variance is given by

$$V[\langle X \rangle - x(t)] \equiv S^2 = \frac{1}{M_\infty - 1} \sum_k^{M_\infty} (\langle X \rangle - x(k\,\Delta t))^2 \qquad (2.76)$$

The notation $V[y - z]$ designates the operation that computes the variance of z with respect to (wrt) y.

Sample Statistics. We want to estimate the trajectory average $\langle X \rangle$, but we do not want to compute the full trajectory of M_∞ points. Instead, we intend to compute only a piece of the trajectory, a segment of M points, where $M \ll M_\infty$. The collection of M consecutive points constitutes *one* sample from the full trajectory. This sample has its own average, given by (2.74),

$$\langle x \rangle = \frac{1}{M} \sum_k^{M} x(k\,\Delta t)$$

and any one value $x(t)$ taken from the computed trajectory has variance about this average given by

$$V[\langle x \rangle - x(t)] \equiv \sigma^2 = \frac{1}{M-1} \sum_k^M (\langle x \rangle - x(k \Delta t))^2 \qquad (2.77)$$

The central question posed at the start of this section can now be recast in this form: How well does the sample average $\langle x \rangle$ approximate the trajectory average $\langle X \rangle$?

Statistics of the Sampling Distribution. Now note that there are many different samples, each containing M points, that could be drawn from the trajectory of M_∞ points. Let the number of distinct samples be M_s, a number whose value depends on how we choose to draw samples. Any one sample, say the kth, has its own average $\langle x \rangle_k$, and therefore a total of M_s possible averages could be computed from one trajectory. These M_s averages form their own distribution, the *sampling distribution*, about their own average $E[\langle x \rangle]$,

$$E[\langle x \rangle] = \frac{1}{M_s} \sum_k^{M_s} \langle x \rangle_k \qquad (2.78)$$

The notation $E[z]$ represents the expected value of z obtained by averaging over all possible samples, each containing M points. For any one sample average relative to the expected average, the variance is

$$V[E[\langle x \rangle] - \langle x \rangle] \equiv \frac{1}{M_s} \sum_k^{M_s} (E[\langle x \rangle] - \langle x \rangle_k)^2 \qquad (2.79)$$

At this point we have introduced three kinds of statistical measures; the notation is summarized in Table 2.1.

TABLE 2.1   Summary of Jargon and Notation Used to Discuss Statistical Sampling

Entity	Number of Components in Entity	Statistical Measures Average	Variance
Full phase-space trajectory	M_∞ discrete points	$\langle X \rangle$	S^2
Sampling distribution	M_s samples from trajectory	$E[\langle x \rangle]$	$V[E[\langle x \rangle] - \langle x \rangle]$
One sample	M discrete points	$\langle x \rangle$	σ^2

Note that $E[\ ]$ and $V[\ ]$ represent operators: the expected-value operator and variance operator, respectively. To help distinguish these from functions, we place their arguments in brackets, rather than in parentheses.

2.8.2   The Nature of Uncertainties

To relate one sample average $\langle x \rangle$ to the trajectory average $\langle X \rangle$, we first assume that one sample average approximates the expected average,

$$E[\langle x \rangle] \approx \langle x \rangle \tag{2.80}$$

and, likewise, one sample variance approximates the expected sample variance,

$$E[\sigma^2] \approx \sigma^2 \tag{2.81}$$

Our problem now is to relate the expected average to the trajectory average and relate the expected sample variance to the trajectory variance.

Recall that the expected average $E[\langle x \rangle]$ is formed from M_s sample averages. Uncertainties in M_s sample averages can be measured by a mean-square error (mse),

$$\text{mse} = \frac{1}{M_s} \sum_k^{M_s} (\langle X \rangle - \langle x \rangle_k)^2 \tag{2.82}$$

If in the quadratic we add and subtract the expected value $E[\langle x \rangle]$, expand the quadratic, and use the definition (2.78) for the expected value, then (2.82) separates into two terms,

$$\text{mse} = (\langle X \rangle - E[\langle x \rangle])^2 + V[E[\langle x \rangle] - \langle x \rangle] \tag{2.83}$$

that is,

$$\text{mse} = (\text{bias})^2 + \text{variance of average wrt expected average} \tag{2.84}$$

Thus, just as in laboratory experiments, (2.83) shows that errors are of two types: *bias or systematic error*, which measures the deviation of the expected value from the trajectory average, and *variance or statistical error*, which measures the distribution of the averages about their average.

Systematic error consistently displaces the computed average from the trajectory average. It may be caused by approximations used in solving the equations of motion, by effects of periodic boundary conditions, by round-off errors, and even by the way the trajectory is sampled for values of $x(t)$. Systematic error will not be reduced by extending or repeating a run. In contrast, statistical error is a consequence of using a finite sum to approximate the time-average integral (2.41). Statistical error is reduced by extending or repeating a run.

We define bias to be systematic error that may include a component from the sampling method used to draw samples,

$$\begin{aligned} \text{Bias} = &\text{systematic error caused by sampling method} \\ &+ \text{systematic error caused by everything else} \end{aligned} \tag{2.85}$$

The sampling component depends on *how* we sample the trajectory, on the number of points M we draw, and perhaps, on the trajectory length M_∞. However, we have some control over the sampling component because we are free to choose how to draw samples; typically, we want a sampling procedure that eliminates the sampling component from the bias. Such sampling methods are said to be *statistically unbiased*. In an unbiased method all sampling errors reside in the variance and (2.83) becomes

$$\text{mse} = (\text{systematic error})^2 + (\text{statistical error})^2 \qquad (2.86)$$

Thus, we presume that, in an unbiased sampling of the trajectory, the remaining systematic and statistical errors are independent and therefore they can be combined in quadrature to obtain an estimate of the total error. *For the remainder of Section 2.8 we ignore systematic errors and concentrate on statistical errors.* Systematic error will claim much of our attention through the remainder of the book.

To estimate a trajectory average and assess the quality of that estimate, we take two steps: (i) We identify an unbiased estimator for the trajectory average, so that (ignoring systematic errors) the first term in the mse (2.83) vanishes, that is,

$$E[\langle x \rangle] = \langle X \rangle \qquad (2.87)$$

Then by using the expected value from (2.80) in (2.87), we approximate the trajectory average as one sample average,

$$\langle X \rangle \approx \langle x \rangle \qquad (2.88)$$

(ii) We identify an estimate of the reliability of (2.88), that is, with what error does one sample average $\langle x \rangle$ approximate a trajectory average $\langle X \rangle$? To do this, we need the variance $V[\langle X \rangle - \langle x \rangle]$. We want to compute only one variance, and so analogous to (2.80) and (2.81), we assume that the variance of one sample average approximates the expected variance

$$V[\langle X \rangle - \langle x \rangle] \approx E\left[(\langle X \rangle - \langle x \rangle)^2\right] \qquad (2.89)$$

Now for an unbiased estimator, it is straightforward to show (Exercise 2.21) that

$$E\left[(\langle X \rangle - \langle x \rangle)^2\right] = V[E[\langle x \rangle] - \langle x \rangle] \qquad (2.90)$$

which is the second term in the mse (2.83). So we are left with the job of devising an estimate for the expected variance, which, according to (2.90), is the variance of one average wrt the expected average.

2.8.3 Simple Random Sampling

As one method for obtaining an unbiased estimator of the trajectory average, we consider simple random sampling: for the moment, ignore the temporal order of points along the trajectory and imagine sampling points at random, letting each point have the same chance of selection. For a trajectory composed of a total of M_∞ points, we can possibly draw M_s different samples, each containing M points (see Section 2.8.1). In simple random sampling the number of samples is given by

$$M_s = \frac{M_\infty!}{M!(M_\infty - M)!} \tag{2.91}$$

That is, we divide M_∞ points into two sets: the set of M selected points and the set of $M_\infty - M$ ignored points. There are M_s unique ways of so dividing the points.

For samples drawn by this random procedure, a simple exercise in combinatorial algebra shows that the sampling is unbiased [24]; that is, the statistical component of the expected average equals the full trajectory average as we require by (2.87). Moreover, an addition exercise in combinatorial algebra shows that the expected variance is proportional to the trajectory variance [24],

$$E\left[(\langle X \rangle - \langle x \rangle)^2\right] = \left(1 - \frac{M}{M_\infty}\right)\frac{S^2}{M} \tag{2.92}$$

This is one of the most important results in sampling theory. Note two bounds. First, if we completely sample the trajectory, so $M = M_\infty$, then the variance vanishes: the expected average is the trajectory average, as claimed in (2.87). Second, if the total trajectory is large compared to the computed segment, so $M_\infty \gg M$, then (2.92) reduces to

$$E\left[(\langle X \rangle - \langle x \rangle)^2\right] = \frac{S^2}{M} \tag{2.93}$$

That is, in an estimate of $\langle X \rangle$ from one sample the precision attained depends on the number of sampled points M, but *it is independent of the total length of the trajectory M_∞.*

This is surprising. Intuitively we feel that a sample average will more closely approximate the trajectory average only as the length of the computed trajectory approaches that of the full trajectory, that is, as $M \to M_\infty$. However, (2.93) says that when M_∞ is very large, the ratio M/M_∞ is unimportant. Equation (2.93) is the statistical basis on which a pollster will randomly sample 1500 people and then claim to know the opinions of the entire U.S. population (a number approaching 300,000,000).

Likewise, (2.93) is the statistical basis on which we claim that molecular dynamics provides reliable estimates to macroscopic properties. In molecular dynamics (2.93) usually applies in two ways because many properties are averages over both time and number of molecules. We want estimates for properties over macroscopic durations (approaching the order of seconds) and macroscopic sizes (of the order of 10^{26} molecules), but we compute only over picoseconds and using only a few hundred molecules. Since $M_{\infty}\Delta t \approx$ 1 sec $\gg M\Delta t \approx 10$ psec, and since $10^{26} \gg N \approx 864$ molecules, (2.93) applies —if the sampling is done randomly.

At this point, we have accomplished step (i) of the procedure outlined at the end of Section 2.8.2: we have identified simple random sampling as one method for generating an unbiased estimate of the trajectory average, so

$$\langle X \rangle \approx \langle x \rangle \qquad (2.88)$$

To accomplish step (ii), we need an estimate for the expected variance. We would like to use (2.93), but to do so we need the trajectory variance S^2, which would require complete sampling of the full trajectory. Can we estimate S^2? Yes, a final exercise in combinatorial algebra [24] shows that for an unbiased estimator the expected sample variance is the trajectory variance,

$$E[\sigma^2] = S^2 \qquad (2.94)$$

We now have all the pieces we need. Combining (2.81), (2.89), (2.93), and (2.94), we obtain a useful approximation for the variance of the sample average wrt the trajectory average,

$$V[\langle X \rangle - \langle x \rangle] \approx \frac{\sigma^2}{M} \qquad (2.95)$$

or in terms of the standard deviation,

$$\boxed{\sigma_{\langle x \rangle} \approx \frac{\sigma}{\sqrt{M}}} \qquad (2.96)$$

It is important to keep clear the distinction between the standard deviation σ of one value $x(t)$ wrt the average $\langle x \rangle$, defined by (2.77), and the standard deviation $\sigma_{\langle x \rangle}$ of the sample average $\langle x \rangle$ wrt the trajectory average $\langle X \rangle$, given in (2.96). Equation (2.96) says that if we wish to decrease the standard deviation of a computed average by a factor of 2, we must increase the length of the simulation by a factor of 4.

Equations (2.88) and (2.96) provide methods for approximating a time average and for quantifying the statistical precision of the approximation. We can actually go further and ask, how confident can we be about our estimates? To answer this, we must know the sampling distribution of the M_s possible averages. In simple random sampling that distribution is known from the *central limit theorem*: No matter how the values $x(t)$ are distributed along the trajectory, if the randomly sampled values are independent, then the M_s averages will tend toward a Gaussian. Moreover, if the $x(t)$ are themselves distributed as a Gaussian along the trajectory (as, in fact, they usually are in molecular dynamics), then the sampling distribution will be exactly Gaussian. Proofs of the central limit theorem are given by Ma [25], van Kampen [26], and Mathews and Walker [27]. Thus, we can express the reliability of a computed sample average in terms of the confidence levels for the Gauss distribution:

$\langle x \rangle$ lies within 0.67 $\sigma_{\langle x \rangle}$ of $\langle X \rangle$ for 50 of every 100 samples

$\langle x \rangle$ lies within 1 $\sigma_{\langle x \rangle}$ of $\langle X \rangle$ for 68 of every 100 samples

$\langle x \rangle$ lies within 2 $\sigma_{\langle x \rangle}$ of $\langle X \rangle$ for 95 of every 100 samples

$\langle x \rangle$ lies within 2.6 $\sigma_{\langle x \rangle}$ of $\langle X \rangle$ for 99 of every 100 samples

2.8.4 Systematic Sampling

The results in Section 2.8.3 are valuable, but they apply to simple random sampling: points from the trajectory are to be chosen arbitrarily with equal probability. But this is not what we do in molecular dynamics. Instead, we generate an arbitrary initial point on the trajectory and thereafter compute successive points systematically. Such systematic sampling does not generate a random sample—or does it?

Note that systematic sampling can produce a random sample if the sampled objects are randomly organized [28]. For example, if we want to produce a random sample of 5 cards from a deck of 52, we could fan the cards face down and arbitrarily select 5. This would be random sampling. Alternatively, we could shuffle the deck to randomly organize it, select an arbitrary starting point (say the twenty-third card), and then systematically select the next five cards. This is analogous to what we do in molecular dynamics: the sampling of the cards would be systematic, but the resulting sample would still be random. So, systematic sampling does not necessarily preclude a random sample. But are points on a phase-space trajectory randomly organized? Do all points on the trajectory have the same chance of being selected? Are sampled points independent?

It is clear that at least over some duration, nearby points on a trajectory are *not* independent. Rather, they are causally connected via the algorithm used to solve the equations of motion. Neighboring points are said to be *serially correlated*; so if we compute a time average $\langle x \rangle$ from a small number of adjacent points, the computed sample is not a random one, the sampling is not statistically unbiased, and the material in Section 2.8.3 must be modified.

However, the situation is salvable for those (many) quantities whose serial correlation times are relatively short. In molecular dynamics, serial correlations are disrupted by the chaotic nature of the phase-space trajectory. That is the theme of Section 2.5: the trajectory, in effect, randomly samples the constant-energy surface and so it is possible to draw, in a systematic way, a random sample. Exceptions occur in metastable situations, such as near phase transitions. So the final question is: Are serial correlation times short enough for most points on a computed trajectory to be uncorrelated?

For a particular property $x(t)$, the serial correlation time can be obtained from a correlation coefficient,

$$C(t) = \frac{\sum\limits_{k}^{M}(x(t_k)-\langle x\rangle)(x(t_k+t)-\langle x\rangle)}{\sum\limits_{k}^{M}(x(t_k)-\langle x\rangle)^2} \qquad (2.97)$$

This coefficient measures how the value of x at time t_k+t is correlated with its value at an earlier time t_k. The function $C(t)$ is exactly that quantity often used in statistical analyses to determine the strength of a correlation between two variables (see, e.g., [29]). Here, however, the two variables are actually the same quantity sampled at two different times. In statistical mechanics, $C(t)$ is called a time correlation function; properties of such functions are discussed in Chapter 7.

The denominator in (2.97) is merely a normalization factor: It forces $C(t) = 1$ when $t = 0$, that is, when there is no elapsed time. Further, the time average $\langle x\rangle$ is subtracted from each $x(t)$ to force $C(t) \to 0$ when the elapsed time grows long. That is, after some sufficiently long time, we expect $x(t_k+t)$ to become uncorrelated with $x(t_k)$. Thus, we expect $C(t)$ to be initially equal to unity (full correlation) and to decay to zero as the serial correlation weakens. Computing a time correlation function enables us to identify a *relaxation time*: a time beyond which, on the average, serial correlations are disrupted.

Figure 2.12 shows correlation coefficients $C(t)$ for the instantaneous temperature and pressure obtained from a simulation of 108 Lennard-Jones atoms. The figure shows that serial correlations have died away in about 0.5 psec or 50 integration time-steps, which is roughly the average collision interval. That is, at the density and temperature of the figure, 50 integration steps is roughly the time required for an atom to travel one mean free path (Exercise 2.17). For times $t > 0.5$ psec in Figure 2.12, $C(t)$ deviates from zero because of statistical fluctuations whose magnitudes depend on the number of atoms and on the number of time origins used in forming the averages. A relaxation time of 50 time-steps is typical of properties that are easily estimated from molecular dynamics, but we caution that each property will have its own relaxation time and that time has to be determined in preliminary runs before production runs are started.

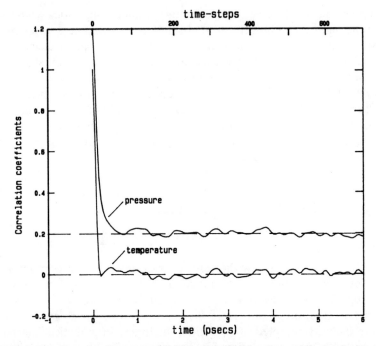

FIGURE 2.12 For simple properties of atomic fluids, the serial correlation time is about 50 integration time-steps. Here we show the correlation coefficient (2.97) for the instantaneous temperature (lower curve) and pressure (upper curve) of 108 Lennard-Jones atoms at $\rho\sigma^3 = 0.6$, $kT/\varepsilon = 1.542$. Each curve is an average over 80,000 time origins. For clarity, the curve for the pressure has been shifted vertically upward by 0.2 units.

A strategy for avoiding serial correlations is to divide the computed trajectory of length $M\Delta t$ into m segments, each of duration $n\Delta t$, where $n\Delta t$ is greater than the relaxation time for $x(t)$, the property of interest. Then to obtain m-values of $x(t)$ from which a sample average $\langle x \rangle$ may be computed, any of the following may be used:

(a) *stratified systematic sampling*, in which one value of $x(t)$ is systematically drawn from each segment, or

(b) *stratified random sampling*, in which one value of $x(t)$ is randomly drawn from each segment, or

(c) *two-stage sampling* (aka coarse graining), in which the n-values in each segment are averaged according to

$$\bar{x} = \frac{1}{n} \sum_{k}^{n} x(k\Delta t) \qquad (2.98)$$

Then $\langle x \rangle$ is estimated by averaging the m coarse-grain averages.

When the number of segments m is reasonably large, these three procedures give comparable results, within the statistical uncertainties provided by the standard deviation (2.96). Which procedure to use depends partly on the property being estimated. For static quantities, such as thermodynamic properties, coarse graining is usually preferred since it often gives smaller standard deviations than the other two methods. For static structural properties, such as the radial distribution function, either of methods (a) or (b) should be used. For these static properties, if you insist on simply averaging all computed values of $x(t)$ and ignore serial correlations, then the standard deviation of the average is no longer given by (2.96). Instead, the uncertainty is increased by a factor that involves the correlation coefficient (2.97) [30]. In computing time correlation functions, then, of course, the correlations are themselves the objective of the study and method (a) should be used with sampling at intervals smaller than one relaxation time.

2.9 PERIODIC BOUNDARY CONDITIONS

Molecular dynamics is typically applied to systems containing several hundred or a few thousand atoms. Such small systems are dominated by surface effects—interactions of the atoms with the container walls. For example, to hold 500 atoms at a liquid density, a cube must have an edge length of about 8.5 atomic diameters; however, wall–fluid interactions extend 4 to 10 atomic diameters from each wall. A simulation of this system would provide information on the behavior of the liquid near a solid surface, not information on the bulk liquid. In simulations in which these surface effects are not of interest, they are removed by using periodic boundary conditions (pbc).

To use pbc in a simulation of N atoms confined to a volume V, we imagine that volume V is only a small portion of the bulk material. The volume V is called the *primary cell*; it is representative of the bulk material to the extent that the bulk is assumed to be composed of the primary cell surrounded by exact replicas of itself. These replicas are called *image cells*. The image cells are each the same size and shape as the primary cell and each image cell contains N atoms, which are *images* of the atoms in the primary cell. Thus the primary cell is imagined to be periodically replicated in all directions to form a macroscopic sample of the substance of interest. This periodicity extends to the positions and momenta of the images in image cells.

To discuss how the positions and momenta of image atoms are related to those of atoms in the primary cell, we introduce the following notation. To each cell we assign a reference frame, located in a corner of the cell, as in Figure 2.13. The frame in the primary cell is called the *primary frame*; the others are image frames. In D dimensions, each cell can be identified by a D-dimensional vector $\boldsymbol{\alpha}$—the cell translation vector—whose components are either signed integers or zero. For example, in three dimensions the primary cell has $\boldsymbol{\alpha} = \{0, 0, 0\}$. Then we write any vector $\mathbf{r}$ that has its head in cell $\boldsymbol{\alpha}_1$

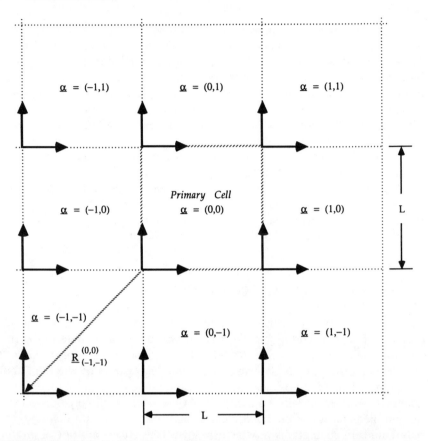

FIGURE 2.13 Periodic boundary conditions in two dimensions. The primary cell is surrounded by eight image cells and each cell is identified by a cell translation vector $\boldsymbol{\alpha}$. Each cell has a reference frame whose origin is the lower left corner of the cell. Each image frame is located wrt the primary frame by a vector **R**, such as the one shown here for cell $\boldsymbol{\alpha} = (-1, -1)$.

but originates at the origin of the reference frame in cell $\boldsymbol{\alpha}_2$ as $\mathbf{r}_{(\boldsymbol{\alpha}_1)}^{(\boldsymbol{\alpha}_2)}$. For cubic cells in three dimensions, the origin of a reference frame in any image cell $\boldsymbol{\alpha}$ can be located wrt the primary frame by a vector $\mathbf{R}_{(\boldsymbol{\alpha})}^{(0,0,0)}$, where

$$\mathbf{R}_{(\boldsymbol{\alpha})}^{(0,0,0)} = L\boldsymbol{\alpha} \tag{2.99}$$

and L is the scalar length of one edge of a cubic cell. A sample **R**-vector is shown, for two dimensions, in Figure 2.13.

Now each image of an atom i occupies the same position wrt its image frame as atom i occupies wrt the primary frame; that is,

$$\mathbf{r}_{i(0,0,0)}^{(0,0,0)} = \mathbf{r}_{i(\boldsymbol{\alpha})}^{(\boldsymbol{\alpha})} \qquad \text{for all image cells } \boldsymbol{\alpha} \tag{2.100}$$

Furthermore, each image of i has the same momentum as i,

$$\mathbf{p}_{i(0,0,0)}^{(0,0,0)} = \mathbf{p}_{i(\alpha)}^{(\alpha)} \qquad \text{for all image cells } \boldsymbol{\alpha} \qquad (2.101)$$

Cells are separated by open boundaries, so atoms and images can freely enter or leave any cell. Nevertheless, the number of atoms in each cell is the constant N because when an atom i leaves the primary cell, an image of i simultaneously enters the primary cell through an opposite face, as illustrated in Figure 2.14. For this to work properly, the shapes of the cells must be space filling.

During a simulation, we need store only the positions of the N atoms in the primary cell. Positions of images can be computed, when needed, by

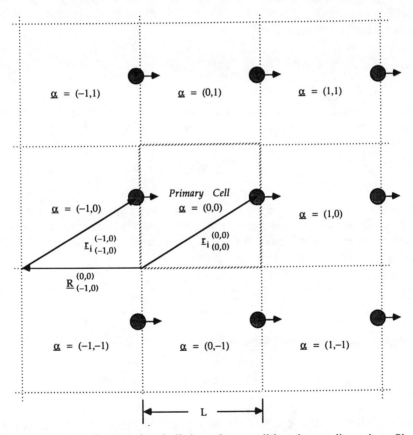

FIGURE 2.14 Application of periodic boundary conditions in two dimensions. Since the events in the primary cell are duplicated in each image cell, there is conservation of mass in each cell. As atom i leaves the primary cell, an image of i enters the primary cell from an adjacent cell. At the instant at which i crosses the cell boundary, its location vector is transformed by the two-dimensional version of (2.114).

coordinate transformations. In three dimensions the simplest such transformations occur for cubic cells. Thus, for a cube of edge L, an image of atom i in cell $\boldsymbol{\alpha}$ is located wrt the primary frame by

$$\mathbf{r}_{i(\alpha)}^{(0,0,0)} = \mathbf{r}_{i(\alpha)}^{(\alpha)} + \mathbf{R}_{(\alpha)}^{(0,0,0)} \tag{2.102}$$

This vector addition is illustrated in Figure 2.14. Using (2.99) and (2.100) on the rhs of (2.102) gives

$$\mathbf{r}_{i(\alpha)}^{(0,0,0)} = \mathbf{r}_{i(0,0,0)}^{(0,0,0)} + L\boldsymbol{\alpha} \tag{2.103}$$

Thus, to compute the position of the image, we need only the position of atom i in the primary cell plus the cell translation vector $\boldsymbol{\alpha}$.

We now consider calculation of pairwise additive forces and potential energies in periodic systems. For just the N atoms in the primary cell, the total potential energy, call it $\mathcal{U}_{pc}$, is

$$\mathcal{U}_{pc} = \frac{1}{2} \sum_{i \neq j} \sum u\left(\mathbf{r}_{ij(0,0,0)}^{(0,0,0)}\right) \tag{2.104}$$

where u is the two-body potential. However, in periodic systems, besides atoms in the primary cell, image atoms also contribute to $\mathcal{U}$; thus,

$$\mathcal{U} = \frac{1}{2} \sum_{\alpha} \sum_{i \neq j} \sum u\left(\mathbf{r}_{ij(0,0,0)}^{(0,0,0)} - \boldsymbol{\alpha}L\right) \tag{2.105}$$

Here $\mathbf{r}_{ij} - \boldsymbol{\alpha}L = \mathbf{r}_i - (\mathbf{r}_j + \boldsymbol{\alpha}L)$. In a D-dimensional system the sum over the cell translation vector $\boldsymbol{\alpha}$ represents a D-fold sum that accumulates contributions from the primary cell ($\boldsymbol{\alpha} = \{0,0,0\}$) and from all image cells. For example, in three dimensions,

$$\sum_{\alpha} (\cdots) = \sum_{\alpha_x = -\infty}^{\infty} \sum_{\alpha_y = -\infty}^{\infty} \sum_{\alpha_z = -\infty}^{\infty} (\cdots) \tag{2.106}$$

Now consider any one atom, m, in the primary cell. Its potential energy is

$$\mathcal{U}_m = \frac{1}{2} \sum_{\alpha} \sum_{j \neq m} u\left(\mathbf{r}_{mj(0,0,0)}^{(0,0,0)} - \boldsymbol{\alpha}L\right) \tag{2.107}$$

and note that each image of atom m also has the potential energy $\mathcal{U}_m$. For example, consider the image having $\boldsymbol{\alpha} = \{1,1,1\}$ and label it m'. Then

$$\mathcal{U}_{m'} = \frac{1}{2} \sum_{\alpha} \sum_{j \neq m'} u\left(\mathbf{r}_{m'j(1,1,1)}^{(1,1,1)} - \boldsymbol{\alpha}L\right) \tag{2.108}$$

But according to (2.100)

$$\mathbf{r}_{mj(0,0,0)}^{(0,0,0)} = \mathbf{r}_{m'j(1,1,1)}^{(1,1,1)} \tag{2.109}$$

Therefore,

$$\mathcal{U}_m = \mathcal{U}_{m'} \tag{2.110}$$

where, in general, m' is any image of atom m. Because of (2.109) and (2.110), the force felt by any image m' is exactly the same as that felt by the real atom m,

$$\frac{\partial \mathcal{U}_m}{\partial \mathbf{r}_{m(0,0,0)}^{(0,0,0)}} = \frac{\partial \mathcal{U}_{m'}}{\partial \mathbf{r}_{m'(1,1,1)}^{(1,1,1)}} \tag{2.111}$$

or

$$\mathbf{F}_m = \mathbf{F}_{m'} \tag{2.112}$$

where, in general, m' is any image of m. Thus, image atoms in image cells follow trajectories that are exact duplicates of those followed by the atoms in the primary cell. Analogous to (2.105), the force on atom m is computed by

$$\mathbf{F}_m = \sum_{\alpha} \sum_{j \neq m} \mathbf{f}(\mathbf{r}_{mj} - \alpha L) \tag{2.113}$$

where $\mathbf{f}$ represents the two-body force, $\mathbf{r}_{mj}$ is the vector between atoms m and j in the primary cell, and the sum over α is as in (2.106).

Boundary conditions are defined by telling what happens to an atom when it encounters a border of the system. In a system having periodic boundaries, any atom that leaves the primary cell is replaced by the image that simultaneously enters the primary cell. For example, if atom i moves from the primary cell to cell α an image is brought into the primary cell from image cell $-\alpha$. Thus, the position of atom i is transformed by

$$\mathbf{r}_{i(0,0,0)}^{(0,0,0)} \Rightarrow \mathbf{r}_{i(-\alpha)}^{(0,0,0)} = \mathbf{r}_{i(0,0,0)}^{(0,0,0)} - L\alpha \tag{2.114a}$$

According to (2.101) the image from cell $-\alpha$ has the same velocity as the real atom i; therefore, the momentum $\mathbf{p}_i$ is unaffected by the translation in position:

$$\mathbf{p}_{i(0,0,0)}^{(0,0,0)} \Rightarrow \mathbf{p}_{i(-\alpha)}^{(0,0,0)} = \mathbf{p}_{i(0,0,0)}^{(0,0,0)} \tag{2.114b}$$

The transformations (2.114) define pbc.

How many image cells do we need? It depends on the range of intermolecular forces. Formally, each atom in the primary cell interacts with all $N-1$ other atoms and all their images. When the forces are sufficiently short ranged (as, e.g., they are in the Lennard-Jones model), then we need only those image cells that adjoin the primary cell. For squares in two dimensions, there are eight adjacent image cells; for cubes in three dimensions, the number of adjacent images is 26. For space-filling shapes other than the cube, transformations analogous to (2.114a) have been derived [31], but in most applications there appears to be little justification for the added complexity introduced by those geometries.

Use of pbc removes unwanted surface effects at the expense of introducing artificial periodicity into the simulated system. In Section 2.10 we discuss how this artificial periodicity affects conserved quantities. In general, use of pbc restricts molecular dynamics to studies of short-range and short-lived phenomena. However, these are not new limitations; they are already imposed on us by the small system size ($N < 1000$) and the small integration time-step needed to solve the equations of motion. For equilibrium properties, particularly thermodynamics and local structure, available evidence [32] suggests that the effects of pbc are small, or at least are overshadowed by other kinds of systematic error. Studies of pbc effects are hampered by the difficulty of isolating the effects of periodicity from other effects, particularly system size. Repeating runs with increasing numbers of atoms tests both effects and is usually an adequate test.

Effects of periodicity are more pronounced on dynamic properties, particularly time correlation functions. The imposed periodicity in space causes a suppressed but finite periodicity in time; for example, in simulations of small systems, the tails of time correlation functions exhibit artifacts caused by periodic boundaries [33]. These unrealistic correlations occur when the delay time for sampling the correlation function exceeds the time τ_{pbc} needed for spatial translations to become periodic. The periodic correlation time τ_{pbc} is equivalent to the time required for a longitudinal wave to traverse the simulation cell of side L and since sound propagates by longitudinal waves, the sonic velocity w provides a convenient means for estimating τ_{pbc},

$$\tau_{pbc} = \frac{L}{w} = \frac{1}{w}\left(\frac{N}{\rho}\right)^{1/3} \tag{2.115}$$

Recall that the sonic velocity is simply related to the adiabatic compressibility κ_s by

$$w = \frac{1}{\sqrt{\rho m \kappa_s}} \tag{2.116}$$

where m is the atomic mass. At a specified system density ρ, we manipulate the periodic correlation time by changing the system size; thus in three dimensions, τ_{pbc} increases as the cube root of the number of atoms. Representative values of τ_{pbc} are given in Table 2.2.

TABLE 2.2   Representative Values for Periodic Correlation Time τ_{pbc} for Lennard-Jones Fluid at $\rho\sigma^3 = 0.72$, $kT/\varepsilon = 1$

N	Container Edge L (Å)	τ_{pbc} (psec)
108	18.1	2.7
256	24.1	3.6
500	30.1	4.5
864	36.2	5.5
1372	42.2	6.4
2048	48.2	7.3
4000	60.3	9.1

The value of the sonic velocity $w\sqrt{m/\varepsilon} = 4.19 \Rightarrow w = 6.62$ Å/psec was computed from the Nicolas et al. [34] equation of state. Property values were made dimensional using potential parameters ($\sigma = 3.405$ Å, $\varepsilon/k = 120$ K) and molecular weight characteristic of argon.

2.10   CONSERVATION PRINCIPLES

In 1918 Noether [35] published theorems stating that in dynamical systems conservation laws are consequences of inherent symmetries. Thus, if we can identify symmetries—operations that leave the system invariant—then we can deduce the corresponding conservation laws [36]. For N-body systems maintained in isolation, neither mass nor energy can be exchanged with the surroundings, and the conserved quantities are mass, energy, linear momentum, and angular momentum. However, when isolated systems are simulated by molecular dynamics, use of pbc may disrupt symmetries and prevent these quantities from being conserved. In this section we consider the effects of periodic boundaries on conservation principles.

Mass. Isolated systems, by definition, have a constant number of atoms N, and in Section 2.9 we noted that pbc do not affect the number of atoms in the primary cell. Whenever an atom leaves the primary cell, an image atom enters through an opposite face.

Total Energy. Invariance of a system under translations in time leads, via Noether's theorem, to conservation of energy. But do pbc affect the energy? For hard bodies, such as hard spheres, the question can be immediately answered. Hard bodies have no potential energy except at the time of collisions, and so, between collisions, the total energy is purely kinetic. Since, according to (2.114b), pbc do not affect atomic momenta, periodic boundaries cannot disturb conservation of energy in hard-body simulations.

In the general case, we can answer the question by considering an atom n about to leave the primary cell. At the instant before leaving, the system's total energy is

$$E_{\text{old}} = \frac{m}{2} \sum_{i \neq n}^{N} v_i^2 + \mathscr{U}' + \frac{m}{2} v_{n,\text{old}}^2 + \mathscr{U}_{n,\text{old}} \tag{2.117}$$

where $\mathscr{U}_n$ is the potential energy of atom n,

$$\mathscr{U}_n = \frac{1}{2} \sum_{\alpha} \sum_{j \neq n} u(\mathbf{r}_{nj} - \alpha L) \tag{2.118}$$

and $\mathscr{U}'$ is the potential energy of all other atoms, $\mathscr{U}' = \mathscr{U} - \mathscr{U}_n$. When atom n leaves the primary cell, the periodic boundary transformations (2.114) are applied and the total energy becomes

$$E_{\text{new}} = \frac{m}{2} \sum_{i \neq n}^{N} v_i^2 + \mathscr{U}' + \frac{m}{2} v_{n,\text{new}}^2 + \mathscr{U}_{n,\text{new}} \tag{2.119}$$

Now the first two terms on the rhs of (2.119) do not involve atom n and are therefore unaffected by the boundary transformation. The "new" terms in (2.119) are the kinetic and potential energies of an image of atom n. But according to (2.101) each image of n has the same velocity as n itself, and according to (2.110) each image of n has the same potential energy as n. Hence

$$E_{\text{new}} = E_{\text{old}} \tag{2.120}$$

Thus, when the potential energy is computed using real and image atoms, as in (2.105), the total system energy is unaffected by pbc.

Linear Momentum. Invariance of a system under translations in space leads to conservation of linear momentum. Since no external forces act on isolated systems, they are invariant under spatial translations, and therefore

$$\mathscr{P} = \sum_{i}^{N} \mathbf{p}_i(t) = \sum_{i}^{N} \mathbf{p}_i(0) = \text{const} \tag{2.121}$$

This conservation principle extends to periodic systems because, according to (2.114b), all images of an atom i have the same momenta as i itself. Thus, $\mathscr{P}$ can be computed from just the N atoms in the primary cell.

Conserved linear momentum poses a delicate problem in the statistical mechanics of isolated systems. The concern is whether there are constants of the motion other than the total energy. If there are, then the trajectory in phase space is confined, not to a constant-energy surface, but to a smaller dimensional surface that is common to all conserved quantities. The full constant-energy surface is not sampled. The possibility of conserved linear momentum is ignored in many developments of statistical mechanics; or if the issue is raised, it is quickly dismissed. One careful discussion is given by Fowler and Guggenheim [37], who point out that real isolated systems are confined to vessels. Such confinement, which provides an interaction between system and container, disrupts linear momentum, so that the phase point is not constrained to a surface of constant $\mathscr{P}$.

However, in the usual molecular dynamics simulation, container walls are replaced by periodic boundaries and linear momentum is conserved. Formally, then, the simulation of an isolated system with periodic boundaries is not quite a realization of the usual microcanonical ensemble of statistical mechanics. The phase point does not sample the full constant-energy surface; instead, it is confined to the surface of intersection between the constant-E and the constant-$\mathscr{P}$ surfaces. In general, this additional constraint has two effects on property values computed from simulation. The first is a loss of three degrees of freedom because $\mathscr{P}_x$, $\mathscr{P}_y$, and $\mathscr{P}_z$ are each individually constants of the motion. Therefore those properties that specifically involve the number of degrees of freedom will be affected. For example, (2.43), which relates the average kinetic energy to the temperature, becomes

$$\tfrac{3}{2}kT = \frac{\langle E_k \rangle}{N-1} + O(N^{-1}) \tag{2.122}$$

Here $\langle E_k \rangle$ is the average translational kinetic energy calculated as in (2.43). The loss of one atomic degree of freedom means that the temperature is a factor of $N/(N-1)$ larger than the value given by (2.43). The second term in (2.122) occurs because a constant-$\mathscr{P}$ ensemble does not strictly mimic the microcanonical ensemble; it is the usual $1/N$ correction for differences in ensembles [38].

Both effects in (2.122) become smaller as the system size, measured by N, increases. As a rule of thumb, both effects can usually be neglected for systems having, say, $N > 100$ atoms. We will assume $N > 100$ for the simulations considered in this book and not worry further about corrections for conservation of linear momentum. If, however, you choose to study a system having $N < 100$, you will have to compute these corrections for each property of interest and decide whether their contribution is important to your application. The effects of conserved linear momentum are discussed in more detail elsewhere [39–41].

Angular Momentum. Invariance under spatial rotations leads to conservation of angular momentum. Since no external torques act on isolated systems, we expect

$$\mathscr{Q} = \sum_i^N \mathbf{r}_i \times \mathbf{p}_i = \text{const} \tag{2.123}$$

where $\times$ represents the vector cross product; however, for the N atoms in the primary cell, periodic boundaries disrupt conservation of $\mathscr{D}$. Consider an atom m that leaves the primary cell during the interval between t and $t + \Delta t$. At time t

$$\mathscr{D}(t) = \sum_{i \neq m}^{N} \{ \mathbf{r}_i(t) \times \mathbf{p}_i(t) \} + \mathbf{r}_m(t) \times \mathbf{p}_m(t) \tag{2.124}$$

However, at $t + \Delta t$ atom m has been replaced in the primary cell by an image having position $\mathbf{r}_m + \alpha L$, so the total angular momentum is now

$$\mathscr{D}(t + \Delta t) = \sum_{i \neq m}^{N} \{ \mathbf{r}_i(t + \Delta t) \times \mathbf{p}_i(t + \Delta t) \} + (\mathbf{r}_m + \alpha L) \times \mathbf{p}_m \tag{2.125}$$

Hence, the value of $\mathscr{D}$ undergoes a step change between t and $t + \Delta t$. However, over a finite time the numbers of atoms exchanged with images will

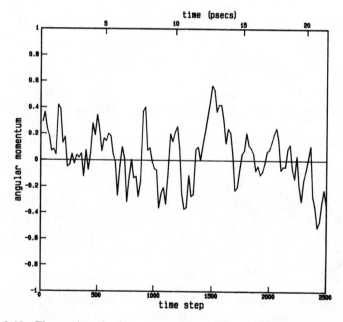

FIGURE 2.15   Fluctuations in the y-component of the angular momentum $\mathscr{D}_y(t) = \sum_i^N (z_i \dot{x}_i - \dot{z}_i x_i)$, computed during a molecular dynamics simulation of 108 Lennard-Jones atoms at $\rho \sigma^3 = 0.7$, $kT/\varepsilon = 1$. Units are σ for position and $(m/\varepsilon)^{1/2}$ for velocity. Values are plotted at intervals of 20 integration time-steps; using potential parameters for argon, 20 time-steps = 0.172 psec.

be nearly equal in all spatial directions so that, as shown in Figure 2.15, components of the angular momentum fluctuate about essentially constant values.

2.11 SUMMARY

In this chapter we have reviewed aspects of multibody dynamics that are fundamental to molecular dynamics simulation. The discussion focused on the interpretation of dynamics as the unfolding of a trajectory in phase space. The unfolding is governed by classical equations of motion in either their Newtonian or Hamiltonian forms. For systems at equilibrium, the trajectory is confined to the hypersurface of a constant Hamiltonian, and if the system is isolated, the Hamiltonian is the total energy. Rudimentary features of phase-space trajectories were illustrated using simple one- and two-dimensional motion. Those simple periodic and quasiperiodic motions should help you become comfortable with the concept of phase space, but we emphasize that the trajectories produced by multibody systems are necessarily more complex. Periodic and quasiperiodic motions are characteristic of systems having few degrees of freedom; when a system has many degrees of freedom that are coupled nonlinearly, its natural tendency is to follow unstable trajectories [42]. Molecular dynamics is a numerical realization of those trajectories.

Once a molecular dynamics simulation has generated a phase-space trajectory, the trajectory serves as raw data for obtaining time averages of properties. We introduced some of the concerns that arise in computing reliable approximations to time averages. It is this computation of time averages that distinguishes molecular dynamics from Monte Carlo simulations and from formal statistical mechanics.

We emphasized that to provide reliable time averages, molecular dynamics must generate phase-space trajectories that are unstable to small perturbations. In effect, the sampling of the constant-Hamiltonian surface should be apparently random. How can *random* generation of phase points be reconciled with the *deterministic* calculation of those points from classical equations of motion? The calculational demonstration in Section 2.5 illustrates that the deterministic is made chaotic because of molecular interactions. After a few interactions positions and velocities are essentially unrelated to earlier positions and velocities: the correlation time is short and the trajectory is unstable. Consequently, over a relatively short duration, the Maxwell distribution of velocities develops and time averages for properties can be computed.

Molecular interactions, then, are both the bane and blessing of molecular dynamics. Molecular interactions cause a simulation to use great amounts of computer time; but without interactions, the simulation would be meaningless.

REFERENCES

[1] H. Goldstein, *Classical Mechanics*, 2nd ed., Addison-Wesley, Reading, MA, 1980, Chapter 8.

[2] R. Penrose, *The Emperor's New Mind*, Oxford University Press, New York, 1989, pp. 176–184.

[3] Hao Bai-Lin, Ed., *Chaos*, World Scientific, Singapore, 1984, Chapter 2.

[4] J. L. Lebowitz and O. Penrose, "Modern Ergodic Theory," *Phys. Today*, **26**(2), 23 (1973).

[5] H. Poincaré, "Sur le Problème des Trois Corps et les Équations de la Dynamique," *Acta Math.*, **13**, 1 (1890); see also R. Kurth, *Axiomatics of Classical Statistical Mechanics*, Pergamon, Oxford, 1960.

[6] W. G. Hoover and W. T. Ashurst, "Nonequilibrium Molecular Dynamics" in *Theoretical Chemistry. Advances and Perspectives*, H. Eyring and D. Henderson, Eds., Academic, New York, 1975, p. 5.

[7] A. N. Kolmogorov, "On the Conservation of Quasi-Periodic Motions for a Small Change in the Hamiltonian Function," *Akad. Nauk. SSSR Doklady*, **98**, 527 (1954); reprinted in *Chaos*, Hao Bai-Lin, Ed., World Scientific, Singapore, 1984.

[8] V. I. Arnol'd, "Proof of a Theorem of A. N. Kolmogorov on the Invariance of Quasi-Periodic Motions under Small Perturbations of the Hamiltonian," *Russ. Math. Surv.*, **18**(5), 9 (1963).

[9] J. Moser, "On Invariant Curves of Area-preserving Mappings on an Annulus," *Nachr. Akad. Wiss. Göttingen Math. Physik*, **K1**, 1 (1962).

[10] E. Fermi, J. Pasta, and S. M. Ulam, "Studies of Non Linear Problems" in *Enrico Fermi: Collected Papers*, Vol. II, University of Chicago Press, Chicago, 1965, p. 978.

[11] S. M. Ulam, *Adventures of a Mathematician*, Scribner's Sons, New York, 1976, pp. 226–228.

[12] S. G. Brush, *The Kind of Motion We Call Heat*, Vol. 2, North-Holland, Amsterdam, 1986, pp. 363–385.

[13] D. Ruelle, *Chaotic Evolution and Strange Attractors*, Cambridge University Press, Cambridge, 1989.

[14] W. G. Hoover, *Molecular Dynamics*, Lecture Notes in Physics, Vol. 258, Springer-Verlag, Berlin, 1986, pp. 6–9.

[15] W. Yourgrau, A. van der Merwe, and G. Raw, *Treatise on Irreversible and Statistical Thermophysics*, Dover, New York, 1982, p. 63f.

[16] S. Chapman and T. G. Cowling, *The Mathematical Theory of Non-Uniform Gases*, Cambridge University Press, London, 1961, pp. 26, 37, 40–41.

[17] G. D. Birkhoff, "Proof of a Recurrence Theorem for Strongly Transitive Systems," *Proc. Nat. Acad. Sci.*, **17**, 650 (1931); "Proof of the Ergodic Theorem," *Proc. Nat. Acad. Sci.*, **17**, 656 (1931).

[18] J. von Neumann, "Proof of the Quasi-Ergodic Hypothesis," *Proc. Nat. Acad. Sci.*, **18**, 70 (1932); "Physical Applications of the Ergodic Hypothesis," *Proc. Nat. Acad. Sci.*, **18**, 263 (1932).

[19] S. K. Ma, *Statistical Mechanics*, World Scientific, Philadelphia, PA, 1985, p. 454.

[20] S. Chapman and T. G. Cowling, *The Mathematical Theory of Non-Uniform Gases*, Cambridge University Press, London, 1961, Chapter 4.

[21] S. K. Ma, *Statistical Mechanics*, World Scientific, Philadelphia, PA, 1985, Chapter 6.

[22] L. D. Landau and E. M. Lifshitz, *Statistical Physics*, 2nd ed., Pergamon, Oxford, 1977, Chapter 12.

[23] B. Williams, *A Sampler on Sampling*, Wiley, New York, 1978.

[24] W. G. Cochran, *Sampling Techniques*, 2nd ed., Wiley, New York, 1963, Chapter 2.

[25] S. K. Ma, *Statistical Mechanics*, World Scientific, Philadelphia, PA, 1985, Chapter 12.

[26] N. G. van Kampen, *Stochastic Processes in Physics and Chemistry*, North-Holland, Amsterdam, 1981, p. 27.

[27] J. Mathews and R. L. Walker, *Mathematical Methods of Physics*, 2nd ed., Addison-Wesley, New York, 1970, Chapter 14.

[28] B. Williams, *A Sampler on Sampling*, Wiley, New York, 1978, p. 135.

[29] J. R. Taylor, *An Introduction to Error Analysis*, University Science Books, Mill Valley, CA, 1982, p. 180.

[30] W. G. Cochran, *Sampling Techniques*, 2nd ed., Wiley, New York, 1963, p. 210.

[31] D. N. Theodorou and U. W. Suter, "Geometrical Considerations in Model Systems with Periodic Boundaries," *J. Chem. Phys.*, **82**, 955 (1985).

[32] L. R. Pratt and S. W. Haan, "Effects of Periodic Boundary Conditions on Equilibrium Properties of Computer Simulated Fluids. I. Theory," *J. Chem. Phys.*, **74**, 1864 (1981); "Effects of Periodic Boundary Conditions on Equilibrium Properties of Computer Simulated Fluids. II. Application to Simple Liquids," *J. Chem. Phys.*, **74**, 1873 (1981).

[33] B. J. Berne and G. D. Harp, "On the Calculation of Time Correlation Functions," *Adv. Chem. Phys.*, **17**, 63, 130–132 (1970).

[34] J. J. Nicolas, K. E. Gubbins, W. B. Streett, and D. J. Tildesley, "Equation of State for the Lennard-Jones Fluid," *Mol. Phys.*, **37**, 1429 (1979).

[35] E. Noether, "Invariant Variation Problems," English translation by M. A. Tavel, in *Transport Theory Stat. Phys.*, **1**, 186 (1971).

[36] P. J. Olver, *Applications of Lie Groups to Differential Equations*, Springer-Verlag, New York, 1986, p. 277ff.

[37] R. H. Fowler and E. A. Guggenheim, *Statistical Thermodynamics*, Cambridge University Press, London, 1939, pp. 10–12.

[38] J. L. Lebowitz, J. K. Percus, and L. Verlet, "Ensemble Dependence of Fluctuations with Application to Machine Computations," *Phys. Rev.*, **153**, 250 (1967).

[39] W. W. Wood in *Fundamental Problems in Statistical Mechanics*, E. G. D. Cohen, Ed., North-Holland, Amsterdam, 1975.

[40] F. Lado, "Some Topics in the Molecular Dynamics Ensemble," *J. Chem. Phys.*, **75**, 5461 (1981).

[41] D. C. Wallace and G. K. Straub, "Ensemble Corrections for the Molecular Dynamics Ensemble," *Phys. Rev. A*, **27**, 2201 (1983).

[42] J. Ford, "How Random Is a Coin Toss?" *Phys. Today*, **36**(4), 40 (1983).

EXERCISES

2.1   Mechanical work is defined by

$$W = \int \mathbf{F} \cdot d\mathbf{r}$$

Show that kinetic energy is the work required to move a body of mass m from rest to velocity $\mathbf{v}$.

2.2   The analytic solutions of (2.26) for the instantaneous position and momentum of the one-dimensional harmonic oscillator are

$$x(t) = x(0)\cos \omega t + p(0)(\omega/\gamma)\sin \omega t$$

$$p(t) = -x(0)m\omega \sin \omega t + p(0)\cos \omega t$$

where $\omega = (\gamma/m)^{1/2}$. Here $x(0)$ and $p(0)$ are the initial position and momentum, respectively.

(a) Using these results, compute the total energy from (2.25) and show that it is constant.

(b) Compute the time-average displacement $\langle x \rangle$ over one period, $T = 2\pi/\omega$, of the motion.

(c) Compute the root-mean-square (rms) displacement $(\langle x^2 \rangle)^{1/2}$ over one period.

(d) Show that $\langle U \rangle = \langle E_k \rangle = \frac{1}{2}E$ over one period; that is, the time-average potential and kinetic energies are equal.

2.3   Consider the trajectory of the ODHO given analytically in the previous exercise. The oscillator has frequency ω (radians/sec) and period $T = 2\pi/\omega$. You intend to compute time averages numerically by sampling the trajectory at discrete increments $\Delta t = T/n$.

(a) If the increment Δt is rational, show that on subsequent cycles your sampling will be periodic; that is, you will repeatedly sample the same points.

(b) Conversely, if the increment is irrational, show that on subsequent cycles your sampling will eventually cover the constant energy surface. How might these two procedures affect your estimates for time averages?

2.4   Using the analytic expressions for the trajectory from Exercise 2.2, show that the ODHO produces stable trajectories; that is, if the initial conditions are perturbed by small amounts, ε_x and ε_p, so

$$x'(0) = x(0) + \varepsilon_x$$

$$p'(0) = p(0) + \varepsilon_p$$

the deviations $\delta x(t) = x'(t) - x(t)$ and $\delta p(t) = p'(t) - p(t)$ do not grow as the oscillator repeats its cycle in the phase plane.

2.5 Consider a one-dimensional oscillator that has potential $u(x) = 0.25\gamma x^4$.

(a) Determine the general expression for the phase-plane trajectory.

(b) Derive Hamilton's equations of motion for this situation.

(c) Derive Newton's equation of motion.

(d) Determine how the total energy divides between the average kinetic and potential energies. To do this, start with the quantity $I = mx^2$. Take two time-derivatives of I, form its time average, and then argue that this average second derivative vanishes because the trajectory is bounded. The rest is algebra to relate the average kinetic and potential energies.

(e) For the special case of $m = 1$ and $\gamma = 1$ (in arbitrary units), plot the phase-plane trajectory that results from initial conditions $x(0) = 2$, $p(0) = 0$.

(f) Compare the plot obtained in (e) with that obtained for the ODHO that also has $m = 1$, $\gamma = 1$ and that starts from the same initial conditions as in (e).

2.6 Consider a one-dimensional oscillator that has potential energy

$$u(x) = -\tfrac{1}{2}x^2 + \tfrac{1}{4}x^4$$

(a) Sketch the potential.

(b) Derive Newton's and Hamilton's equations of motion.

(c) For what values of the total energy will the motion be confined to the region of either potential well and, therefore, not sample the full energy surface?

(d) Qualitatively sketch the family of phase-plane trajectories for various energy values.

2.7 A one-dimensional oscillator having $m = 1$ exhibits the phase-plane trajectory shown in Figure E2.7. Assuming $u = 0$ when $x = 0$, deduce the form of the potential energy function $u(x)$.

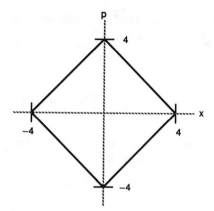

FIGURE E2.7

2.8 Consider a ball of mass m connected to two springs, as in Figure E2.8. The spring on the left is harmonic, has spring constant γ_1, and has its natural length when the ball is at r_1. The other spring is anharmonic, having the potential from Exercise 2.5 with spring constant γ_2 and its natural length when the ball is at r_2. Assume the system is isolated from its surroundings and that the ball can only move in one dimension. The displacement of the ball from the equilibrium position of spring 1 is measured by x, and its displacement from the equilibrium position of spring 2 is measured by y; hence,

$$x + y = \text{const} = c$$

Thus, only one distance variable, say x, is needed to locate the ball. The potential energy can then be written as

$$u(x) = \tfrac{1}{2}\gamma_1 x^2 + \tfrac{1}{4}\gamma_2(c - x)^4$$

(a) Obtain an expression for the value of x at which the forces on the ball are exactly balanced.

(b) Determine the equation for the phase-plane trajectory.

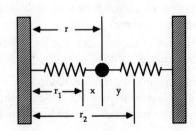

FIGURE E2.8

(c) Derive Hamilton's equations of motion for this system.

(d) Derive Newton's equations of motion.

(e) Study the behavior of the phase-plane trajectory by plotting the trajectory for various values of the spring constants γ_1 and γ_2, and the distance c.

2.9 A ball of mass m rolls on a floor between two walls, as in Figure E2.9. At the point of contact with each wall the ball encounters a harmonic spring as shown in the figure. Each spring has spring constant γ, and when not in contact with the ball, each lies at its natural length, which is a distance l from the midpoint between the two walls. Assuming the ball's motion is one dimensional and neglecting rolling friction, the potential energy can be written in terms of x, the distance from one wall, as

$$u(x) = \begin{cases} \frac{1}{2}\gamma(x-l)^2 & \text{if } x > l \\ 0 & \text{if } -l < x < l \\ \frac{1}{2}\gamma(x+l)^2 & \text{if } x < -l \end{cases}$$

Sketch the phase-plane trajectory for the motion of the ball.

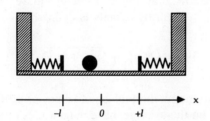

FIGURE E2.9

2.10 The solutions to the equations of motion for the two-dimensional uncoupled harmonic oscillator of Figure 2.4 are

$$x(t) = x(0)\cos\omega_x t + p_x(0)(\omega_x/\gamma_x)\sin\omega_x t$$

$$p_x(t) = -x(0)m\omega_x \sin\omega_x t + p_x(0)\cos\omega_x t$$

$$y(t) = y(0)\cos\omega_y t + p_y(0)(\omega_y/\gamma_y)\sin\omega_y t$$

$$p_y(t) = -y(0)m\omega_y \sin\omega_y t + p_y(0)\cos\omega_y t$$

where $\omega_x = (\gamma_x/m)^{1/2}$ and $\omega_y = (\gamma_y/m)^{1/2}$. Write a computer program that computes $x(t)$, $y(t)$, $p_x(t)$, and $p_y(t)$ as functions of time. From computed values, plot $\{y(t) \text{ vs. } x(t)\}$, $\{p_y(t) \text{ vs. } p_x(t)\}$, $\{p_x(t) \text{ vs. } x(t)\}$, and $\{p_y(t) \text{ vs. } x(t)\}$ and study how these plots behave for various initial conditions (a) when $(\gamma_x/\gamma_y)^{1/2}$ is rational and (b) when $(\gamma_x/\gamma_y)^{1/2}$ is irrational.

2.11 Consider the two-dimensional harmonic oscillator shown in Figure 2.4 and adopt the same parameter values ($m = 1$, $\gamma_x = 4$, $\gamma_y = 10$) and initial conditions [$x(0) = y(0) = 2$, $p_x(0) = p_y(0) = 0$] as in the figure (all in arbitrary units).

(a) In Section 2.4.1 we noted that the phase space is four dimensional and the constant-energy surface is three dimensional but the motion of the ball is restricted to a torus in two dimensions. Find a position (x, y) and momentum (p_x, p_y) that lies on the constant-energy surface but is not accessible to the ball, that is, a phase point that does *not* lie on the two-dimensional torus.

(b) Is the total linear momentum conserved for either situation depicted in Figure 2.4?

2.12 Contrive arguments based on the central limit theorem to explain (a) why molecular velocities describe a Maxwell distribution and (b) why instantaneous values of dynamic quantities describe Gaussian distributions.

2.13 (a) A system of N atoms, each of mass m, is in thermal equilibrium at temperature T. Show that the root-mean-square (rms) velocity equals $(3kT/m)^{1/2}$. The rms velocity is defined by

$$v_{rms} = \sqrt{\langle v_x^2 \rangle + \langle v_y^2 \rangle + \langle v_z^2 \rangle}$$

☒ (b) Of 1000 argon atoms in thermal equilibrium at 27°C, determine the number that, on the average, have magnitudes of their x-velocities greater than 1000 km/hr.

2.14 From the temperature distribution for 108 atoms given in Figure 2.11, ☒ determine the number of atoms, on the average, that would have kinetic energies corresponding to temperatures greater than $kT/\varepsilon = 1.6$.

2.15 We define an ideal gas as a system of noninteracting point particles—the points do not collide, they merely pass through one another. Explain why it would be fruitless to attempt a molecular dynamics simulation of an ideal gas. For a system of N ideal-gas particles, how many constants of the motion are there?

2.16 Prove that in an ideal gas the velocity of sound is proportional to the square root of absolute temperature and is independent of density.

2.17 Consider N atoms, each of diameter σ and mass m, occupying volume V at temperature T. We want to estimate the mean free path l, that is, the average distance an atom travels between collisions. To do so, we first estimate the collision rate N_c, which is the average number of collisions per unit time:

$N_c \approx$ average number of collisions experienced by one
atom of *radius* σ moving in a field of N points

This approximation to N_c is equal to the volume cut by a circle of radius σ moving at a velocity $\bar{v}_r$ through the substance of number density $\rho = N/V$. The velocity $\bar{v}_r$ is an average relative velocity of the atoms, which should be proportional to v_{rms}:

$$N_c \approx \left(\begin{array}{c} \text{area of circle} \\ \text{of radius } \sigma \end{array} \right) \times \left(\begin{array}{c} \text{velocity} \\ \text{of circle} \end{array} \right) \times \left(\begin{array}{c} \text{number density} \\ \text{of field} \end{array} \right)$$

$$= (\pi \sigma^2) \bar{v}_r \rho$$

We must now estimate $\bar{v}_r$. Assume all atoms are moving at the same speed—that of the rms velocity. When two atoms approach one another, at some angle θ, their relative velocity is $v_r = 2v_{rms} \sin(\theta/2)$. (See Figure E2.17.) The average is then

$$\bar{v}_r = \overline{2v_{rms} \sin \frac{\theta}{2}}$$

where

$$\overline{\sin \frac{\theta}{2}} = \frac{1}{2} \int_0^\pi d\theta \sin \frac{\theta}{2} \sin \theta = \frac{2}{3}$$

(a) Now using the expression for v_{rms} from Exercise 2.13, show that the dimensionless mean free path is given by

$$\frac{l}{\sigma} = \frac{3}{4\pi\rho\sigma^3}$$

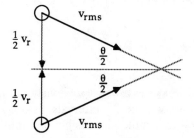

FIGURE E2.17

(b) Show that the dimensionless collision interval—the mean time between collisions—is

$$\frac{t_c}{\sigma\sqrt{m/\varepsilon}} = \frac{3}{4\pi(\rho\sigma^3)\sqrt{3kT/\varepsilon}}$$

(c) For an integration time step $\Delta t = 0.004(\sigma\sqrt{m/\varepsilon})$, show that at $\rho\sigma^3 = 0.7$ and $kT/\varepsilon = 1.2$, the collision interval is just under 50 time steps.

2.18 For a system of 100 hard spheres at density $\rho\sigma^3 = 0.5$, use the following procedure to estimate a lower bound for the Poincaré recurrence time. (This scheme is from S. Chandrasekhar, *Rev. Mod. Phys.*, **15**, 1 (1943), App. VI.)

(a) Between collisions, hard spheres travel along straight lines with constant velocities. So consider just the recurrence in momentum space, which is $3N$ dimensional. Now we must choose within what tolerance we will be satisfied that the initial velocities have been repeated. Since we want only a lower bound, we agree to be satisfied if each sphere regains each component of its initial velocity within 1% of 5σ, where σ is the standard deviation from the Maxwell velocity distribution. Thus, divide the range of attainable velocities ($\langle v \rangle \pm 2.5\sigma$) on each axis of momentum space into 100 equal segments. We therefore have subdivided momentum space into

$$N_s = 100^{3N} \text{ velocity intervals}$$

The number of ways W of distributing $3N$ velocity components among N_s intervals is given by (2.91); that is,

$$W \approx \frac{N_s!}{(3N)!(N_s - 3N)!}$$

Use Stirling's approximation (see Appendix M) for large x: $\ln x! \approx x \ln x - x$ to show that when $N_s \gg 3N$, W is approximated by

$$W \approx \frac{(N_s)^{3N}}{(3N)!}$$

Assuming each collision gives a new (previously unsampled) distribution of velocities among the N_s intervals, estimate the number of collisions required to sample all possible distributions.

(b) Use the estimate for the mean collision interval obtained in Exercise 2.17 to estimate the time required for the Poincaré recurrence estimated in (a). Note that the rate N_c in Exercise 2.17 is for one atom, so the total number of collisions per unit time is NN_c. If we use values for the parameters ε, σ, and m applicable to argon, the time unit in Exercise 2.17 is $\sigma/(m/\varepsilon)^{1/2} = 2$ psec. How does your estimate of the recurrence time compare with the age of the universe, which is thought to be about 15 billion years? (Recall we have ignored the time needed for a recurrence in position space, which must increase your estimate by at least a factor of N.)

2.19 Use a combination of words and equations to distinguish among the following:

(a) The variance of one number with respect to the sample average

(b) The variance of one sample average with respect to its expected average

(c) The variance of one sample average with respect to the trajectory average

(d) The variance of the expected sample average with respect to the trajectory average

2.20 A room contains five people who have the following amounts of money in their pockets (in the local coin of the realm):

Person	1	2	3	4	5
Amount	27	11	6	35	19

(a) How many unique samples (M_s) of two people can be drawn from this population of five?

(b) Do a calculation to demonstrate that the average of all M_s possible samples of two people equals the true average; that is, demonstrate that

$$E[\langle x \rangle] = \langle X \rangle$$

(c) Now prove for the general case of simple random sampling that $E[\langle x \rangle] = \langle X \rangle$ and therefore that the average $\langle x \rangle$ from one sample is an unbiased estimate of $\langle X \rangle$.

(d) Compute the variance S^2 of each person's pocket money relative to the true average; then compute the expected variance for your M_s two-person samples. Now show that these two variances are related by (2.92). What happens to the relation if you normalize S^2 by M_∞ rather than by $M_\infty - 1$?

2.21 For an unbiased estimator of $\langle X \rangle$, prove (2.90), which states that the expected mean-square error in $\langle x \rangle$ relative to $\langle X \rangle$ is the same as the variance of $\langle x \rangle$ relative to its expected value.

2.22 Carry out the manipulations that lead from (2.82) to (2.83).

2.23 Write out a table of the vectors $\boldsymbol{\alpha}$ used to identify the 26 periodic image cells that pertain to a three-dimensional cube.

2.24 In a cube of edge $L = 5$, an atom center is located at $\mathbf{r} = 2\hat{\mathbf{i}} + 3\hat{\mathbf{j}} + 4\hat{\mathbf{k}}$, with all distances measured in units of the atomic diameter σ. List the position vectors of the 26 images of the atom, with all vectors measured relative to the same origin as is $\mathbf{r}$.

2.25 In a cube of edge $L = 5$, atoms 1 and 2 are located at positions

$$\mathbf{r}_1 = 0.5\hat{\mathbf{i}} + 3\hat{\mathbf{j}} + \hat{\mathbf{k}} \qquad \text{and} \qquad \mathbf{r}_2 = 4\hat{\mathbf{i}} + 0.4\hat{\mathbf{j}} + 4\hat{\mathbf{k}}.$$

All distances are in units of one atomic diameter. Devise an algorithm that finds the scalar distance between atom 1 and the image of 2 that is closest to atom 1.

2.26 For quantities evaluated along periodic trajectories, the correlation coefficient (2.97) will not decay to zero at long times; rather, it will also be periodic. Determine the expression for the correlation $C(t)$ of the position $x(t)$ of the ODHO when the motion is started from $x(0) = x_0$ and $p(0) = 0$.

2.27 Some simulators claim that if time averages for properties converge to stable values, then the simulation is ergodic. Discuss, with specific examples, whether the following situations satisfy the premise and, if so, whether the motion could nevertheless be nonergodic:

   (a) Systematic sampling of periodic motion,

   (b) Random sampling of quasiperiodic motion,

   (c) Any sampling of metastable motion.

2.28 The volume of an n-dimensional sphere of radius r is given by

$$\Psi(r) = \frac{\pi^{n/2} r^n}{\Gamma\left(\dfrac{n}{2} + 1\right)}$$

where $\Gamma(x)$ is the gamma function: $\Gamma(x + 1) = x\Gamma(x) = x!$ For large x, we have Stirling's approximation (see Appendix M), $\ln x! \approx x \ln x - x$.

   (a) For an isolated system of volume V containing N ideal gas atoms at total energy E, show that the accessible volume Ψ in phase space is

$$\Psi \approx \left[Ve^{3/2} \left(\frac{4\pi mE}{3N} \right)^{3/2} \right]^N$$

(b) Now the entropy of an isolated system is related to Ψ by

$$S(N,V,E) = k \ln(\Psi/N!)$$

The factor of $N!$ arises because the atoms are indistinguishable: there are $N!$ equivalent ways of distributing N atoms among N locations in phase space. Using this, the result from (a), and the fundamental equation from thermodynamics

$$dE = TdS - PdV + \mu\,dN$$

where μ is the chemical potential, (i) derive the relation between E and temperature, thus showing that the ideal gas internal energy is independent of density, and (ii) derive the ideal gas equation of state.

3

HARD SPHERES

Molecular dynamics algorithms divide into two broad classes: those for soft bodies, for which the intermolecular forces are continuous functions of the distance between molecules (Chapters 4 and 5), and those for hard bodies, for which the forces are discontinuous. For hard bodies the discontinuity in the force extends to the intermolecular potential; in particular, hard spheres of diameter σ interact through a potential energy function $u(r)$ of the form

$$u(r) = \begin{cases} \infty & r \leq \sigma \\ 0 & r > \sigma \end{cases} \qquad (3.1)$$

which is shown in Figure 3.1. Thus, hard spheres exert forces on one another only when they collide.[†] Between collisions the spheres travel along straight lines at constant velocities, and so, rather than compute trajectories, the simulation algorithm [1] computes the times of collisions. The calculation is purely algebraic because collisions are taken to be perfectly elastic: during a collision no energy is transferred either to deform a sphere or to change its internal state. Hence, collisions disrupt neither conservation of linear momentum nor conservation of kinetic energy, and these two principles enable us to determine collision times (Sections 3.1–3.3).

As do all molecular dynamics methods, the hard-sphere algorithm divides into three tasks: initialization (Section 3.4), equilibration (Section 3.5), and production. In spite of its algebraic determinism, the algorithm produces phase-space trajectories that are chaotic, chaotic not only because of colli-

[†]W. Thomson, Lecture XX, p. 211: "Imagine what is the force of the collision between molecules. Take two billiard balls..."

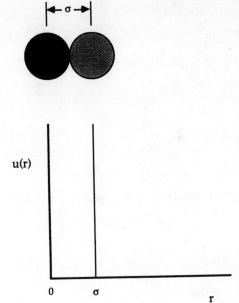

FIGURE 3.1 The pair potential energy function $u(r)$ for hard spheres of diameter σ. Two spheres exert no force on one another unless they collide.

sions (Section 2.5), but also because of a delicate coupling among the algorithm, software, and hardware (Section 3.6). The hard-sphere substance may exist as either fluid or solid, and therefore the algorithm must be used with care near the phase transition (Section 3.7). Otherwise, the most challenging part of hard-sphere molecular dynamics comes at the end of a run, when we attempt to assess the reliability of its results (Section 3.8).

3.1 KINEMATICS OF HARD-SPHERE COLLISIONS

Consider two hard spheres having equal masses ($m_1 = m_2 \equiv m$) and equal diameters ($\sigma_1 = \sigma_2$), with their motion constrained to one dimension. At some time t_1 these spheres are separated by a distance x and have velocities v_1 and v_2 such that they will collide at some later time $t_2 > t_1$. Now at time t_1 no forces are acting on the spheres; hence, by Newton's first law, v_1 and v_2 are constants. Our immediate objective is to determine the velocities v_1 and v_2 after the collision at t_2.

Conservation of linear momentum gives

$$mv_1 + mv_2 = mv_1' + mv_2' \tag{3.2}$$

and conservation of energy gives

$$\tfrac{1}{2}mv_1^2 + \tfrac{1}{2}mv_2^2 = \tfrac{1}{2}mv_1'^2 + \tfrac{1}{2}mv_2'^2 \tag{3.3}$$

Since the masses are the same, these reduce to

$$v_1 + v_2 = v_1 + v_2 \tag{3.4}$$

$$v_1^2 + v_2^2 = v_1^2 + v_2^2 \tag{3.5}$$

Now (3.5) can be rearranged to

$$v_1^2 - v_1^2 = v_2^2 - v_2^2 \tag{3.6}$$

which factors into

$$(v_1 + v_1)(v_1 - v_1) = (v_2 + v_2)(v_2 - v_2) \tag{3.7}$$

Likewise (3.4) rearranges to

$$v_1 - v_1 = v_2 - v_2 \tag{3.8}$$

Using (3.8) to eliminate $v_1 - v_1$ from the lhs of (3.7) leaves

$$v_1 + v_1 = v_2 + v_2 \tag{3.9}$$

Now adding (3.8) and (3.9) gives

$$v_1 = v_2 \qquad v_2 = v_1 \tag{3.10}$$

Thus, for purely one-dimensional motion, a perfectly elastic collision between spheres of equal masses causes the two spheres to exchange velocities. In a molecular dynamics simulation of one-dimensional hard spheres the initially assigned velocity distribution is preserved, the simulation is not ergodic, and property values from such a simulation are erroneous. However, if the Maxwell velocity distribution is initially assigned, that distribution is maintained, and property values are in good agreement with theoretical results [2].

Our interest here is in using the one-dimensional result (3.10) to deduce the kinematics of hard-sphere collisions in two and three dimensions. In any space—one, two, or three dimensions—when hard spheres collide, the repulsive force is exerted along the line of sphere centers $\mathbf{r}_{12} = \mathbf{r}_2 - \mathbf{r}_1$: only components of the velocities parallel to $\mathbf{r}_{12}$ are affected by a perfectly elastic collision. Thus, we define a set of mutually orthogonal axes $\{\hat{\mathbf{r}}_{12}, \hat{\mathbf{j}}, \hat{\mathbf{k}}\}$, where the carets denote unit vectors: $\hat{\mathbf{r}}_{12} = \mathbf{r}_{12} / |\mathbf{r}_{12}|$. The precollision velocity of sphere i can be written as

$$\mathbf{v}_i = (\mathbf{v}_i \cdot \hat{\mathbf{r}}_{12})\hat{\mathbf{r}}_{12} + (\mathbf{v}_i \cdot \hat{\mathbf{j}})\hat{\mathbf{j}} + (\mathbf{v}_i \cdot \hat{\mathbf{k}})\hat{\mathbf{k}} \tag{3.11}$$

and its postcollision velocity written as

$$\mathbf{v}_i = (\mathbf{v}_i \cdot \hat{\mathbf{r}}_{12})\hat{\mathbf{r}}_{12} + (\mathbf{v}_i \cdot \hat{\mathbf{j}})\hat{\mathbf{j}} + (\mathbf{v}_i \cdot \hat{\mathbf{k}})\hat{\mathbf{k}} \qquad (3.12)$$

Now, from our one-dimensional results (3.10), we expect that, on collision of spheres 1 and 2, components of $\mathbf{v}_1$ and $\mathbf{v}_2$ parallel to $\mathbf{r}_{12}$ will be exchanged:

$$\mathbf{v}_1 \cdot \hat{\mathbf{r}}_{12} = \mathbf{v}_2 \cdot \hat{\mathbf{r}}_{12} \qquad (3.13a)$$

$$\mathbf{v}_2 \cdot \hat{\mathbf{r}}_{12} = \mathbf{v}_1 \cdot \hat{\mathbf{r}}_{12} \qquad (3.13b)$$

while components perpendicular to $\mathbf{r}_{12}$ are unaffected:

$$\mathbf{v}_1 \cdot \hat{\mathbf{j}} = \mathbf{v}_1 \cdot \hat{\mathbf{j}} \qquad (3.14a)$$

$$\mathbf{v}_2 \cdot \hat{\mathbf{j}} = \mathbf{v}_2 \cdot \hat{\mathbf{j}} \qquad (3.14b)$$

$$\mathbf{v}_1 \cdot \hat{\mathbf{k}} = \mathbf{v}_1 \cdot \hat{\mathbf{k}} \qquad (3.14c)$$

$$\mathbf{v}_2 \cdot \hat{\mathbf{k}} = \mathbf{v}_2 \cdot \hat{\mathbf{k}} \qquad (3.14d)$$

Equations (3.13) and (3.14) identify *specular reflection*: a perfectly elastic collision in which the velocity component normal to the collision surface is changed, while components parallel to the surface are unaffected.

Using (3.13a), (3.14a), and (3.14c) in (3.12) for sphere 1 gives

$$\mathbf{v}_1 = (\mathbf{v}_2 \cdot \hat{\mathbf{r}}_{12})\hat{\mathbf{r}}_{12} + (\mathbf{v}_1 \cdot \hat{\mathbf{j}})\hat{\mathbf{j}} + (\mathbf{v}_1 \cdot \hat{\mathbf{k}})\hat{\mathbf{k}} \qquad (3.15)$$

Adding and subtracting the $\hat{\mathbf{r}}_{12}$-component of $\mathbf{v}_1$ on the rhs of (3.15) gives

$$\mathbf{v}_1 = (\mathbf{v}_1 \cdot \hat{\mathbf{r}}_{12})\hat{\mathbf{r}}_{12} + (\mathbf{v}_1 \cdot \hat{\mathbf{j}})\hat{\mathbf{j}} + (\mathbf{v}_1 \cdot \hat{\mathbf{k}})\hat{\mathbf{k}} + [(\mathbf{v}_2 - \mathbf{v}_1) \cdot \hat{\mathbf{r}}_{12}]\hat{\mathbf{r}}_{12} \qquad (3.16)$$

and using (3.11) for $\mathbf{v}_1$ in (3.16) leaves

$$\mathbf{v}_1 = \mathbf{v}_1 - [(\mathbf{v}_1 - \mathbf{v}_2) \cdot \hat{\mathbf{r}}_{12}]\hat{\mathbf{r}}_{12} \qquad (3.17)$$

The analogous procedure for sphere 2 yields

$$\mathbf{v}_2 = \mathbf{v}_2 + [(\mathbf{v}_1 - \mathbf{v}_2) \cdot \hat{\mathbf{r}}_{12}]\hat{\mathbf{r}}_{12} \qquad (3.18)$$

Equations (3.17) and (3.18) give the postcollision velocities $\mathbf{v}_i$ in terms of the (known) precollision velocities $\mathbf{v}_i$ while avoiding explicit determination of the components appearing in (3.14). Equations (3.17) and (3.18) are used in the hard-sphere simulation program contained in Appendix K. Note that these equations only apply for perfectly elastic collisions between two spheres of equal masses; however, the sphere diameters need not be equal.

The ergodicity of simulations in two and three dimensions should not be affected by the velocity exchange represented by (3.13) because each collision will, in general, occur at a different orientation of the line of centers $\hat{\mathbf{r}}_{12}$. Over time, collisions promote sampling of molecular velocities from an equilibrium distribution.

3.2 COLLISION TIMES

Before applying (3.17) and (3.18), we must determine whether spheres 1 and 2 will collide and, if so, when. In this section we first describe the calculation of collision times for any two spheres and then discuss the tabulation of those times to reduce the computational expense of a simulation.

3.2.1 Collision Time Calculations

Consider two spheres, 1 and 2, each having diameter σ. When they collide, the centers of 1 and 2 will be separated by distance σ; that is,

$$|\mathbf{r}_1(t_c) - \mathbf{r}_2(t_c)| = \sigma \tag{3.19}$$

or

$$[\mathbf{r}_1(t_c) - \mathbf{r}_2(t_c)]^2 = \sigma^2 \tag{3.20}$$

Between collisions, the spheres transverse straight lines at constant velocities; so from the known positions $\mathbf{r}_i(t_0)$, the position of sphere i at any later time t is

$$\mathbf{r}_i(t) = \mathbf{r}_i(t_0) + (t - t_0)\mathbf{v}_i(t_0) \tag{3.21}$$

Using (3.21) in (3.20) for $\mathbf{r}_1$ and $\mathbf{r}_2$ produces a quadratic for the collision time t_c,

$$[\mathbf{r}_{12} + (t_c - t_0)\mathbf{v}_{12}]^2 = \sigma^2 \tag{3.22}$$

where

$$\mathbf{r}_{12} \equiv \mathbf{r}_1(t_0) - \mathbf{r}_2(t_0) \quad \text{and} \quad \mathbf{v}_{12} \equiv \mathbf{v}_1(t_0) - \mathbf{v}_2(t_0) \tag{3.23}$$

or

$$t_c = t_0 + \frac{(-\mathbf{v}_{12} \cdot \mathbf{r}_{12}) \pm \sqrt{(\mathbf{v}_{12} \cdot \mathbf{r}_{12})^2 - v_{12}^2(r_{12}^2 - \sigma^2)}}{v_{12}^2} \tag{3.24}$$

This equation contains all possibilities regarding an encounter between spheres 1 and 2. Specifically, it allows us to distinguish three important situations.

$$\text{Situation } A: \quad \mathbf{v}_{12} \cdot \mathbf{r}_{12} < 0 \quad \text{for collision} \tag{3.25}$$

The product $\mathbf{v}_{12} \cdot \mathbf{r}_{12}$ is proportional to the component of the velocity difference along the line of centers. If this component is negative, then the spheres are approaching one another and a collision may occur. If it is positive, the spheres are moving away from one another and they will definitely not collide.

$$\text{Situation } B: \quad (\mathbf{v}_{12} \cdot \mathbf{r}_{12})^2 - v_{12}^2(r_{12}^2 - \sigma^2) \geq 0 \quad \text{for collision} \tag{3.26}$$

Equation (3.25) is only a necessary, not a sufficient, condition for a collision; that is, the spheres may be moving toward one another but still not collide. A sufficient condition for collision is provided by (3.26), for which a geometric interpretation is given in Figure 3.2. To prove (3.26), we proceed as follows. In Figure 3.2, the relative velocity $\mathbf{v}_{12}$ prescribes the trajectory of sphere 1 relative to 2. When spheres 1 and 2 collide at time t_c, then the law of cosines gives

$$\sigma^2 = r_{12}^2 + v_{12}^2 - 2r_{12}v_{12}\cos\theta \tag{3.27}$$

So for given relative separation $r_{12} = |\mathbf{r}_{12}|$ and relative velocity $v_{12} = |\mathbf{v}_{12}|$, whether or not a collision occurs depends on angle θ. At given r_{12} and v_{12}, the largest value of θ, call it θ_L, for collision occurs when sphere 1 just grazes sphere 2, that is, when the triangle formed by r_{12}, v_{12}, and σ is right; then

$$r_{12}^2 = \sigma^2 + v_{12}^2 \tag{3.28}$$

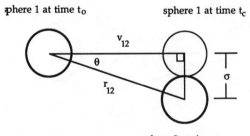

sphere 1 at time t_0 sphere 1 at time t_c

sphere 2 at time t_c

FIGURE 3.2 Geometric interpretation of the discriminant in (3.24) showing the largest value of angle θ for which spheres 1 and 2 collide. This limiting value of the angle of approach occurs when the triangle $\{v_{12}, r_{12}, \sigma\}$ is right.

Collisions occur whenever the angle θ is less than the limiting value θ_L, or equivalently, when

$$2r_{12}v_{12}\cos\theta \geq 2r_{12}v_{12}\cos\theta_L \tag{3.29}$$

For the limiting right triangle we have

$$\cos\theta_L = \frac{v_{12}}{r_{12}} \tag{3.30}$$

and from (3.28)

$$v_{12}^2 = r_{12}^2 - \sigma^2 \tag{3.31}$$

Using (3.30) in (3.29) gives

$$r_{12}v_{12}\cos\theta \geq v_{12}^2 \tag{3.32}$$

Squaring (3.32) yields

$$r_{12}^2 v_{12}^2 \cos^2\theta = (\mathbf{r}_{12}\cdot\mathbf{v}_{12})^2 \geq v_{12}^4 \tag{3.33}$$

Using (3.31) for one v_{12}^2 on the rhs of (3.33) gives, for a collision, (3.26). QED

Situation C: When the condition (3.26) is satisfied, the quadratic (3.24) has two roots. These roots imply that, for a given separation $\mathbf{r}_{12}$ and relative velocity $\mathbf{v}_{12}$ leading to a collision, two geometric arrangements have the spheres in contact, as in Figure 3.3. Since the spheres are impenetrable, only the short-time root (negative sign in 3.24) is physically possible. In the case of the grazing contact, the discriminant is zero, the equality in (3.26) applies, and the quadratic (3.24) has a single root.

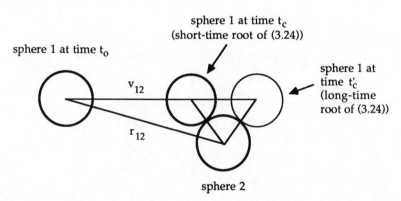

FIGURE 3.3 Geometric interpretation of the two roots of the quadratic (3.24). For given $\mathbf{r}_{12}$ and $\mathbf{v}_{12}$ there are two geometric arrangements that have spheres 1 and 2 in contact, but for impenetrable spheres only the short-time root is physically possible.

3.2.2 Collision Time Tabulation

The heart of the hard-sphere simulation problem is this: from the known positions and velocities of N spheres at time t_0, determine which two spheres will collide next and when the collision will occur. Our first approach to this problem would probably be to test the conditions (3.25) and (3.26) and, when applicable, solve the quadratic (3.24) for each of the $\frac{1}{2}N(N-1)$ pairs of spheres. However, much of this computation can be avoided by maintaining a table of collision times.

The table is constructed during initialization prior to entering the main simulation loop. From the initially assigned positions and velocities we determine the collision times for all $\frac{1}{2}N(N-1)$ pairs. For each sphere i the collision time and its collision partner j are entered into the table. Then in the main simulation loop, the identification of colliding pairs involves searching the table for the next collision time. Following each collision, we revise all table entries that involve either of the colliding pair; this allows either of the two to be properly identified as participants in subsequent collisions along their new trajectories.

When we compute collision times for systems having periodic boundaries, we must consider not only whether a sphere i will collide with sphere j, but also whether i will collide with an image of j. We limit sampling to the 26

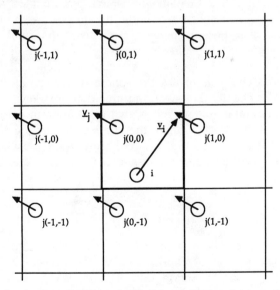

FIGURE 3.4 An instant during a hard-disk simulation in two dimensions with periodic boundary conditions. The positions and velocities of disks i and j are arranged to show that disk i will collide with an image of j but not necessarily with j itself nor with image $j(0, -1)$, which is currently the j-image closest to i. The cell translation vectors $\boldsymbol{\alpha} = (m, n)$ are described in Figure 2.13.

image cells that adjoin the primary cell (see Section 2.9). Since j and all its images have the same velocity vector, their trajectories are all parallel; therefore, several images of j (including j itself) may possibly satisfy the collision conditions (3.25) and (3.26) (at different times, of course), but we need find only the image of j that *first* collides with i. All other images will be simultaneously colliding with some image of i. Note that, as shown in Figure 3.4, neither j nor the image of j closest to i will necessarily be first to collide with i.

3.3   SIMULATION ALGORITHM

With the results from the previous sections we can now discuss the algorithm for simulating pure hard spheres in three dimensions. We divide the algorithm into three parts: (a) initialization, (b) equilibration, and (c) production. Before discussing these steps, we describe a set of units that can be used in the code.

3.3.1   System of Units

Following the ideas of corresponding states, molecular dynamics programs are conventionally written in a system of units in which dimensional quantities are made unitless. The fundamental dimensions are mass, length, energy, and time. For a pure substance we choose the unit of mass to be that of one sphere, for then mass does not appear explicitly in the simulation program.

For the unit of length some practitioners use the sphere diameter σ while others use the length of one edge of the cubic container, L. In theoretical work the obvious choice is to use the sphere diameter and we follow that convention in this chapter. However, in simulation programs L offers a slight computational advantage, so we follow that convention in the code provided in Appendix K. In units of σ, then, the number density $\rho = N/V$ becomes $\rho^* = N\sigma^3/V$. Besides ρ we can measure density with the packing fraction η, which is the ratio of the volume of N spheres (V_{spheres}) to the volume of their container (V),

$$\eta = \frac{V_{\text{spheres}}}{V} = \frac{N\sigma^3 \pi}{V6} = \frac{\pi \rho^*}{6} \qquad (3.34)$$

Because the hard-sphere potential (3.1) is either zero or infinite, it provides no natural unit of energy, and so the convention is to use the thermal energy kT as the energy unit. This choice forces all dimensional isotherms to collapse onto the same reduced isotherm, and the reduced total energy always obeys

$$\frac{\langle E \rangle}{NkT} = \frac{3}{2} \qquad (3.35)$$

where E is both the total energy and the kinetic energy. With this choice, the reduced time is $t^* = t/[\sigma(m/kT)^{1/2}]$, the reduced velocity is $v^* = v(m/kT)^{1/2}$, and therefore changing the velocities cannot affect results for properties. For example, if all velocities are doubled, the unit of time is halved so that the times between collisions are unchanged; consequently, the total energy still obeys (3.35) and values for other reduced properties are unchanged. Since we have only one reduced isotherm, assigning a value to one other reduced intensive variable—say, the packing fraction—fixes the thermodynamic state.

3.3.2 Steps in the Algorithm

Several hard-sphere algorithms have been used, but one of the simplest is the original, devised by Alder and Wainwright [1] and discussed by Erpenbeck and Wood [3]. We summarize it here and use it in the code in Appendix K.

Initialization. Initialization establishes the thermodynamic state and assigns initial positions and velocities to the spheres. From those initial conditions we can then build the table of collision times.

1. Choose a number of spheres N and packing fraction η.
2. Compute the volume of the primary cube, $V/\sigma^3 = (N\pi/6\eta)$ and adopt the length of one edge of the cube as the unit of distance, that is, set $L = 1$. Then in these units, compute the sphere diameter $\sigma = (N\pi/6\eta)^{-1/3}$.
3. Assign initial positions $\mathbf{r}_i(0)$, $i = 1, \ldots, N$. Typically the initial positions are the sites on a face-centered-cubic (fcc) lattice; alternatively, positions can be taken from the end of an earlier simulation. The construction of an fcc lattice is described in Section 3.4.
4. Assign initial velocities $\mathbf{v}_i(0)$, $i = 1, \ldots, N$. The program in Appendix K randomly assigns the components of each $\mathbf{v}_i$ from a uniform distribution on $[-1, +1]$. A slight gain in computer time can be realized by assigning the velocities directly from a Maxwell distribution. If the initial velocities are not Maxwellian, then they are scaled to yield zero total linear momentum. Even if velocities are assigned from a Maxwell distribution, you should scale them to zero linear momentum because otherwise, for a small number of atoms, the Maxwellian velocities will not sum to exactly zero. Again, velocities could be taken from a previous simulation.
5. Construct the table of collision times by solving (3.24) for each of $\frac{1}{2}N(N-1)$ pairs of spheres. This completes the initialization phase.

Equilibration. Time averages for equilibrium properties must be independent of initial conditions and so, in a crude sense, equilibration develops the dynamics over a time sufficient for the system to "forget" how it was prepared.

6. Now execute the simulation loop (steps 7–11) over a prescribed number of collisions to allow the system to relax from the initial conditions. The duration of this equilibration phase is determined by monitoring parameters that measure order in both position and momentum space. These are discussed in detail in Section 3.5.

Production. After equilibration the program enters the production phase: the main simulation loop that generates the equilibrium phase-space trajectory from which properties will be calculated.

7. From the table of collision times, determine the duration Δt until the next collision and identify the colliding spheres i and j.
8. From their current positions $\mathbf{r}_k(t_0)$, advance all spheres through time Δt to the collision,

$$\mathbf{r}_k(t_0 + \Delta t) = \mathbf{r}_k(t_0) + \mathbf{v}_k(t_0)\,\Delta t \qquad k = 1, \ldots, N \qquad (3.36)$$

9. Apply periodic boundary conditions to any sphere leaving the primary cell during Δt.
10. From (3.17) and (3.18) obtain the postcollision velocities of i and j.
11. Compute new entries for the table of collision times by applying the quadratic (3.24) to i, to j, and to any other sphere that would have collided with i or j had i and j not collided.
12. Compute contributions to equilibrium properties.
13. Iterate steps 7–12 over the desired number of collisions.

This algorithm differs in two ways from that discussed by Erpenbeck and Wood [3]. First, we do not tabulate the time when a sphere crosses a boundary and is replaced by a periodic image; instead, in step 9 we check whether each sphere has crossed a boundary since the last collision. The second difference is that when the collision time table is built in step 5 and updated in step 11, we do not explicitly identify which image of sphere j will collide with i. Instead, in step 10, we compute the postcollision velocities using the minimum image distance between spheres i and j; that is, we apply the following transformations:

$$\text{If } x_{ij} > \tfrac{1}{2} \quad \text{then } x_{ij} \rightarrow x_{ij} - 1$$

$$\text{If } x_{ij} < -\tfrac{1}{2} \quad \text{then } x_{ij} \rightarrow x_{ij} + 1$$

where x_{ij} is the x-component of the vector between centers of spheres i and

j. Similar transformations are applied to the y- and z-components. Here distances are measured in units of one edge of the cubic container ($L = 1$), so if a component of $\mathbf{r}_{ij}$ has magnitude greater than half the box width, then the transformation is applied. At a collision the magnitude of $\mathbf{r}_{ij}$ must be one sphere diameter σ and σ must necessarily be smaller than the box width L (see step 2). The minimum image criterion successfully identifies the image involved in a collision because, at a specified time, sphere i can possibly collide only with j or *one* of its images. The image closest to i is necessarily the one involved in the collision.

There is probably little advantage, either way, in these differences in algorithms, but since in the soft-sphere programs we use these devices for dealing with images, there is pedagogical value in using them here as well.

3.4 INITIAL POSITIONS AND VELOCITIES

An efficient way to start a simulation is to take the initial positions from the end of an earlier run. But this possibility may not be viable if we are changing the number of molecules from one run to the next or if otherwise we do not have positions from a previous run (after all, one has to start somewhere). We may be tempted to try to assign positions randomly, but unless the density is fairly small, random assignment proves time consuming because hard spheres cannot overlap. The usual alternative is to initially locate atoms on a regular lattice structure, preferably the structure into which the substance of interest crystallizes. Since argon crystallizes into the fcc structure, simulations of monatomic substances are conventionally started from an fcc lattice. In fluid simulations, then, part of the objective of equilibration is to melt the lattice.

A space lattice, such as fcc, is constructed by replicating a fundamental arrangement of atoms called the *unit cell*. The unit cell prescribes the locations of the smallest number n of atoms needed to define the symmetry of the lattice. For cubic structures, each side of the unit cell has length a and the density of the cell is the density of the full lattice. Therefore, for a cubic unit cell composed of identical atoms, the density is merely

$$\rho = \frac{n}{a^3} \tag{3.37}$$

The fcc unit cell contains four atoms having location vectors

$$\mathbf{r}_1 = (0,0,0) \quad \mathbf{r}_2 = \left(0, \tfrac{1}{2}a, \tfrac{1}{2}a\right) \quad \mathbf{r}_3 = \left(\tfrac{1}{2}a, 0, \tfrac{1}{2}a\right) \quad \mathbf{r}_4 = \left(\tfrac{1}{2}a, \tfrac{1}{2}a, 0\right)$$

$$\tag{3.38}$$

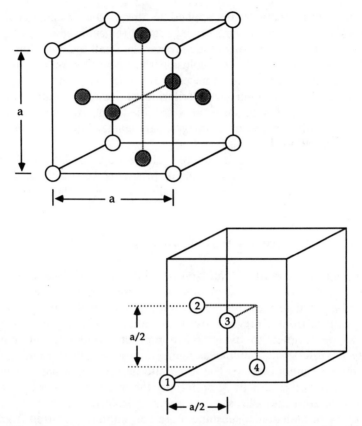

FIGURE 3.5 Top shows one cube from an fcc lattice having side of length a. A shaded atom lies in each face of the cube. Bottom shows the locations of the four atoms that define the fcc unit cell.

where the origin, and hence the first sphere, is located at one corner of the cell, as shown in Figure 3.5. A full fcc lattice can be built from these four vectors by repeatedly translating them through the distance a in each Cartesian direction. An fcc lattice is completely filled by N atoms, where $N = 4I^3$ and I is an integer. Thus, simulations of simple fluids are often done using $32, 108, 256, 500, 864, \ldots$ atoms.

The largest number of hard spheres that can be loaded into a system of fixed volume defines a maximum packing fraction. For hard spheres, the maximum occurs when the spheres are arranged in either a hexagonal-close-packed or a cubic-close-packed lattice. The cubic-close-packed structure has the fcc unit cell, so we can immediately compute the maximum packing fraction for hard spheres. When close packed, the four spheres of the unit cell are in contact, any two sphere centers are separated by a sphere diameter σ, and so, by Pythagoras, $\sigma^2 = 2(a/2)^2$ or $a = \sigma\sqrt{2}$. Then

$$\eta_{max} = \rho\sigma^3 \frac{\pi}{6} = \frac{n\sigma^3}{a^3} \frac{\pi}{6} = \frac{\pi}{6}\sqrt{2} = 0.74048\ldots \qquad (3.39)$$

"Many mathematicians believe, and all physicists know" [4] that (3.39) gives the largest possible packing density for uniform hard spheres [5].

Initial velocities can also be taken from a previous simulation; otherwise, they can be assigned from a uniform velocity distribution on some interval, say $[-1, +1]$. These velocities are immediately translated by a factor so the total linear momentum is zero; thus

$$v_{xi}^{new} = v_{xi}^{old} - \frac{1}{N}\sum_i^N v_{xi}^{old} \qquad (3.40)$$

and similarly for the y- and z-components. From these initially assigned velocities the Maxwell velocity distribution (2.54) should develop during equilibration. Alternatively, the Maxwell distribution could be assigned directly using, say, the Box–Muller procedure [6], but it is edifying to monitor the development of the Maxwell distribution from an arbitrary set of initial conditions.

3.5 MONITORING EQUILIBRATION

Equilibration, the approach to equilibrium, allows the phase-space trajectory to move from arbitrarily or pathologically assigned initial conditions to the region of phase space that is most accessible—the region of equilibrium states. To monitor equilibration, we would like to have two numbers: one that tracks the development of disorder in atomic positions and a second that tracks the development of the Maxwell distribution of velocities. When we start from a regular lattice, we can use a translational order parameter to monitor positional disorder. To monitor development of the Maxwell velocity distribution, we compute the kinetic part of Boltzmann's H-function.

3.5.1 Positional Disorder

To monitor the dissolution of an fcc lattice composed of N atoms, Verlet [7] introduced a translational order parameter λ, which we define as

$$\lambda = \tfrac{1}{3}\left[\lambda_x + \lambda_y + \lambda_z\right] \qquad (3.41)$$

where

$$\lambda_x = \frac{1}{N} \sum_i^N \cos\left(\frac{4\pi x_i}{a}\right) \tag{3.42}$$

and a is the length of one edge of the fcc unit cell. Initially, when all atoms occupy fcc lattice sites, $\lambda = 1$ because the positional components x_i, y_i, and z_i are all integer multiples of $\frac{1}{2}a$. When the lattice is completely dissolved, λ fluctuates about zero because then the atoms are distributed randomly about the original lattice sites; the fluctuations are of magnitude $\sqrt{N}/N$. Thus, as the fcc lattice dissolves, λ decays from unity to zero. An order parameter analogous to (3.42) can be defined when a simulation is started from some regular structure other than the fcc.

The time required for λ to exhibit stable fluctuations about zero depends on the packing fraction and on how the initial velocities are assigned. If

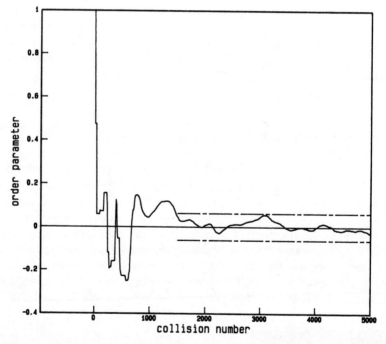

FIGURE 3.6 Disintegration of the initial fcc lattice, monitored by the translational order parameter λ (3.41) in a simulation of 256 hard spheres at packing fraction $\eta = 0.4$. Instantaneous values of λ are plotted here at intervals of 10 collisions. The run was started by assigning the same velocity (v_x, v_y, v_z) to 128 spheres in one-half of the container; those in the other half were assigned the velocity $(-v_x, -v_y, -v_z)$. The broken horizontal lines for collision numbers over 1500 mark the expected bounds on fluctuations in λ, namely, $\sqrt{N}/N = \pm 0.0625$.

initial velocities are taken randomly from a uniform distribution, then about 1000 collisions for 256 atoms are enough to completely dissolve the lattice. If initial velocities are not random but instead are assigned by some regular procedure, then a longer equilibration period is required. For example, Figure 3.6 shows the order parameter through the equilibration of a simulation of 256 hard spheres at packing fraction $\eta = 0.4$. Instead of random velocities, 128 of the spheres (1–128) in one-half of the container were each given the same velocity (v_x, v_y, v_z). The spheres in the other half (129–256) were given velocities of the same magnitude but oriented in the opposite direction $(-v_x, -v_y, -v_z)$. Figure 3.6 shows that the initial fcc lattice is quickly destroyed, but remnants persist over the first 1000 collisions; after 1500 collisions, the remnants are dissolved and λ is fluctuating about zero.

The order parameter is useful not only for monitoring the dissolution of initial conditions but also for monitoring the presence of metastable structures. Figure 3.7 shows an example taken from another simulation of 256 hard spheres at $\eta = 0.4$. For the run in Figure 3.7 the initial velocities were

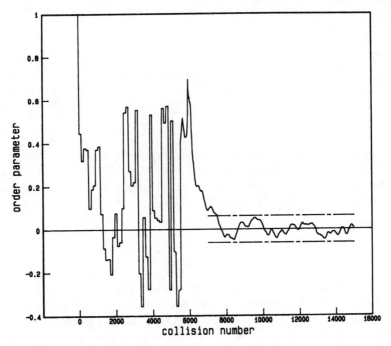

FIGURE 3.7   Same as Figure 3.6, except the initial assignment of velocities differs. This run was started by assigning the same velocity (v_x, v_y, v_z) to every other sphere; the remaining 128 spheres were assigned velocity $(-v_x, -v_y, -v_z)$. Total kinetic energy is the same as in Figure 3.6. Here, instantaneous values of λ are plotted at intervals of 20 collisions. The broken horizontal lines for collision numbers over 7000 mark the expected bounds on fluctuations in λ; $\sqrt{N}/N = \pm 0.0625$.

assigned in much the same way as for Figure 3.6, except that now the velocity (v_x, v_y, v_z) was assigned to alternate spheres $(1, 3, 5, \ldots, 255)$, while the opposite velocity $(-v_x, -v_y, -v_z)$ was assigned to the rest $(2, 4, 6, \ldots, 256)$. Figure 3.7 shows that about 7500 collisions are needed to completely disrupt the initial fcc lattice. From collision 1000 to 6000 metastable remnants of the lattice persist and the system is not at equilibrium.

3.5.2 Velocity Distribution

To monitor the development of Maxwell's velocity distribution, we could compute the velocity distribution $f(v)$ itself and watch how it evolves through equilibration. It is cumbersome, however, to monitor the temporal evolution of a function; we would prefer to monitor a single number. Fortunately, a single value is provided by Boltzmann's H-function.

To compute the kinetic part of the H-function, we first need to evaluate the velocity distribution. According to (2.51), the distribution measures the fraction of N atoms having velocity within an interval $d\mathbf{v}$ of a particular value $\mathbf{v}$,

$$f(\mathbf{v})\,d\mathbf{v} = \frac{1}{N} N(\mathbf{v})\,d\mathbf{v} \qquad (2.51)$$

For one Cartesian component, say v_x, (2.51) becomes

$$f(v_x)\,dv_x = \frac{1}{N} N(v_x)\,dv_x \qquad (3.43)$$

We can determine $f(v_x)$ during a simulation by selecting a finite interval width, Δv_x, and counting the number of atoms having velocities in $v_x \pm \frac{1}{2}\Delta v_x$. The counting operation can be represented by a δ-symbol, as discussed in Appendix A, so the working form of (3.43) is

$$f(v_x)\,\Delta v_x = \frac{1}{N} \sum_i^N \delta(v_x - v_{xi})\,\Delta v_x \qquad (3.44)$$

For one component of the velocity (3.44) gives the instantaneous velocity distribution—the distribution at one moment during the simulation. The corresponding value of the instantaneous (kinetic) H-function is then, from (2.57),

$$H_x = \sum_{\Delta v_x} f(v_x)\ln f(v_x)\,\Delta v_x \qquad (3.45)$$

Although this finite approximation to the integral in (2.57) is crude, (3.45) is accurate because Δv_x is small. Using expressions analogous to (3.45) for the

y- and z-components, we can write

$$H = \tfrac{1}{3}\left(H_x + H_y + H_z\right) \tag{3.46}$$

Figure 3.8 shows the instantaneous H-function (3.45) computed during the equilibration of the 256-hard-sphere simulation described earlier for Figure 3.6. Recall in that simulation atoms in one half of the container were each given the velocity (v_x, v_y, v_z) and those in the other half were given $(-v_x, -v_y, -v_z)$. In the figure, the horizontal line at $H = -0.163$ is the value of the H-function for the Maxwell distribution at the kinetic energy of the run. In these calculations, the increment used for sampling the velocities was $\Delta v_x = 0.05$. Figure 3.8(a) shows that the Maxwell distribution develops within about 1200 collisions. The simulation value for H is slightly displaced from Maxwell's value because of number dependence: repeating the run using more atoms causes H to converge closer to Maxwell's value. The broken line in Figure 3.8(a) is an average H-function computed from (3.46) but using running time averages for the velocity distributions (3.44). This H-function decays in nearly monotone fashion and asymptotically approaches the Maxwell value.

To verify that the Maxwell distribution has indeed developed, we show in Figure 3.8(b) the velocity distribution $f(v_x)$ time averaged over the 5000 collisions of Figure 3.8(a) and compare it with the Maxwell distribution at the same kinetic energy. The agreement is good; the two erroneous points (at $v = \pm 0.2$) were caused by the nonequilibrium distribution that dominates the first 1000 collisions used in computing the average.

Should we use the instantaneous or the running average H-function to monitor equilibration? The running average, it turns out, can mislead and so we should use the instantaneous H-function. This point is illustrated in Figure 3.9 which shows the development of the velocity distribution for the 256-hard-sphere simulation discussed in Figure 3.7. In Figure 3.9(a) the broken line, representing the H-function from the running average distribution, immediately decays toward the Maxwell value. However, over the first 5000 collisions, the instantaneous H-function does not decay at all; rather, it fluctuates about its initial value. (Recall the broken line is not an average of H, but rather H for an average f, so the broken line need not be bounded by the instantaneous values of the solid line.) Because of the way the initial velocities were assigned, Maxwell's distribution does not develop during the first 5000 collisions of the run; see Figure 3.9(b). This metastable, nonequilibrium behavior is consistent with the behavior of the order parameter shown in Figure 3.7 for the same run.

At about collision 5500, the instantaneous H-function in Figure 3.9(a) begins to decay toward its equilibrium value and then the Maxwell distribution develops. The velocity distribution averaged over collisions 6000–11,000 is in good agreement with Maxwell, as shown in Figure 3.9(c). Thus, the instantaneous H-function can help identify the presence of metastabilities in momentum space as well as help monitor the approach to equilibrium.

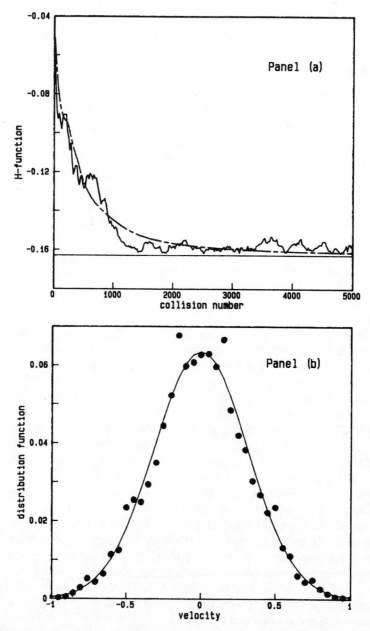

FIGURE 3.8 Approach to equilibrium as monitored by the H-function in the simulation of 256 hard spheres described in Figure 3.6. (a) Solid line is the instantaneous function H_x from (3.45); broken line is H from (3.46) but using running average velocity distributions $\langle f(v_x) \rangle$, $\langle f(v_y) \rangle$, and $\langle f(v_z) \rangle$; horizontal line is $H = -0.163$ computed using the Maxwell distribution. (b) The velocity distribution $f(v_x)\,dv_x$ averaged over the first 5000 collisions of the run (points) compared to the Maxwell distribution (line) at the same kinetic energy.

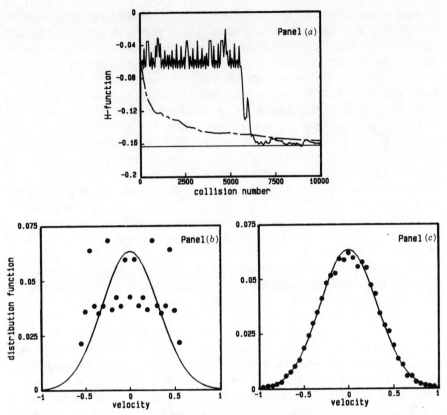

FIGURE 3.9 Approach to equilibrium for the 256-hard-sphere simulation described in Figure 3.7. (a) Same as Figure 3.8(a). (b) The velocity distribution $f(v_x)dv_x$ averaged over the first 5000 collisions of the run (points) compared to the Maxwell distribution (line) at the same kinetic energy. (c) The velocity distribution $f(v_x)dv_x$ averaged over collisions 6000–11,000 of the run (points) compared to the Maxwell distribution (line).

3.6 UNPREDICTABILITY

In Chapter 1 we distinguished between deterministic processes that are calculable and those that are predictable: *calculable* means an algorithm can be prescribed for connecting system outputs to system inputs while *predictable* means the algorithm can be implemented numerically. Tacit in the idea of predictability is the idea of reproducibility. If we repeatedly provide the same inputs to the algorithm, we expect a predictable process to consistently provide the same outputs. However, trajectories generated in hard-sphere simulations are sensitive to initial conditions: if we repeat a run with minute changes in initial conditions, those small changes will eventually lead to trajectories completely different from those obtained in the original run. In this section we address the extent to which the hard-sphere simulation algorithm is unpredictable because misconceptions about predictability influence our expectations about the performance of the algorithm.

One way in which unpredictability enters a simulation is through multibody and near-multibody collisions. Note that the hard-sphere algorithm considers only binary collisions: when a collision time is computed from (3.24), we presume only two spheres are involved. Ignoring multibody collisions does not in itself introduce unpredictability: a system allowing only binary collisions is a well-defined and legitimate model (its realism is a separate issue altogether). Instead, unpredictability arises from sensitivity to initial conditions, the finite precision of the calculations, and near-multibody collisions which, from time to time, occur.

To illustrate that the output from the algorithm can be unpredictable even though the inputs are known, consider a simple three-body collision among two-dimensional hard disks, as shown in Figure 3.10. The three disks were given initial positions and velocities so as to cause a three-body collision; hand calculations of the postcollision velocities result in the solid-line vectors shown in Figure 3.10. Then we applied the simulation algorithm of Section 3.3 to the same situation. The algorithm correctly finds that the collision time of disks 1 and 2 is the same as that for disks 2 and 3, but in step 7 the algorithm must find *the* next collision—the algorithm presumes no simultaneous collisions involving one disk. The algorithm makes an *arbitrary* choice as to whether collision 1–2 occurs before or after 2–3. The arbitrariness is introduced in the way the algorithm is implemented, that is, on the logic of the program. For example, if the code searches for the next collision by first comparing collision times in the order disk 1, disk 2, disk 3, then the program resolves the three-body encounter into three two-body collisions: disks 1 and 2 collide, then disks 2 and 3 collide, then 1 and 2 collide again. After these three collisions, the velocities are as shown by the broken-line vectors in Figure 3.10(*a*). The postcollision velocities obtained by the algorithm are very different from the hand-calculated velocities. If this three-body encounter had occurred in an *N*-body simulation, the phase-space trajectories originating from the algorithm's interpretation of the three-body encounter would differ significantly from the trajectories originating from the hand calculation.

We can carry this example a step further by using it to illustrate that the output of the algorithm shown in Figure 3.10(*a*) is arbitrary. We redo the simulation using the same algorithm and exactly the same inputs as before: same initial positions and velocities of the three disks, same computer, same precision arithmetic. We make only one change: the order of searching for the next collision is reversed from disk 1, then 2, then 3 to disk 3, then disk 2, then 1. This is not a change from one algorithm to another—step 7 of the algorithm merely says find the next collision; the algorithm does not specify *how* we are to find the next collision. The algorithm computes the same collision times as before and again it resolves the three-body encounter into

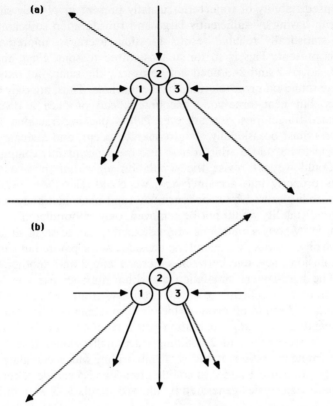

FIGURE 3.10 Unpredictability resulting from a near-three-body collision in a two-dimensional hard-disk simulation. Vectors with heads directed toward (away from) disks are precollision (postcollision) velocities. Solid-lined postcollision velocities are from an analytic calculation of the three-body encounter. Broken-lined velocities are from a simulation program that identifies only two-body collisions. (a) Order of the search for the shortest collision time was disk 1, disk 2, then disk 3. (b) Order of the search was reversed to disk 3, disk 2, then 1. Otherwise, the two simulations were identical. Lengths of the drawn vectors are proportional to the magnitudes of the velocities.

three two-body collisions. But now the three collisions are first 2–3, then 1–2, then 2–3 again. After these three collisions the velocities are as shown by the broken-line vectors in Figure 3.10(b). These postcollision velocities differ not only from the hand calculations but also from the outputs of the same algorithm and same inputs shown in Figure 3.10(a). Note that we cannot even do a hand calculation that reproduces the algorithm's results without knowing both the logic of implementation and the round-off error that enables the algorithm to distinguish collision 1–2 from 2–3: the output from the algorithm is calculable but unpredictable.

Is unpredictability of trajectories usually present in N-body simulations? In any run having N sufficiently large and run duration sufficiently long to provide statistically reliable results for time averages, unpredictability, it seems, is present. This is so for at least three reasons. First, the collision times of, say, 1–2 and 2–3 need not be exactly the same, but only the same within the round-off error tolerated by the machine. Thus, not only three-body collisions, but near-three-body encounters, can produce a divergence in phase-space trajectories. Second, even at low packing fractions, three-body encounters must occasionally, if infrequently, occur, and a single encounter is sufficient to introduce arbitrariness into the subsequently computed trajectories. Could we not revise the simulation algorithm to take three-body collisions properly into account? Yes, we could. But there would still be unpredictability caused by near-four-body (and near-multibody) encounters. The unpredictability would not be removed, only reformulated.

Third, in N-body simulations unpredictability can arise in situations that do not involve multibody encounters: consider two separate but simultaneous binary collisions, say, one between spheres 1 and 2 and another between 3 and 4. The hard-sphere simulation algorithm must choose one collision to occur first, and the scheduling will be arbitrary, arbitrary because it depends on accumulated round-off errors, which in turn depend on how the algorithm is implemented. One implementation will accumulate round-off in one way and lead to spheres 1 and 2 colliding fractionally sooner than 3 and 4. A different implementation of the same algorithm will accumulate round-off differently and cause 1 and 2 to collide after 3 and 4 collide. Thereafter, the phase-space trajectories generated by the two simulations *may* differ.

3.7 PHASE DIAGRAM

After equilibrium a simulation enters its main phase: computation of the equilibrium phase-space trajectory from which properties are evaluated. In Chapters 6 and 7 expressions are given for computing static and dynamic properties. Those expressions generally apply to systems of both hard and soft bodies, with one important exception. For hard bodies, if a property explicitly involves forces acting among atoms, then an indirect means for computing the property must be found to avoid the discontinuity in the force that occurs when hard bodies collide. This problem arises in computing elements of the stress tensor, which contribute to the pressure, the shear viscosity, and the bulk viscosity. To illustrate, we consider the pressure and then we describe the phase diagram.

3.7.1 Pressure

In statistical mechanics an expression for the pressure is usually derived [8] starting from the virial theorem of classical dynamics [9]; however, in systems

confined by periodic boundaries we may question the validity of the derivation of the virial theorem [3]. This uncertainty can be avoided by deriving the
pressure in kinetic theory [3]; a derivation is given in Appendix B. Both
derivations—that from the virial theorem and that in Appendix B—produce
the same form for the equation of state. For spheres interacting with pairwise
intermolecular forces the result is

$$Z \equiv \frac{PV}{NkT} = 1 + \frac{1}{3NkT} \left\langle \sum_{i<j} \sum \mathbf{F}_{ij} \cdot \mathbf{r}_{ij} \right\rangle \tag{3.47}$$

In this expression Z is the compressibility factor for N atoms contained in
volume V at temperature T, while $\mathbf{F}_{ij}$ is the force acting between atoms i and
j, $\mathbf{r}_{ij}$ is (at one instant) the vector distance between atoms i and j, and the
angle brackets represent the usual time average. The time-averaged quantity
in (3.47) is the virial.

Since hard spheres exert forces only at collisions, we do not need to sum
over all pairs of spheres; instead we need to identify when any two spheres i
and j collide. Therefore, we can rewrite the time average in (3.47) to
explicitly accumulate the virial from all collisions that occur during an
interval t,

$$Z = 1 + \frac{1}{3NkT} \frac{1}{t} \int_0^t (\mathbf{F}_{ij} \cdot \mathbf{r}_{ij}) \, \delta(\tau - t_c) \, d\tau \tag{3.48}$$

Here, t_c is the collision time for atoms i and j and the δ-symbol selects from
the duration t all collision times for any two atoms i and j. Now, Newton's
second law (2.1) interprets the force as the time rate of change of momentum,

$$\mathbf{F}_{ij} \approx m \frac{\Delta \mathbf{v}_{ij}}{\Delta t} \tag{3.49}$$

where $\Delta \mathbf{v}_{ij}$ is the change in velocity on collision and Δt is the interval
between successive collisions. Putting (3.49) into (3.48) and writing the
integral in (3.48) as a finite sum, we obtain

$$Z = 1 + \frac{1}{3NkT} \frac{1}{t} \sum_{c=1}^{N_c} m \, \Delta \mathbf{v}_{ij}(t_c) \cdot \mathbf{r}_{ij}(t_c) \tag{3.50}$$

In (3.50) the index c runs over a total of N_c collisions that occur during
interval t. The vector $\mathbf{r}_{ij}$ lies between the center of atoms i and j, and since it
is evaluated only at a collision (t_c), its magnitude is always the hard-sphere
diameter σ.

For an isolated hard-sphere system, the temperature is constant and proportional to the kinetic energy E_k,

$$E_k = \frac{m}{2} \sum_{i=1}^{N} v_i^2 = \frac{3}{2} NkT \qquad (2.43)$$

so (3.50) can be written as

$$Z = 1 + \frac{m}{2E_k} \frac{1}{t} \sum_{c=1}^{N_c} \Delta \mathbf{v}_{ij}(t_c) \cdot \mathbf{r}_{ij}(t_c) \qquad (3.51)$$

For a pure fluid of hard spheres each having diameter σ, (3.51) simplifies because the spheres undergo specular reflection, (3.13) and (3.14). Consequently, at a collision between spheres i and j, $\Delta \mathbf{v}_{ij}$ lies along the line of centers $\mathbf{r}_{ij}$ and the magnitude of $\mathbf{r}_{ij}$ is necessarily σ. Thus we have

$$\boxed{Z = 1 + \frac{m\sigma}{2E_k} \frac{1}{t} \sum_{c=1}^{N_c} |\Delta \mathbf{v}_{ij}(t_c)|} \qquad (3.52)$$

The evaluation of Z from (3.52) is computationally easy because, in a hard-sphere simulation, the quantity $\Delta \mathbf{v}_{ij}$ at collision is already being computed to carry the simulation forward.

Note that in (3.52) we do not actually evaluate the full pressure; rather, we compute only a portion—the nonideal-gas part that arises from intermolecular forces. The ideal-gas pressure is of no interest; it is the nonideal part that is so difficult to characterize theoretically and that can be studied so effectively by simulation. This nonideal pressure (sometimes called the residual pressure, other times the excess pressure) is present and calculable whether or not hard walls bound the system. In fact, then, the calculation of the pressure is something of a misnomer in simulation work; we really calculate the intermolecular virial, $Z - 1$. It is this quantity that should be statistically analyzed and reported as a result from a molecular simulation.

3.7.2 Phase Transition

The hard-sphere substance is the simplest material to exhibit melting. At high densities hard-sphere systems display characteristics of solids: long-range order and small values of the self-diffusion coefficient. At low densities they display characteristics of fluids: a lack of long-range order plus moderate values of the diffusion coefficient. In the absence of attractive forces, solidification must be a geometric consequence of packing nondeformable objects into a relatively small volume. However, hard-sphere systems are limited to

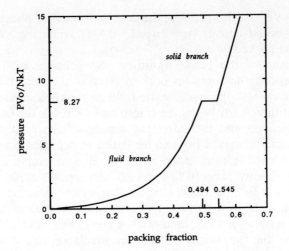

FIGURE 3.11 Pressure–density phase diagram for pure hard spheres. V_0 is the volume of the close-packed substance ($\eta = 0.7405$). The fluid branch is from the Boublik–Nezbeda [13] equation of state (3.54); the solid branch is from molecular dynamics data of Alder et al. [15]. The tie line at $PV_0/NkT = 8.27$ connects solid ($\eta = 0.545$) and fluid ($\eta = 0.494$) states having the same value for the chemical potential, as computed by Hoover and Ree [14].

solid and fluid phases (see Figure 3.11); evidently, hard spheres are too simple to exhibit a gas–liquid transition.

In the fluid phase, simulation data for the PVT equation of state have traditionally been represented by the Carnahan–Starling [10] equation

$$Z = \frac{1 + \eta + \eta^2 - \eta^3}{(1-\eta)^3} \qquad (3.53)$$

where $\eta = N\sigma^3\pi/6V$ is the packing fraction. This equation applies for $0 \le \eta \le 0.494$. By fitting to recent simulation data [11, 12], Boublik and Nezbeda [13] have refined the correlation (3.53); their representation, again for the fluid phase, is

$$Z = \frac{1 + \eta + \eta^2 - b_1\eta^3 - b_2\eta^4 - b_3\eta^5}{(1-\eta)^3} \qquad (3.54)$$

with $b_1 = 0.764314$, $b_2 = 0.151532$, and $b_3 = 0.654551$. This equation was used in constructing the hard-sphere phase diagram shown in Figure 3.11.

Hoover and Ree [14] have precisely located the phase transition by equating the chemical potentials of the solid and fluid phases. They locate the tie line at $P/\rho_o kT = 8.27$, where ρ_o is the density at close packing [from

(3.39), $\rho_o \sigma^3 = \sqrt{2}$]. At this pressure the density of the fluid phase is two-thirds of the close-packed density ($\eta = 2\eta_o/3 = 0.494$), while the solid density is 73.6% of close packed ($\eta = 0.736 \ \eta_o = 0.545$).

In the region of the phase transition, the behavior of a hard-sphere simulation depends not only on packing fraction (i.e., the thermodynamic state) but also on how the run is started. When runs are started from an fcc lattice at densities clearly below the transition ($\eta < 0.43$), the lattice will melt during equilibration and thereafter the system will be fluid. At the other extreme, if runs are started from an fcc lattice at densities $\eta > 0.5$, then the lattice will be a stable equilibrium structure. However, runs started from a lattice in the density range $0.43 < \eta < 0.5$ may exhibit metastabilities that alternate between solid and fluid [16]; the lifetimes of these metastable states will depend strongly on system size (N). Furthermore, if a run is started from a disordered state at $\eta > 0.5$, then often a metastable fluid will persist over the duration of the run, simply because the simulation time is insufficient to allow crystallization to occur.

3.8   ASSESSING RELIABILITY OF RESULTS

Those having little computational experience often assume computers are infallible, and therefore they fail to question results obtained from a simulation. This attitude is akin to the expectation that by merely viewing the scene of the crime, Holmes will be able to identify the culprit. In fact, the reality is far less mechanical and far more engaging. Once we have a working code and can generate some results, we face the task of verification: how do we convince ourselves that the code is running properly and that its results are reliable? The safe approach is to assume that the code is wrong until sufficient evidence is obtained to the contrary (guilty until proven innocent). To gather that evidence, the simulator must subject the code to every reasonable test that can be devised. Often the evidence is provided by the quantities we routinely monitor during a run: the difficulty is in training the eye to recognize what is, in fact, visible. We tend to see what we expect to see —if we expect a code to be correct, then we will tend to interpret its output as confirming that expectation. Training the eye implies developing a skill, and skills are developed, in large part, by trial and error—in other words, by experience.

This section is intended to start you toward gaining the experience needed for sound verification. First (Section 3.8.1) we list simple checks that can be used to test whether a code is running correctly. Then we discuss the uncertainties that always accompany computed time averages: statistical errors (Section 3.8.2) and systematic errors (Section 3.8.3). Ultimately, though, you cannot learn simulation by only reading a book; to become proficient, you must put code into a machine and *think* about its results.

3.8.1 Checking Reliability of the Code

Here we cite several tests that can be used to verify a hard-sphere simulation code. These are listed without comment, as they have already been discussed earlier in the book. The intention is merely to provide a listing for easy reference.

Equilibration

1. The positional order parameter should be fluctuating about zero, with fluctuations having magnitudes $\leq \sqrt{N}/N$ (Section 3.5.1).
2. The kinetic H-function should approach a value consistent with the Maxwell distribution (Sections 2.7.1 and 3.5.2).

Conservation Principles (Section 2.10)

3. The number of spheres in the primary cell should be constant with time.
4. Each Cartesian component of the total linear momentum should be zero.
5. The total kinetic energy should be constant with time; moreover, the kinetic energy should be equally partitioned among its three Cartesian components (Section 2.7.1).

Values of Properties

6. Instantaneous property values should be stable, fluctuating about constant values.
7. Averages should be reproducible, within statistical uncertainties, when runs are repeated from different initial conditions.

We caution that this list cites *necessary* conditions for reliable performance; none of these, separately or together, provides *sufficient* evidence that a code is correct. The exercises at the end of the chapter suggest ways of implementing and adding to the list.

3.8.2 Statistical Error

After a code has been tested, we can begin production runs to accumulate averages for properties. To have a concrete example for the following discussion, we adopt the hard-sphere virial $Z - 1$ as the property of interest. In planning production runs, we are immediately faced with the questions of (i) how to compute averages and (ii) how long the run should be. The answers depend on the amount of statistical error we are willing to tolerate.

Time Averages by Coarse Graining. In the discussion of statistical sampling in Section 2.8.4, we pointed out that a common strategy for avoiding serial correlations is to divide a computed phase-space trajectory into segments whose durations are longer than the relaxation time for the property of interest. So we need to estimate the relaxation time for the hard-sphere virial. For the Lennard-Jones substance, we found (Figure 2.12) that the relaxation time for the instantaneous pressure is about one mean collision interval, and so we might expect similar behavior for hard spheres. Indeed, this turns out to be the case: on computing the time correlation function (2.97) for the hard-sphere virial $Z - 1$, we find that the relaxation time is, on the average, one collision time. This seems reasonable because successive collisions in a hard-sphere simulation occur predominantly between pairs of spheres whose positions and velocities are, at most, weakly correlated. So, at least for the virial, we need not worry about serial correlations, and consequently, the duration of segments is dictated by the target set for statistical uncertainties.

Of the three methods suggested in Section 2.8.4 for sampling segments, coarse graining is preferred for static properties such as the virial. There are at least a couple of reasons for this choice. One is that coarse graining tends to provide smaller statistical uncertainties than the other methods. Other methods use only one instantaneous $Z(t)$-value from each segment, but coarse graining uses many values, thereby smoothing large variations among segment samples. The second reason is a practical one. In coarse graining we can choose the duration of segments *after* the run, without storing instantaneous values of the property $Z(t)$. During a run, all we need do is compute the running average $\langle Z \rangle_R$ and store it (on disk or tape) at intervals. Let the running average at collision number R be

$$\langle Z \rangle_R = \frac{1}{R} \sum_{c=1}^{R} Z(t_c) \qquad R = 1, 2, \ldots, M \tag{3.55}$$

After the run, we want to divide the M collisions into m segments of n collisions each (see Figure 3.12) and then form m coarse-grain averages,

$$\overline{Z}_i = \frac{1}{n} \sum_{c=(i-1)n+1}^{ni} Z(t_c) \qquad i = 1, 2, \ldots, m \tag{3.56}$$

Now note that at the nth collision the running average is the first coarse-grain average. Thereafter (Exercise 3.7), at any collision number R that is an integer multiple of n ($r = R/n$), the rth coarse-grain average can be computed from running averages by

$$\overline{Z}_r = \frac{R}{n} \langle Z \rangle_R - \frac{R-n}{n} \langle Z \rangle_{R-n} \tag{3.57}$$

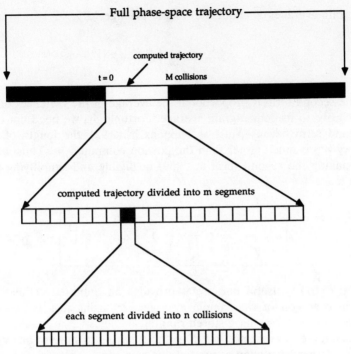

FIGURE 3.12   Schematic representation of dividing a computed trajectory for determining time averages by coarse graining. The M collisions computed during a simulation are divided into m segments, each containing n collisions.

Not only is (3.57) convenient, but it also allows us to study the effects of changing the length n of the segments.

When M collisions are divided into segments having the same size n, then the sample average $\langle Z \rangle$ computed from the coarse-grain averages

$$\langle Z \rangle = \frac{1}{m} \sum_{k=1}^{m} \overline{Z}_k \qquad (3.58)$$

is independent of the length n of each coarse-graining segment. Moreover, the statistical variance in $\langle Z \rangle$, as an unbiased estimator of the (unknown) trajectory average, is given by [17]

$$\sigma_{\langle Z \rangle}^2 = \frac{1}{m}\left(1 - \frac{m}{m_\infty}\right)\sigma_1^2 + \frac{1}{mn}\left(\frac{m}{m_\infty}\right)\left(1 - \frac{n}{n_\infty}\right)\sigma_2^2 \qquad (3.59)$$

Here m_∞ is the total number of segments that occupy the full phase-space trajectory and n_∞ is the total number of Z-values that fall in one segment. The first term on the rhs of (3.59) is proportional to the variance of the coarse-grain averages $\overline{Z}$ about the sample average $\langle Z \rangle$,

$$\sigma_1^2 = \frac{1}{m-1} \sum_{k=1}^{m} \left[\overline{Z}_k - \langle Z \rangle \right]^2 \qquad (3.60)$$

while the second term is proportional the average of the variances of each Z value relative to its coarse-grain average. Fortunately, we need not compute the second term because, just as happens in (2.92), the length of the full trajectory m_∞ is much larger than the portion computed m. Thus, m / m_∞ is small, making the second term in (3.59) negligible and simplifying the first term to leave

$$\boxed{\sigma_{\langle Z \rangle} = \frac{1}{\sqrt{m}} \sqrt{\frac{1}{m-1} \sum_{k=1}^{m} \left[\overline{Z}_k - \langle Z \rangle \right]^2}} \qquad (3.61)$$

Equation (3.61) is useful because it provides an unbiased estimate of the statistical precision of the sample average $\langle Z \rangle$ using just m coarse-grain averages. Further, if we have stored running averages $\langle Z \rangle_R$ at intervals, then we can compute the variance in $\langle Z \rangle$ without having to store or analyze all M values of Z that have been computed during a run.

Equation (3.61) offers still another advantage in that, for a computed trajectory of fixed length $M = mn$, the standard deviation from (3.61) is often insensitive to the size m of segments into which the trajectory is divided [17]. This insensitivity is illustrated in Table 3.1. Often, then, we can choose the number of coarse-grain segments based on considerations other than statisti-

TABLE 3.1   Insensitivity of Statistical Uncertainty (3.61) to Size of Segments Used for Coarse Graining[a]

Number of Coarse-Grain Segments m	Number of Collisions per Segment n	$\sigma_{\langle Z \rangle}$	χ^2
50	10,000	0.0079	2.2[b]
100	5,000	0.0088	1.4[b]
250	2,000	0.0085	0.8[c]
500	1,000	0.0082	0.4[c]
1250	400	0.0084	1.6[c]

[a] Here the statistical uncertainty is shown for the compressibility factor from a simulation of 500 hard spheres at packing fraction $\eta = 0.30$. The total computed trajectory length was $M = 500,000$ collisions, with the instantaneous virial computed every tenth collision. The average from this run would be reported at the 95% confidence level as $\langle Z \rangle - 1 = 2.98 \pm 2\sigma_{\langle Z \rangle} = 2.98 \pm 0.02$.
[b] Using 8 bins (see Section 3.8.3).
[c] Using 18 bins.

cal uncertainty. For example, a lower bound must be placed on m to have enough segments for testing that the distribution of coarse-grain averages is Gaussian. As a rule of thumb, we should divide the trajectory so that we have at least $m = 40$ segments.

Length of Run. The principal reason for performing long runs is not to improve accuracy but rather to decrease statistical uncertainties. This point is illustrated in Figure 3.13, which shows how the statistical error in the average virial decreases with run length in a simulation of 108 hard spheres. The error bars in the figure are at the 95% confidence level; that is, they represent $\pm 2\sigma_{\langle Z\rangle}$ with the standard deviation computed by coarse graining (3.61). In the figure, as the number of coarse-grain averages increases, the size of each coarse-graining segment remains fixed at 5000 collisions. Thus, the last point on the right in the figure represents an average over 10^6 collisions since it was computed from 200 segments.

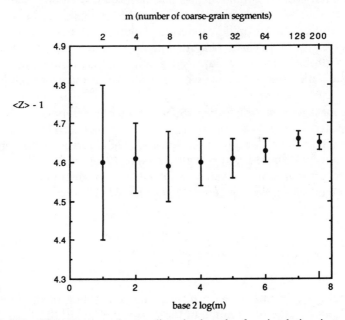

FIGURE 3.13 The purpose of extending the length of a simulation is to decrease statistical uncertainties. Here, for example, increasing the run length by a factor of 100 changes the average virial by only 11% (from 4.60 to 4.65), while the statistical uncertainties have decreased by a factor of 10. The simulation used 108 hard spheres at packing fraction $\eta = 0.3633$. The error bars are $\pm 2\sigma_{\langle Z\rangle}$, where $\sigma_{\langle Z\rangle}$ is the standard deviation of the average computed by coarse graining (3.61). Each coarse-grain segment extended over 5000 collisions, with the instantaneous virial Z computed every tenth collision.

Equation (3.61) shows that uncertainties decrease only as the root of the number of coarse-grain averages m. This behavior is confirmed in Figure 3.13. Thus, using $m = 2$ coarse-grain averages the uncertainty is 0.4; to decrease this by a factor of 10, to 0.04, the run had to be extended by a factor of 100, to $m = 200$. However, increasing the run length by a factor of 100 changed the average itself by only 11%, from 4.60 to 4.65.

To choose the run length, then, we first perform a short pilot run to estimate property averages and their statistical uncertainties. Next we identify the magnitude of the uncertainty we are willing to tolerate, and finally, we estimate the necessary run length by multiplying the length of the pilot by the square of the ratio of pilot uncertainties to desired uncertainties.

3.8.3   Systematic Error

Systematic errors are any of many possible kinds of bias that consistently displace the computed average from the complete trajectory average. It is certainly possible to have small statistical error and large systematic error, so the computed average can be precise but, nevertheless, wrong. Some systematic errors readily reveal themselves; for example, blunders in coding may quickly produce catastrophic numerical failure of the calculation. Other systematic errors may remain obscure and undetected except to thorough and persistent probing. No procedure can guarantee that results have tolerably small systematic errors: systematic error is an ever-present possibility.

Checking for a Normal Distribution. Some systematic errors not only displace the computed average from the true average, but they also bias the distribution of values about their average. Since simple thermodynamic properties are normally distributed about their averages (Section 2.7.2), one test for systematic error is whether values exhibit Gaussian distributions. To make the test quantitative, we can compute *chi squared* [18, 19]:

$$\chi^2 = \frac{1}{n_b - 3} \sum_{k=1}^{n_b} \frac{[No_k - Ne_k]^2}{Ne_k} \tag{3.62}$$

To compute χ^2, assume we have m instantaneous values of the property A. We first compute the distribution of these m values about their average $\langle A \rangle$ and compute their standard deviation σ (not the standard deviation of the average). We want to judge how this distribution compares with the Gaussian that has the same average and same standard deviation. To do so, divide the observed range of A-values into n_b bins. The width of the bins need not be uniform; rather, the number of A-values, as predicted by the Gaussian, should be about the same for each bin. The predicted number of A-values in bin k is the *expected* number Ne_k. If bin k has width ΔA and is centered at

A_k, then Ne_k can be calculated from

$$Ne_k = \frac{m}{\sigma\sqrt{2\pi}} \int_{A_k - \Delta A/2}^{A_k + \Delta A/2} \exp\left[\frac{-(A_k - \langle A \rangle)^2}{2\sigma^2}\right] dA \qquad (3.63)$$

From the computed distribution of A-values, we now count the number of values actually found in each bin; these are the observed numbers No_k. The measure χ^2 compares the observed and expected numbers, according to (3.62). In (3.62) the normalization factor is $n_b - 3$ because, in solving (3.63) for the expected numbers Ne_k, we have used three pieces of information from the observed distribution: the total number of observations m, the average $\langle A \rangle$, and the standard deviation σ. In a sense, the expected numbers have been obtained from a three-parameter fit to the observed distribution.

If χ^2 is zero, then the observed distribution agrees perfectly with a Gaussian. This does not prove that the observations were actually drawn from a Gaussian but merely that the observations are consistent with a Gaussian. If χ^2 is large, then the observed distribution is not Gaussian. We will be satisfied if χ^2 is of order unity, for then the observed distribution is consistent with a Gaussian, within unavoidable sampling errors. To obtain a statistically meaningful value for χ^2, the bin width should provide an expected number equal to at least five A-values in each bin. Thus, for eight bins, we need a sample of at least 40 A-values. Representative values of χ^2 are shown in Table 3.1. In that table, χ^2 is used to test whether m coarse grain averages are consistent with a Gaussian. The calculations used either 8 or 18 bins and the resulting values for χ^2 are consistently of order 1.

Thermodynamic Limit. In simulations that are intended to provide properties for a system of macroscopic size (one that contains of the order of Avogadro's number of molecules), there will be systematic error because of the finite number of molecules used in the simulation. To remove this error, we extrapolate simulation results to large particle number N and large system volume V, with the density N/V held fixed; this extrapolation defines the thermodynamic limit

$$\lim_{\substack{N \to \text{large} \\ V \to \text{large}}} (\text{simulation results})_{N/V \text{ fixed}} \approx (\text{macroscopic results}) \qquad (3.64)$$

Simulations may be affected by two kinds of finite-size effects: (a) simple N dependence [20, 21], which can be expressed as a power series in $1/N$, and (b) anomalous N dependence [22], which is caused by periodic boundary conditions and cannot be expressed as a regular function of N. The relative importance of these two effects depends on the property under investigation, on state condition, on the intermolecular force law, and on the environmen-

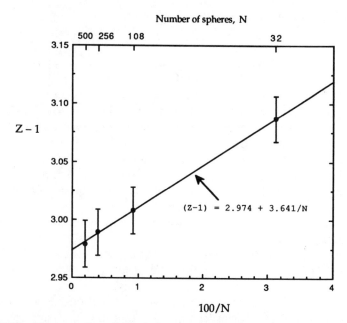

FIGURE 3.14 Extrapolation of simulation results to the thermodynamic limit for the hard-sphere compressibility factor $(Z-1)$ at packing fraction $\eta = 0.30$. The runs for 108, 256, and 500 spheres each ran for 500,000 collisions; that for $N = 32$ ran for 200,000 collisions. The averages and standard deviations from the average were obtained by coarse graining (3.61).

tal variables used to fix the thermodynamic state (i.e., the ensemble). For example, if the properties under study are the virial coefficients, then when N, V, and E (microcanonical) or N, V, and T (canonical ensemble) are used to fix the state, both effects contribute to the N dependence [22]. When μ, V, and T (grand ensemble) are used, only the anomalous dependence affects the virial coefficients, and when N, P, and T (isothermal–isobaric ensemble) are used, only the simple N dependence affects them [23]. Here μ is the chemical potential.

For hard spheres, simple N dependence always causes the compressibility factor to increase with increasing N. However, in isolated (NVE) fluid states having $N > 32$, the anomalous effect dominates the simple effect and Z decreases with increasing N [16]. This behavior is illustrated in Figure 3.14, which shows simulation results for the hard sphere Z at packing fraction $\eta = 0.3$. Those data are well represented by a simple least-squares fit to a straight line in $1/N$.

Independent Tests. Recall that simply repeating runs probes only statistical errors; reproducibility implies nothing about systematic errors. To probe for systematic errors we must repeat runs in different computing environments. For example, we may

(a) repeat runs using the same code on different computers,
(b) repeat runs using the same code but different compilers,
(c) repeat runs using the same algorithm but different code (say, written by different programmers),
(d) repeat runs using different algorithms, and
(e) repeat runs using different simulation methods (say, molecular dynamics and Monte Carlo).

The extent to which these tests are performed reflects the care of the simulator and the importance the simulator attaches to the results. In Chapter 4 we will compare results obtained from different machines (a), different compilers (b), and different algorithms (d); so here, we are content to illustrate only (e).

Table 3.2 shows hard-sphere compressibility factors obtained from NVT–Monte Carlo and from NVE–molecular dynamics. The Monte Carlo data are taken from Labik and Malijevsky [11], whose simulations used 864 spheres and runs of up to 10^7 configurations per state point. The molecular dynamics runs used 256 spheres and the average Z was accumulated by coarse graining using segments of 5000 collisions each. Using this coarse-graining procedure, preliminary pilot runs indicated that production runs of 500,000 collisions would yield statistical uncertainties comparable to those quoted by Labik and Malijevsky. The extrapolation in Figure 3.14 suggests that with $N = 256$, these uncertainties probably capture the thermodynamic

TABLE 3.2 Test for Systematic Errors: NVE–Molecular Dynamics and NVT–Monte Carlo Results Compared for Compressibility Factors of Hard Spheres

	$\langle Z \rangle - 1$	
Packing Fraction, η	Molecular Dynamics	Monte Carlo[a]
0.4504	8.40 ± 0.04	8.41 ± 0.04
0.3965	5.80 ± 0.03	5.81 ± 0.02
0.3670	4.76 ± 0.02	4.72 ± 0.02
0.3149	3.31 ± 0.02	3.30 ± 0.01
0.2610	2.26 ± 0.01	2.26 ± 0.01
0.2097	1.525 ± 0.008	1.526 ± 0.008
0.1572	0.971 ± 0.005	0.964 ± 0.006

[a] From Labik and Malijevsky [11].

limit. For the molecular dynamics results in Table 3.2, the uncertainties are standard deviations (3.61) of the average at the 95% confidence level.

When results from completely independent simulations are in good agreement, such as in Table 3.2, then we have a particularly powerful demonstration that systematic errors have been reduced to tolerable levels. Even if the property of interest is not accessible by different methods, we should still be able to identify some properties that could be realized by two independent algorithms. Agreement between those secondary properties would, at least, provide evidence that the basic simulation methodology was being properly performed.

3.9   SUMMARY

Simulation of hard spheres offers several advantages to those starting to learn molecular dynamics: (a) The equations of motion reduce to algebraic equations, so we are not distracted by problems in numerically solving differential equations. (b) Sphere interactions are short range, the code executes relatively fast, and consequently we can quickly obtain reasonably precise results. (c) Yet despite its simplicity, the hard-sphere system poses most of the computational difficulties that arise in more complex models. These include, for example, equilibration, assessing statistical and systematic errors, number dependence, the possibility of a phase transition, and possibilities for entering metastable states.

Overall, the chapter illustrates how concepts from diverse areas can be organized to yield a properly functioning molecular dynamics code. Thus, from classical analytic dynamics we obtain a description of the kinematics of hard-sphere collisions. From nonlinear dynamics we anticipate that a system started from an arbitrary nonequilibrium condition will evolve to an equilibrium state; moreover, once at equilibrium, the system will span the available constant-energy surface in a random manner. From kinetic theory we learn how to compute properties and that instantaneous values of those properties will fluctuate with specific distributions about average values. From sampling theory we obtain guidance in sampling those fluctuating quantities to obtain statistically precise averages. When these disparate concepts are combined in a programming language and implemented on a machine, the end result must be seen as at least remarkable and perhaps miraculous.

That molecular dynamics averages are actually meaningful seems paradoxical because the simulation procedure demands a certain precision but, at the same time, it appears to forgive lapses in accuracy. Consider again the results in Table 3.2: two very different algorithms—molecular dynamics and Monte Carlo—generate different microscopic behavior, and yet the resulting macroscopic averages are essentially identical. Something of fundamental importance is at work here and it is not clear that a simple understanding of all the parts is sufficient to understand the whole. By a different route, van Kampen has arrived at this same dilemma [24]:

> ...*experience has taught us* that in spite of our ignorance of most of the microscopic variables it is still possible to detect regularities in the macroscopic behavior and formulate them in general laws. It thus *appears* that the precise values of those microscopic variables are not important and *therefore* one might as well average over them. ... It is the task of the physicist to explain how this miracle comes about.... This seems to me the crucial problem of statistical mechanics.

REFERENCES

[1] B. J. Alder and T. E. Wainwright, "Studies in Molecular Dynamics. I. General Method," *J. Chem. Phys.*, **31**, 459 (1959).

[2] J. W. Haus and H. J. Raveché, "Computer Studies of Dynamics in One Dimension: Hard Rods," *J. Chem. Phys.*, **68**, 4969 (1978).

[3] J. J. Erpenbeck and W. W. Wood, "Molecular Dynamics Techniques for Hard-Core Systems," in *Statistical Mechanics, Part B: Time-dependent Processes*, B. J. Berne, Ed., Plenum, New York, 1977, Chapter 1.

[4] C. A. Rogers, "The Packing of Equal Spheres," *Proc. Lond. Math. Soc.*, **8**, 609 (1958).

[5] J. H. Conway and N. J. A. Sloane, *Sphere Packings, Lattices, and Groups*, Springer-Verlag, New York, 1988.

[6] W. H. Press, B. P. Flannery, S. A. Teukolsky, and W. T. Vetterling, *Numerical Recipes*, Cambridge University Press, New York, 1986, p. 202.

[7] L. Verlet, "Computer 'Experiments' on Classical Fluids. I. Thermodynamical Properties of Lennard-Jones Molecules," *Phys. Rev.*, **159**, 98 (1967).

[8] C. G. Gray and K. E. Gubbins, *Theory of Molecular Fluids*, Vol. 1, Oxford University Press, Oxford, 1984, Appendix E.

[9] H. Goldstein, *Classical Mechanics*, 2nd ed., Addison-Wesley, Reading, MA, 1980, p. 82f.

[10] N. F. Carnahan and K. E. Starling, "Equation of State for Nonattracting Rigid Spheres," *J. Chem. Phys.*, **51**, 635 (1969).

[11] S. Labik and A. Malijevsky, "Monte Carlo Simulations of the Radial Distribution Function of Fluid Hard Spheres," *Mol. Phys.*, **42**, 739 (1981).

[12] J. J. Erpenbeck and W. W. Wood, "Molecular Dynamics Calculations of the Hard-Sphere Equation of State," *J. Stat. Phys.*, **35**, 321 (1984).

[13] T. Boublik and I. Nezbeda, "$P-V-T$ Behaviour of Hard Body Fluids. Theory and Experiment," *Collection Czechoslovak Chem. Commun.*, **51**, 2301 (1986).

[14] W. G. Hoover and F. H. Ree, "Melting Transition and Communal Entropy for Hard Spheres," *J. Chem. Phys.*, **49**, 3609 (1968).

[15] B. J. Alder, W. G. Hoover, and D. A. Young, "Studies in Molecular Dynamics. V. High-Density Equation of State and Entropy for Hard Disks and Spheres," *J. Chem. Phys.*, **49**, 3688 (1968).

[16] B. J. Alder and T. E. Wainwright, "Studies in Molecular Dynamics. II. Behavior of a Small Number of Elastic Spheres," *J. Chem. Phys.*, **33**, 1439 (1960).

[17] W. G. Cochran, *Sampling Techniques*, 2nd ed., Wiley, New York, 1963, Chapter 10.

[18] J. R. Taylor, *An Introduction to Error Analysis*, University Science Books, Mill Valley, CA, 1982, Chapter 12.

[19] W. H. Press, B. P. Flannery, S. A. Teukolsky, and W. T. Vetterling, *Numerical Recipes*, Cambridge University Press, New York, 1986, Chapter 13.

[20] I. Oppenheim and P. Mazur, "Density Expansions of Distribution Functions. I. Virial Expansion for Finite Closed Systems; Canonical Ensemble," *Physica*, **23**, 197 (1957).

[21] J. L. Lebowitz and J. K. Percus, "Long-Range Correlations in a Closed System with Applications to Nonuniform Fluids," *Phys. Rev.*, **122**, 1675 (1961).

[22] J. L. Lebowitz and J. K. Percus, "Thermodynamic Properties of Small Systems," *Phys. Rev.*, **124**, 1673 (1961).

[23] Z. W. Salsburg, "Statistical Mechanical Properties of Small Systems in the Isothermal–Isobaric Ensemble," *J. Chem. Phys.*, **44**, 3090 (1966); **45**, 2719 (1966).

[24] N. G. van Kampen, *Stochastic Processes in Physics and Chemistry*, North-Holland, Amsterdam, 1981, p. 57.

EXERCISES

3.1 Show that Equations (3.17) and (3.18), which relate postcollision velocities to precollision velocities of two spheres, satisfy conservation of energy and linear momentum.

3.2 What physical situation is indicated by a negative collision time appearing as the smallest entry in the initial creation of the table of collision times?

3.3 Given initial positions $r_i(0)$ and velocities $v_i(0)$ for two hard spheres, write their vector separation at any time t as

$$r_{12}(t) = r_{12}(0) + t v_{12}(0)$$

Now show that provided the spheres do not collide, the time to the closest point of approach (cpa) is given by

$$t_{cpa} = \frac{-r_{12}(0) \cdot v_{12}(0)}{v_{12}(0) \cdot v_{12}(0)}$$

3.4 Derive the expression for the collision time, analogous to (3.24), for the situation in which the two colliding spheres have different diameters.

3.5 How do you expect the mean time between collisions to vary with packing fraction in the hard-sphere fluid phase?

3.6 Explain why the order parameter λ_x based on an fcc lattice is defined in (3.42) by $\cos(4\pi x_i / a)$ rather than by $\cos(2\pi x_i / a)$.

3.7 Derive (3.57), which enables us to compute coarse-grain averages from a table of running averages.

3.8 Prove (3.58); that is, when coarse-grain segments are the same size, the sample average is an unweighted average of the coarse-grain averages.

3.9 Consider two hard disks in an isolated, two-dimensional system. Write ⊠ a program that for given initial positions and velocities determines whether the two disks will collide.

(a) If they do, obtain the time of collision, the positions at collision, and the postcollision velocities.

(b) If they do not collide, obtain their positions at the closest point of approach.

Use your program to explore the following class of situations: disk 1 is assigned initial position $r_1(0) = 4j$ and initial velocity $v_1(0) = -j$. Disk 2 is given initial position $r_2(0) = -i$ and, in turn, each of the following initial velocities: (a) $6j$, (b) $5j$, (c) $4j$, (d) $3j$, (e) $2j$, (f) j, (g) $-4j$, and (h) $-5j$. Here i and j are unit vectors in the $+x$- and $+y$-directions, respectively.

3.10 One possible and subtle source of systematic error is error in the ⊠ intrinsic functions that are provided by high-level programming languages: SQRT, EXP, ABS in Fortran, for example. Devise and implement a program that exercises all the intrinsic functions used in the hard-sphere program MDHS given in Appendix K. Your program should test (randomly, perhaps?) the full range of argument values that might be encountered during a simulation.

3.11 List at least five errors that could reside in the hard-sphere molecular dynamics code, yet the code would still display conservation of total energy.

3.12 Write a program that uses the Box–Muller procedure [6] to draw ⊠ samples from a Gaussian distribution. Then compute chi squared (3.62) to test the distribution of drawn samples. Explore the sensitivity of chi squared to the number of samples drawn and to the number of bins used.

3.13 This exercise demonstrates that we could use a uniform distribution of ⊠ random numbers to obtain an initial velocity distribution that is Maxwellian. At the end of Section 2.8.3 it was remarked that the central limit theorem guarantees that n sample averages, each formed

from m values randomly drawn from any distribution, will tend to a Gaussian distribution. Test this claim by using the random-number generator ROULET from Appendix H to generate uniformly distributed values. Compute chi squared (3.62) to test whether the averages are in fact Gaussian. Explore how the Gaussian (measured by chi squared) is affected by changes in m and n. In particular, what is the smallest value of m that produces a satisfactory Gaussian?

3.14 Do you expect that the Maxwell distribution of velocities will develop
☒   for hard-sphere simulations in the solid phase? Run a simulation to test your answer.

3.15 Contrive other pathological initial conditions, such as those described
☒   in Section 3.5, use them to start simulations using MDHS from Appendix K, and monitor how the order parameter and H-function respond.

3.16 For atomic fluids the second virial coefficient is related to the intermolecular pair potential by

$$B = 2\pi N_A \int_0^\infty \left[1 - \exp\left(-\frac{u(r)}{kT}\right)\right] r^2 \, dr$$

where N_A is Avogadro's number, k is Boltzmann's constant, T is absolute temperature, and r is the distance between two atoms.

(a) Evaluate the second virial coefficient for a fluid of pure hard spheres.

(b) Show whether the Carnahan–Starling equation of state (3.53) reproduces your expression for B.

(c) Show whether the Boublik–Nezbeda equation (3.54) reproduces your expression.

3.17 Compare the predictions of both the Carnahan–Starling equation of state (3.53) and the Boublik–Nezbeda equation (3.54) with the simulation results for Z given in Table 3.2.

3.18 In the hard-sphere code MDHS in Appendix K, insert lines to compute
☒   the quantity $Q = \Sigma \mathbf{v}_i \cdot \mathbf{r}_i$ for the N spheres in the primary cell. Also form its time average. Then perform runs to confirm that Q is bounded and that its time average tends to zero as the run progresses.

3.19 The equation of state for pure hard spheres can be rigorously expressed in several distinct ways. One rigorous expression is (D. A. McQuarrie, *Statistical Mechanics*, Harper & Row, New York, 1976, p. 280)

$$Z = 1 + \left(\tfrac{2}{3}\pi\sigma^3\right)\rho g_+(\sigma)$$

where $g_+(\sigma)$ is the value of the radial distribution function at the distance of contact of two spheres when approached from separations $r > \sigma$. Why is it difficult to obtain $g_+(\sigma)$ directly from simulation?

3.20 Still another expression for the hard-sphere pressure can be obtained by recognizing that the nonideal-gas contribution to Z is proportional to the hard-sphere collision rate N_c, where N_c is defined in Exercise 2.17. To avoid evaluating the proportionality constant, form the ratio of N_c to its low-density limit N_{c0},

$$\frac{N_c}{N_{c0}} = \frac{Z-1}{\frac{2}{3}\pi\rho\sigma^3}$$

where the low-density limit of $Z - 1$ has been obtained from the expression in Exercise 3.19; that is, $g_+(\sigma) = 1$. Now use the result from Exercise 2.17 as an estimate for N_{c0} and show that

$$Z - 1 \approx \frac{N_c\sigma}{2}\sqrt{\frac{m}{3kT}}$$

Discuss how this expression could be used in a hard-sphere simulation to determine Z.

3.21 Run the hard-sphere program MDHS in Appendix K for (say) 32 ☒ spheres over an extended duration and study the effect of run length on statistical precision by constructing a plot analogous to Figure 3.13 and a table analogous to Table 3.1.

3.22 Perform a series of hard-sphere runs at one packing fraction but with ☒ different numbers of spheres. Then apply the thermodynamic limit, as illustrated in Figure 3.14, to the compressibility factor. Compute the uncertainty in your limiting value as suggested in Appendix E. Compare your limiting value of Z with those predicted by the empirical equations (3.53) and (3.54).

3.23 Test the principle of time-reversal invariance by modifying the pro-☒ gram to perform the following:
■ (a) Store the initial positions of all N spheres.
(b) Run the simulation for MN collisions.
(c) At collision number MN reverse the directions of all velocity vectors.
(d) Continue the run for another MN collisions.
(e) At the end of the total $2MN$ collisions compute the rms deviation between the initial and final positions.
Repeat the exercise for several values of M (e.g., $M = 50, 100, 500, 1000$).

3.24 Modify the code MDHS in Appendix K to print the velocity distribution ☒ $f(v)$ at regular intervals (say every $100N$ collisions). Compare each $f(v)$ with the Maxwell distribution by computing chi squared. Determine how many collisions are required for chi squared to fall within 1 ± 0.2. Repeat at different packing fractions and for different numbers of spheres.

3.25 Run the pure hard-sphere program to evaluate the virial at various ☒ packing fractions. Use your results to test the Carnahan–Starling equation of state (3.53). Repeat this exercise for $N = 32, 108, 256, 500$ spheres and note how the virial evolves over the duration of each run.

3.26 What modifications would be needed to use the hard-sphere program for simulations of hard disks in two dimensions?

3.27 For the same initial positions, packing fraction, number of spheres, ☒ and run duration, run the hard-sphere program three times using different initial velocities (change the seed on the random-number generator). Observe how the kinetic energy and virial respond. Are the different results for the virial consistent within their combined statistical uncertainties?

3.28 What modifications would be needed to use the hard-sphere program for simulations of binary mixtures of spheres of different diameters?

3.29 Test the hard-sphere code MDHS in Appendix K for systematic errors ☒ by executing the code under the same conditions (number of spheres, packing fraction, and seed to random-number generator) on different compilers and then on different machines.

3.30 If the statistical error in a simulation result for the compressibility factor Z is computed to be ± 0.04 and systematic errors are estimated to be ± 0.04, then what is your estimate of the total uncertainty in Z?

3.31 When simulation results and their associated uncertainties are combined to obtain new property values, we are faced with error propagation (see Appendix E). Use the molecular dynamics results for $Z - 1$ in Table 3.2 to estimate the isothermal compressibility κ_T,

$$\kappa_T = \frac{1}{\rho} \left(\frac{\partial \rho}{\partial P} \right)_T$$

of the hard-sphere fluid at $\eta = 0.4235$. Compute the compressibility in the form $\kappa_T / (\kappa_T)_{ig}$, where $(\kappa_T)_{ig}$ is the ideal-gas value. Using the uncertainties quoted in the table, also determine the uncertainty in your result.

3.32 Devise and implement procedures that test separately the principal ☒ subroutines used in the hard-sphere code MDHS given in Appendix K.

3.33 Execute the hard-sphere code MDHS in Appendix K for 2000 collisions
⊠ and 108 spheres to produce a set of baseline data with which to
compare results. Then introduce the following "errors" into the simulation code (one error per run), execute for 2000 collisions, and
compare with the baseline data. After each run describe the nature of
the error introduced and explain its consequences.

    (a) In subroutine HSUPDT, line 13, place a C in column one.
        {IF (I.EQ.NATOMB..

    (b) In subroutine HSMOVE, line 19, place a C in column one.
        {IF (X(I).GT.1...

    (c) In subroutine HSMOVE, line 16, replace X1 with Y1.
        {under ADVANCE ALL SPHERES

    (d) In subroutine HSMOVE, line 9, replace 1 with 3. {DO 400 I=1,...

    (e) In subroutine HSCRSH, line 13, replace 1.DO with 0.DO.
        {RX=RX-1.DO

    (f) In subroutine HSCRSH, line 30, replace -DELVX with +DELVX.

    (g) Contrive two more errors of your own and test their consequences.

3.34 Appendix C contains a derivation of the *NVE* partition function for a
one-dimensional system of hard spheres. Use that result to derive
expressions for the entropy, pressure, constant volume heat capacity,
chemical potential, and total energy. The answers are contained in the
appendix.

3.35 How would you modify the hard-sphere algorithm to simulate spheres
that interact with the square-well potential:

$$u(r) = \begin{cases} \infty & r \le \sigma \\ -\varepsilon & \sigma < r < \lambda\sigma \\ 0 & r \ge \lambda\sigma \end{cases}$$

Here σ is the hard-sphere diameter, λ is a constant factor, say, 1.5,
and ε is a constant well depth.

4

FINITE-DIFFERENCE METHODS

Consider this problem: Determine the location of a car at 10:30, 11:00, and 12:00 given that the car left Salina, Kansas, at 10:00, traveling west on Interstate-70 at a constant 50 mph. The solution of this problem is governed by an ordinary differential equation

$$v = \frac{dx}{dt} = 50 \text{ mph} \qquad (4.1)$$

We are told a rate and an initial condition and, consequently, such problems are called *initial-value problems*. The solution is trivial because the rate is constant; a simple integration of (4.1) yields an algebraic equation that can be solved for the distance from Salina for any time subsequent to 10:00. This problem is analogous to the hard-sphere simulation problem described in the last chapter. In a hard-sphere simulation we need solve only algebraic equations because, between collisions, the spheres follow straight-line trajectories at constant velocities.

But what if either problem were more realistic? In the case of the car, we know, ignoring the use of cruise control, that the car's speed will not be constant: it will vary in response to local traffic, topography of the road, and inattentiveness of the driver. In this more realistic situation not only does the car's speed change with time, but we do not even have an analytic form for *how* the speed changes. Perhaps we have a table of speedometer readings at points during the journey. How do we solve this problem? How do we determine the car's position on I-70 at various points in time? Now we must solve the differential equation and do so numerically—the analytic form for the rate is unknown.

This problem is analogous to a molecular dynamics simulation of molecules whose potential energy varies continuously with distance ("soft bodies"). Because each molecule is simultaneously interacting with many other molecules, soft-body trajectories are not straight lines nor are the velocities constant between collisions. In fact the idea of a collision is not as precise as in the hard-body case. Collisions between soft bodies are not instantaneous; rather, they are strong repulsive interactions that occur over a finite duration. As with the realistic car situation, soft-body simulations require a numerical method for solving differential equations.

The classic tools for attacking initial-value problems are finite-difference methods.[†] These methods replace differentials, such as dx and dt, with finite differences Δx and Δt; they replace differential equations with finite-difference equations; and over a small but finite time Δt, they assume the rate (or some known function of the rate) is constant. Then to solve the car problem, we proceed in the following stepwise fashion: From the known initial position of the car $x(t_0)$, we use the assumed constant rate to approximate the position $x(t_0 + \Delta t)$ after the lapse of the small interval Δt. From this approximate position, with a revised value for the rate (say, from our table of speedometer readings), we step forward another increment to estimate the position at $x(t_0 + 2\Delta t)$. After many such steps we have an approximation to the car's path $x(t)$. This strategy we also use in soft-body molecular dynamics.

As indicated in Figure 1.5, molecular dynamics divides into two great tasks: generate the trajectory in phase space and analyze that trajectory for the properties of interest. Each task incurs uncertainties that detract from the reliability of the results. In generating soft-body trajectories, uncertainties arise from the finite-difference method used (truncation error) and from the way it is implemented (round-off error). Therefore, we must be concerned not merely with finite-difference methods per se, but with the broader problem of how those methods can undermine the reliability of simulation results. We begin by introducing a prototypical finite-difference algorithm (Section 4.1) and use that algorithm to define truncation and round-off errors (Section 4.2). The terms *truncation* and *round-off* refer to sources of errors; however, we must consider not only what causes errors but also how errors propagate—whether they grow as the simulation proceeds. This is the issue of algorithmic stability (Section 4.3). For simple finite-difference algorithms applied to linear differential equations, we can readily perform a stability analysis, but for the nonlinear differential equations used in molecular dynamics (Section 4.4) such an analysis is less direct. Moreover, in soft-body simulations we may find that uncertainties in computed trajectories are

---

[†]W. Thomson, Lecture VI, p. 61: "The problem that I put before you here is given in that work [Lagrange's *Mécanique Analytique*] under the title of vibrations of a linear system of bodies. Lagrange applies what he calls the algorithm of finite differences to the solutions."

compounded by subtle interplay among errors, instabilities, and nonergodic-ity (Section 4.5).

4.1   A PROTOTYPE: EULER'S METHOD

Several finite-difference methods originate from truncated Taylor expansions and the simplest is Euler's method, which is a Taylor expansion truncated after the first-order term

$$x(t + \Delta t) = x(t) + \dot{x}(t)\,\Delta t \tag{4.2}$$

From the known (or estimated) value of x at t, this method estimates x at $t + \Delta t$ by extrapolating from $x(t)$ the straight line that has slope dx/dt, evaluated at t. As a concrete example, consider Euler's method applied to the ODHO of Figure 2.2. To estimate the oscillator's position x and velocity v, Euler's method uses

$$x(t + \Delta t) = x(t) + \dot{x}(t)\,\Delta t \tag{4.3}$$

$$v(t + \Delta t) = v(t) + \dot{v}(t)\,\Delta t \tag{4.4}$$

Now, according to (2.26), the ODHO has

$$\dot{x}(t) = v(t) \tag{4.5}$$

$$\dot{v}(t) = \ddot{x}(t) = -\frac{\gamma}{m}x(t) = -\omega^2 x(t) \tag{4.6}$$

where ω is the frequency of the oscillation. Therefore, by using these expressions in (4.3) and (4.4), Euler's method for the ODHO becomes

$$x(t + \Delta t) = x(t) + v(t)\,\Delta t \tag{4.7}$$

$$v(t + \Delta t) = v(t) - \omega^2 x(t)\,\Delta t \tag{4.8}$$

For specified values of the mass m, spring constant γ, and initial conditions $x(t_0)$ and $v(t_0)$, iterative application of (4.7) and (4.8) generates an approximation to the trajectory $\{x(t), v(t)\}$. This calculated trajectory will have errors associated with it; moreover, those errors may accumulate as the iterative calculation proceeds. These issues are discussed in the next two sections.

4.2 ERRORS

A finite-difference method incurs two types of errors: truncation error and round-off error. *Truncation error* refers to the accuracy with which a finite difference method approximates the true solution to a differential equation. When a finite-difference equation is written in a Taylor series form, truncation error is measured by the first nonzero term that has been omitted from the series. The Taylor series is

$$x(t + \Delta t) = x(t) + \frac{dx(t)}{dt} \Delta t + \frac{1}{2} \frac{d^2 x(t)}{dt^2} \Delta t^2 + \frac{1}{3!} \frac{d^3 x(t)}{dt^3} \Delta t^3 + \cdots$$

(4.9)

and comparing this with (4.2) identifies the truncation error (te) in Euler's method as

$$\text{te} = \frac{1}{2} \frac{d^2 x(t)}{dt^2} \Delta t^2$$

(4.10)

A method whose truncation error varies as $(\Delta t)^{n+1}$ is said to be an nth-order method; hence, Euler's is first order. We typically use a system of units such that $\Delta t < 1$; therefore, for a given time-step size, high-order methods have smaller truncation errors than low-order methods. Truncation error is inherent in the algorithm; its value is the same regardless of how the actual calculations are performed, whether on a computer, on a calculator, or by pencil and paper.

In contrast, *round-off error* encompasses all errors that result from the implementation of the finite-difference algorithm. For example, round-off error is affected by the number of significant figures kept at each stage of the calculation, by the order in which the calculations are actually performed, and by any approximations used in evaluating square roots, exponentials, and so on.

Both round-off and truncation errors can be subdivided into global and local errors. *Local error* is that incurred during one step (Δt) of the algorithm, while *global error* is local error accumulated over the entire calculation. It is local truncation error, such as (4.10), that is used to define the order of a method. But instead of local errors, we should be more concerned with global errors, and in general, global error varies by one factor of Δt *less* than local error. To see this, consider an nth-order algorithm that has local truncation error

$$\text{lte} = kx^{(n+1)} \Delta t^{n+1}$$

(4.11)

where k is a constant and $x^{(n+1)}$ means the $(n+1)$th derivative of x. Over M integration steps, each of size Δt, the global truncation error is

$$\text{gte} = k \sum_{i=1}^{M} x_i^{(n+1)} \Delta t^{n+1} = k \Delta t^{n+1} \sum_{i=1}^{M} x_i^{(n+1)} \qquad (4.12)$$

In general the derivatives under the sum will vary in value from one step to the next, but we can apply a mean-value theorem: there will be an average value of the M-derivatives such that the M-term sum can be replaced by M times the average,

$$\text{gte} = k \Delta t^{n+1} M \overline{x^{(n+1)}} \qquad (4.13)$$

Here the overbar indicates the average. Now the number of steps M is related to the total duration t of the calculation by $M = t/\Delta t$; hence,

$$\text{gte} = k \Delta t^{n+1} \frac{t}{\Delta t} \overline{x^{(n+1)}} = k \Delta t^n t \overline{x^{(n+1)}} \qquad (4.14)$$

thus, gte varies as $(\Delta t)^n$, although lte varies as $(\Delta t)^{(n+1)}$. QED. If, instead of gte, we evaluate the average gte per step, $\langle \text{gte} \rangle$, then we have a global error that behaves with Δt in the same way as the local truncation error,

$$\langle \text{gte} \rangle = \frac{\text{gte}}{M} = k \Delta t^{n+1} \overline{x^{(n+1)}} \qquad (4.15)$$

The results (4.14) and (4.15) apply to one-step methods, which are those that use information only from the current step to estimate x at the next step. In contrast, multistep methods use estimates for x at previous steps, as well as the current step, to estimate x at the next step; in those cases, gte is a more complicated function of the step size.

Global truncation error and global round-off error (gre) both depend on the size of the integration step Δt, so having chosen an algorithm, we must determine the magnitude of Δt that produces acceptably small global errors. Unfortunately, gte and gre are affected differently by changes in the step size. Global truncation error decreases with decreasing Δt, as indicated above. In contrast, gre depends on the number of calculations: increasing the number of calculations increases the opportunities for round-off and produces higher gre. Thus at some point, decreasing Δt for the same duration $\tau = t_{\text{final}} - t_{\text{initial}}$ does not produce more accurate results, as shown in Figure 4.1. Usually the value of Δt having the smallest total error is too small to be useful—too much computer time would be required for a simulation. Some value of Δt larger than that for minimum total error is used; its value is determined empirically in test calculations.

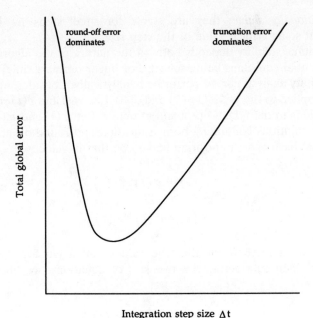

FIGURE 4.1 Schematic of how truncation and round-off errors contribute to the total global error generated by a finite-difference algorithm. At large step sizes the global error is dominated by truncation errors in the algorithm; however, for small step sizes global error is dominated by round-off. For each step size, the behavior shown applies for global error accumulated over a fixed duration of calculation, not for a fixed number of integration steps.

The principal defenses against round-off errors are to write efficient, nonredundant code and to use high-precision arithmetic: double precision rather than single. The principal defense against truncation error is to reduce the step size. However, if Δt must be made unacceptably small, the only recourse is to change finite-difference algorithms.

4.3 ALGORITHMIC STABILITY

In addition to the magnitudes of errors involved in using finite-difference methods, we must also be concerned with how the algorithm propagates those errors. This is the issue of algorithmic stability—a concept that is distinct from the idea of trajectory stability discussed in Sections 2.4 and 2.5. If an algorithm amplifies errors from one step to the next, then the algorithm is *unstable*, and ultimately the calculation will abort in a numerical overflow. Conversely, if the algorithm does not amplify errors from one step to the next, then the method is *stable*. Most algorithms used in molecular dynamics are *conditionally stable*: they are stable for small steps Δt but become unstable at some critical value of the step size.

Algorithmic stability depends both on the nature of the algorithm and on the differential equations being solved. For linear ordinary differential equations, stability analysis can be performed analytically. Consider, again, Euler's method applied to the ODHO (4.7) and (4.8). Let $x(t)$ and $v(t)$ represent the true solutions to the ODHO problem at time t. Let $x'(t)$ and $v'(t)$ represent erroneous solutions containing both truncation and round-off errors. Let e_x be the total local error in position and e_v be that in velocity,

$$e_x(t) = x'(t) - x(t) \tag{4.16}$$

$$e_v(t) = v'(t) - v(t) \tag{4.17}$$

Writing Euler's method for the true values and again for the erroneous values and then subtracting the two sets of equations, we obtain, for the ODHO,

$$e_x(t + \Delta t) = e_x(t) + e_v(t)\,\Delta t \tag{4.18}$$

$$e_v(t + \Delta t) = e_v(t) - \omega^2 e_x(t)\,\Delta t \tag{4.19}$$

which relate the errors at time $t + \Delta t$ to those at time t. These two equations can be written in vector form as

$$e(t + \Delta t) = \mathbf{A}e(t) \tag{4.20}$$

where $e = (e_x e_v)^T$ is the column vector of errors and $\mathbf{A}$ is the stability matrix

$$\mathbf{A} = \begin{bmatrix} 1 & \Delta t \\ -\omega^2 \Delta t & 1 \end{bmatrix} \tag{4.21}$$

Now, if $\mathbf{A}$ has any eigenvalue λ that lies outside the unit circle ($\lambda^2 > 1$) on the complex plane, then $\mathbf{A}$ amplifies errors and the algorithm is unstable. The eigenvalues satisfy the characteristic equation

$$|\mathbf{A} - \lambda \mathbf{I}| = 0 \tag{4.22}$$

where $\mathbf{I}$ is the identity matrix. For $\mathbf{A}$ given by (4.21) this equation is

$$(1 - \lambda)^2 + \omega^2 (\Delta t)^2 = 0 \tag{4.23}$$

which has solutions

$$\lambda = 1 \pm \Delta t \sqrt{-\omega^2} \qquad (4.24)$$

Thus, for any time step Δt, $|\lambda| > 1$ and Euler's method is unstable when applied to the ODHO problem. Note that the stability of the algorithm is independent of the conditions $x(0)$ and $v(0)$ that start the calculations.

To illustrate a conditionally stable algorithm, we modify Euler's method for the ODHO. First write a forward Taylor series, truncated at first order, to estimate the velocity $v(t + \Delta t)$,

$$v(t + \Delta t) = v(t) + \dot{v}(t) \Delta t \qquad (4.25)$$

and then write a backward Taylor series, also truncated at first order, to estimate the position

$$x(t) = x(t + \Delta t) - \dot{x}(t + \Delta t) \Delta t \qquad (4.26)$$

The modified Euler method is then

$$x(t + \Delta t) = x(t) + [v(t) + \dot{v}(t) \Delta t] \Delta t \qquad (4.27)$$

$$v(t + \Delta t) = v(t) + \dot{v}(t) \Delta t \qquad (4.28)$$

For the ODHO the time derivatives are given by (4.5) and (4.6), so the algorithm becomes

$$x(t + \Delta t) = \left[1 - \omega^2 (\Delta t)^2\right] x(t) + v(t) \Delta t \qquad (4.29)$$

$$v(t + \Delta t) = v(t) - \omega^2 x(t) \Delta t \qquad (4.30)$$

The stability matrix for this method is

$$\mathbf{A} = \begin{bmatrix} \left(1 - \omega^2 (\Delta t)^2\right) & \Delta t \\ -\omega^2 \Delta t & 1 \end{bmatrix} \qquad (4.31)$$

Solving the characteristic equation shows that the step size Δt must satisfy

$$\left| 2 - \omega^2 (\Delta t)^2 \pm \omega \Delta t \sqrt{\omega^2 (\Delta t)^2 - 4} \right| < 2 \qquad (4.32)$$

That is, we must have $-2 < \omega \Delta t < 2$ for this modified Euler method to provide stable solutions to the ODHO problem.

It is important to realize that stable solutions are not necessarily accurate solutions. This point is illustrated in Figure 4.2, which shows ODHO phase-

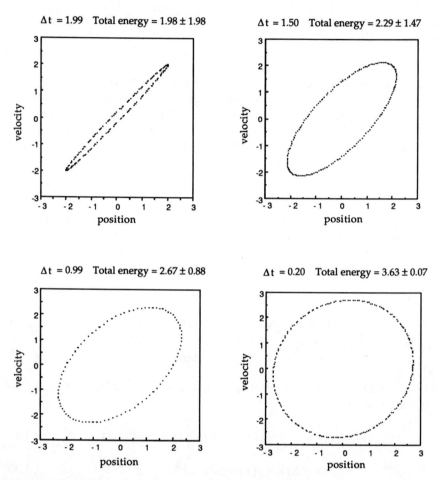

FIGURE 4.2 Stable solutions are not necessarily accurate. Shown here are phase-plane trajectories for the ODHO calculated from the modified Euler algorithm (4.29) and (4.30) using four values of the time-step Δt. Each trajectory shows 200 steps starting from $x(0) = v(0) = 2$, with $m = \gamma = 1$ (arbitrary units). Thus in each case the true solution to the ODHO is a circle with total energy equal to 4. Each of the four values of Δt is below the stability limit $\Delta t = 2$, but the accuracy becomes acceptable only when $\Delta t < 0.02$.

plane trajectories calculated from the modified Euler method (4.29) and (4.30). The figure shows how the shape of the trajectory is affected by the size of the integration time step Δt. Calculations attempted with Δt only marginally greater than the stability limit ($\Delta t = 2$) quickly led to numerical overflows. The values of the mass, spring constant, and initial conditions were set such that the true solution to the ODHO describes a circle in the phase plane, corresponding to a total energy $E = 4$ arbitrary units. The calculations show that although the algorithm is stable for $\Delta t < 2$, the total energy is not properly conserved at its initial value of $E = 4$ until the time step is reduced to $\Delta t < 0.02$, which is a value 100 times smaller than the stability limit.

The lesson from these simple examples is that a finite-difference algorithm applied to a particular set of differential equations yields a stability matrix that may produce inherently stable, unstable, or conditionally stable behavior. Our first response to an unstable algorithm is to decrease the step size; however, if stability can be achieved only by using an intolerably small Δt, then we must either change algorithms or find another problem (i.e., change the differential equations).

Molecular dynamics involves nonlinear ordinary differential equations, so the analytic stability analysis described above cannot be used. Instead, we may do an approximate analysis by linearizing the differential equations or we may attack the stability problem using the methods of Lyapunov [1]. Both strategies are cumbersome when applied to the equations of motion used in molecular dynamics, and what we do instead is a numerical analysis using trial values for Δt. That is, through a series of short test runs, we identify the critical step Δt at which the algorithm becomes unstable and then choose a smaller operational value that establishes conservation of energy.

4.4 ALGORITHMS FOR MOLECULAR DYNAMICS

Of the large number of finite-difference methods that can be devised, the most commonly used are the Runge–Kutta (RK) methods [2]. These methods have the structure of Euler's method (in fact, Euler's is the first-order RK algorithm)

$$x(t + \Delta t) = x(t) + \dot{x}\,\Delta t \tag{4.33}$$

The various RK methods differ from one another in how the slope $\dot{x}$ is estimated. Each RK algorithm estimates the slope at points along the interval Δt and then computes a weighted average to get a single value to be used in (4.33). For low-order RK algorithms, the number of slope evaluations during one step equals the order of the method. Thus the popular fourth-order RK algorithm (see Exercise 4.11) makes four estimates of the slope for each step forward in the solution procedure.

Runge–Kutta methods generally have good stability characteristics; nevertheless, they have been little used in molecular dynamics because, for large numbers of molecules, the RK algorithms are too slow. In molecular dynamics the evaluation of intermolecular forces is by far the most time-consuming calculation. Therefore, since a fourth-order RK algorithm would require four force evaluations per atom per step, such an algorithm would execute almost four times slower than a method that needs only one force evaluation per atom per step. Runge–Kutta methods have been used in simulations of systems involving a very few degrees of freedom [3].

The RK methods suggest that the order of a method can often be increased by using positions and velocities from several points in time rather than from just the current time. We would like to use such information, but we want to avoid the expense of evaluating intermolecular forces more than once per atom per step. These conflicting goals can be met in two general ways: (a) use positions and velocities from previously calculated steps, the principal example of this approach being Verlet's algorithm, or (b) estimate positions and velocities for future steps. The latter are the predictor–corrector algorithms.

4.4.1   Verlet's Algorithm

The simplest finite-difference method that has been widely used in molecular dynamics is a third-order Störmer algorithm, first used by Verlet [4] and known to simulators as Verlet's method. The algorithm is a combination of two Taylor expansions, combined as follows. First write the Taylor series for position from time t forward to $t + \Delta t$:

$$x(t + \Delta t) = x(t) + \frac{dx(t)}{dt}\Delta t + \frac{1}{2}\frac{d^2x(t)}{dt^2}\Delta t^2 + \frac{1}{3!}\frac{d^3x(t)}{dt^3}\Delta t^3 + O(\Delta t^4)$$

$$(4.34)$$

Then write the Taylor series from t backward to $t - \Delta t$:

$$x(t - \Delta t) = x(t) - \frac{dx(t)}{dt}\Delta t + \frac{1}{2}\frac{d^2x(t)}{dt^2}\Delta t^2 - \frac{1}{3!}\frac{d^3x(t)}{dt^3}\Delta t^3 + O(\Delta t^4)$$

$$(4.35)$$

Adding these two expansions eliminates all odd-order terms, leaving

$$x(t + \Delta t) = 2x(t) - x(t - \Delta t) + \frac{d^2x(t)}{dt^2}\Delta t^2 + O(\Delta t^4) \quad (4.36)$$

This is Verlet's algorithm for positions. It has a local truncation error that varies as $(\Delta t)^4$ and hence is third order, even though it contains no third-order derivatives. Nor does (4.36) for positions involve any function of the velocities; the acceleration in (4.36) is, of course, obtained from the intermolecular forces and Newton's second law. To estimate velocities, practitioners have contrived various schemes, one being an estimate for the velocity at the half-step:

$$v\left(t + \tfrac{1}{2}\Delta t\right) \approx \frac{x(t + \Delta t) - x(t)}{\Delta t} \tag{4.37}$$

Verlet himself used the first-order central difference estimator

$$v(t) \approx \frac{x(t + \Delta t) - x(t - \Delta t)}{2\Delta t} \tag{4.38}$$

Verlet's algorithm is a two-step method because it estimates $x(t + \Delta t)$ from the current position $x(t)$ and the previous position $x(t - \Delta t)$. Therefore it is not self-starting: initial positions $x(0)$ and velocities $v(0)$ are not sufficient to begin a calculation, and something special must be done at $t = 0$ (say, a backward Euler method) to get $x(-\Delta t)$.

The Verlet algorithm offers the virtues of simplicity and good stability for moderately large time steps. In its original form it treated molecular velocities as less important than positions—a view in conflict with the attitude that the phase-space trajectory depends equally on positions and velocities. Modern formulations [5–7] of the method often overcome this asymmetric view.

4.4.2   General Predictor–Corrector Algorithms

Predictor–corrector methods are composed of three steps: prediction, evaluation, and correction. In particular, from the current position $x(t)$ and velocity $v(t)$ the steps are as follows:

1. Predict the position $x(t + \Delta t)$ and velocity $v(t + \Delta t)$ at the end of the next step.
2. Evaluate the forces at $t + \Delta t$ using the predicted position.
3. Correct the predictions using some combination of the predicted and previous values of position and velocity.

As a simple example, consider the following predictor–corrector based on Euler's method and applied to the ODHO:

1. *Predict* $x(t + \Delta t)$ and $v(t + \Delta t)$ using Euler's method (4.7) and (4.8):

$$x(t + \Delta t) = x(t) + v(t)\,\Delta t$$

$$v(t + \Delta t) = v(t) - \omega^2 x(t)\,\Delta t$$

2. *Evaluate* the force at $t + \Delta t$:

$$\frac{f(t + \Delta t)}{m} = \frac{dv}{dt} = -\omega^2 x(t + \Delta t) \qquad (4.39)$$

3. *Correct* the predictions. Here we choose the same form as Euler's method but compute the slopes at the end of the step rather than at the beginning:

$$x(t + \Delta t) = x(t) + v(t + \Delta t)\Delta t \qquad (4.40)$$

$$v(t + \Delta t) = v(t) - \omega^2 x(t + \Delta t)\Delta t \qquad (4.41)$$

The force evaluation implicit in (4.8) would actually be done in the previous step $(t - \Delta t)$ and stored for use in the current step, so this algorithm meets the goals of requiring only one force evaluation per step while providing an algorithm of higher order than Euler's method.

Predictor–corrector methods offer great flexibility in that many choices are possible for both the prediction and correction steps. They may be either one-step, in which case they are self-starting, or multistep methods, in which case something special must be done to start the calculation. With judicious combinations of predictor and corrector they offer good stability because the corrector step amounts to a feedback mechanism that can dampen instabilities that might be introduced by the predictor.

Any given predictor–corrector algorithm can be made more elaborate and of higher order by repeating the evaluation and correction steps. Let P be prediction, E be evaluation, and C be correction. Then the procedure described above can be represented as PEC. If the corrected positions and velocities are used as second predictions, we obtain the algorithm $PEC(EC) = P(EC)^2$. Obviously, the E and C steps can be repeated as many times as desired, $P(EC)^n$. This strategy, while simple to implement, is rarely done in molecular dynamics because each E-step requires force calculations; a $P(EC)^n$ simulation will execute nearly n times slower than the PEC simulation.

4.4.3 Gear's Predictor–Corrector Algorithms

Predictor–corrector algorithms were first introduced into molecular dynamics by Rahman [8]. Those commonly used in molecular dynamics are often taken from the collection of methods devised by Gear [9]. The one used in the programs in Appendices I and L consists of the following steps.

Predict molecular positions r_i at time $t + \Delta t$ using a fifth-order Taylor series based on positions and their derivatives at time t. Thus, the derivatives $\dot{r}_i$, $\ddot{r}_i$, $r_i^{(iii)}$, $r_i^{(iv)}$, and $r_i^{(v)}$ are needed at each step; these are also predicted at time $t + \Delta t$ by applying Taylor expansions at t:

$$\mathbf{r}_i(t + \Delta t) = \mathbf{r}_i(t) + \dot{\mathbf{r}}_i(t)\,\Delta t + \ddot{\mathbf{r}}_i(t)\frac{(\Delta t)^2}{2!} + \mathbf{r}_i^{(\text{iii})}(t)\frac{(\Delta t)^3}{3!}$$

$$+\,\mathbf{r}_i^{(\text{iv})}(t)\frac{(\Delta t)^4}{4!} + \mathbf{r}_i^{(\text{v})}(t)\frac{(\Delta t)^5}{5!} \tag{4.42}$$

$$\dot{\mathbf{r}}_i(t + \Delta t) = \dot{\mathbf{r}}_i(t) + \ddot{\mathbf{r}}_i(t)\,\Delta t + \mathbf{r}_i^{(\text{iii})}(t)\frac{(\Delta t)^2}{2!}$$

$$+\,\mathbf{r}_i^{(\text{iv})}(t)\frac{(\Delta t)^3}{3!} + \mathbf{r}_i^{(\text{v})}(t)\frac{(\Delta t)^4}{4!} \tag{4.43}$$

$$\ddot{\mathbf{r}}_i(t + \Delta t) = \ddot{\mathbf{r}}_i(t) + \mathbf{r}_i^{(\text{iii})}(t)\,\Delta t + \mathbf{r}_i^{(\text{iv})}(t)\frac{(\Delta t)^2}{2!} + \mathbf{r}_i^{(\text{v})}(t)\frac{(\Delta t)^3}{3!} \tag{4.44}$$

$$\mathbf{r}_i^{(\text{iii})}(t + \Delta t) = \mathbf{r}_i^{(\text{iii})}(t) + \mathbf{r}_i^{(\text{iv})}(t)\,\Delta t + \mathbf{r}_i^{(\text{v})}(t)\frac{(\Delta t)^2}{2!} \tag{4.45}$$

$$\mathbf{r}_i^{(\text{iv})}(t + \Delta t) = \mathbf{r}_i^{(\text{iv})}(t) + \mathbf{r}_i^{(\text{v})}(t)\,\Delta t \tag{4.46}$$

$$\mathbf{r}_i^{(\text{v})}(t + \Delta t) = \mathbf{r}_i^{(\text{v})}(t) \tag{4.47}$$

Evaluate the intermolecular force $\mathbf{F}_i$ on each molecule at time $t + \Delta t$ using the predicted positions. For continuous potential energy functions $u(r_{ij})$ that act between atoms i and j, the force on each molecule is given by

$$\mathbf{F}_i = -\sum_{j \neq i} \frac{\partial u(r_{ij})}{\partial r_{ij}} \hat{\mathbf{r}}_{ij} \tag{4.48}$$

where $\hat{\mathbf{r}}_{ij}$ is the unit vector in the $\mathbf{r}_{ij}$-direction. Evaluation of forces is time consuming because the sum in (4.48) must be performed for *each* molecule i in the system. However, there are several ways to save time; one is that Newton's third law (2.6) can be applied,

$$\mathbf{F}(\mathbf{r}_{ij}) = -\mathbf{F}(\mathbf{r}_{ji}) \tag{4.49}$$

to decrease the amount of computation by a factor of 2. Other time-saving devices are discussed in Section 5.1.

Correct the predicted positions and their derivatives using the discrepancy $\Delta \ddot{\mathbf{r}}_i$ between the predicted acceleration and that given by the evaluated force $\mathbf{F}_i$. With the forces at $t + \Delta t$ obtained from (4.48), Newton's second law (2.1) can be used to determine the accelerations $\ddot{\mathbf{r}}_i(t + \Delta t)$. The difference between the predicted accelerations and evaluated accelerations is then formed,

$$\Delta \ddot{\mathbf{r}}_i = \left[\ddot{\mathbf{r}}_i(t + \Delta t) - \ddot{\mathbf{r}}_i^P(t + \Delta t) \right] \tag{4.50}$$

In Gear's algorithms for second-order differential equations, this difference term is used to correct all predicted positions and their derivatives; thus,

$$\mathbf{r}_i = \mathbf{r}_i^P + \alpha_0 \Delta \mathbf{R2} \tag{4.51}$$

$$\dot{\mathbf{r}}_i \Delta t = \dot{\mathbf{r}}_i^P \Delta t + \alpha_1 \Delta \mathbf{R2} \tag{4.52}$$

$$\frac{\ddot{\mathbf{r}}_i(\Delta t)^2}{2!} = \frac{\ddot{\mathbf{r}}_i^P(\Delta t)^2}{2!} + \alpha_2 \Delta \mathbf{R2} \tag{4.53}$$

$$\frac{\mathbf{r}_i^{(iii)}(\Delta t)^3}{3!} = \frac{\mathbf{r}_i^{(iii)P}(\Delta t)^3}{3!} + \alpha_3 \Delta \mathbf{R2} \tag{4.54}$$

$$\frac{\mathbf{r}_i^{(iv)}(\Delta t)^4}{4!} = \frac{\mathbf{r}_i^{(iv)P}(\Delta t)^4}{4!} + \alpha_4 \Delta \mathbf{R2} \tag{4.55}$$

$$\frac{\mathbf{r}_i^{(v)}(\Delta t)^5}{5!} = \frac{\mathbf{r}_i^{(v)P}(\Delta t)^5}{5!} + \alpha_5 \Delta \mathbf{R2} \tag{4.56}$$

where

$$\Delta \mathbf{R2} \equiv \frac{\Delta \ddot{\mathbf{r}}_i(\Delta t)^2}{2!} \tag{4.57}$$

TABLE 4.1   Values of α_i Parameters in Gear's Predictor–Corrector Algorithm[a] for Second-Order Differential Equations Using Predictors of Order q

α_i	$q = 3$	$q = 4$	$q = 5$
α_0	$\frac{1}{6}$	$\frac{19}{120}$	$\frac{3}{16}$
α_1	$\frac{5}{6}$	$\frac{3}{4}$	$\frac{251}{360}$
α_2	1	1	1
α_3	$\frac{1}{3}$	$\frac{1}{2}$	$\frac{11}{18}$
α_4	—	$\frac{1}{12}$	$\frac{1}{6}$
α_5	—	—	$\frac{1}{60}$

[a] From ref. 9, except that for $q = 5$, $\alpha_0 = 3/16$ seems to be somewhat better than Gear's original value.

The parameters α_i promote numerical stability of the algorithm. The α_i depend on the order of the differential equations to be solved and on the order of the Taylor series predictor. Gear determined their values by applying each algorithm to linear differential equations and analyzing the resulting stability matrices. For a q-order predictor, the values of the α_i were chosen to make the local truncation error of $O(\Delta t^{q+1})$; then the method will be stable for second-order differential equations with global truncation error of $O(\Delta t^{q+1-2}) = O(\Delta t^{q-1})$. Values of the α_i for third-, fourth-, and fifth-order predictors are given in Table 4.1.

4.4.4 Energy Conservation

Finite-difference algorithms used in molecular dynamics are often judged by their ability to conserve energy, and in this section we illustrate such tests. To compare the Verlet and Gear algorithms, we computed the dynamics of a single collision between two otherwise isolated Lennard-Jones atoms with the motion restricted to one dimension. Initially the atoms were separated by a distance of 2.6σ and each atom was given an initial approach velocity $v^* = v(m/\varepsilon)^{1/2} = 1$. The equations of motion were then integrated through the collision until the atoms were again separated by 2.6σ. To accumulate a global error, we compared the total energy at each step with its value at the first step,

$$ge = \sqrt{\sum_{k=1}^{M} \left[E^*(0) - E^*(k\,\Delta t) \right]^2} \tag{4.58}$$

where $E^* = E/\varepsilon$. This sum was accumulated as a function of step size Δt, and since in each case the initial and final positions of the atoms were the same, the total duration $(M\Delta t)$ of each run was the same.

We show in Figure 4.3 how global error in the Gear algorithm changes with step size in both single- and double-precision arithmetic. Here we use the time step in dimensionless form: $\Delta t^* = \Delta t/\sigma(m/\varepsilon)^{1/2}$, where ε and σ are potential parameters and m is the atomic mass. For values of ε, σ, and m characteristic of argon, $\Delta t^* = 0.005$ corresponds to about 0.011 psec. In Figure 4.3, for time steps larger than $\Delta t^* = 0.005$, the single- and double-precision results coincide. However, for time steps smaller than 0.001, single-precision round-off error dominates truncation error; so as the time step is decreased below 0.001 in the single-precision calculation, the global error in energy increases. This computed behavior confirms the general trend shown schematically in Figure 4.1. In contrast, the double-precision error continues to decrease as Δt^* is decreased below 0.001.

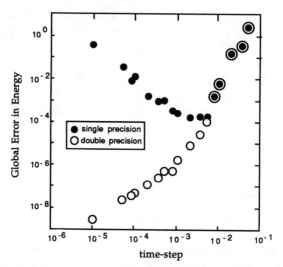

FIGURE 4.3 One way to control round-off error is to increase the precision of the calculation. Shown here are global errors in total energy computed through a one-dimensional collision between two Lennard-Jones atoms. The trajectories were computed using Gear's predictor–corrector algorithm. For time-steps $\Delta t^* = \Delta t / \sigma (m/\varepsilon)^{1/2} > 5 \times 10^{-3}$ truncation error makes the largest contribution to the global error (cf. Fig. 4.1), so single- and double-precision calculations are of the same reliability. For time-steps $\Delta t^* < 10^{-3}$ global errors in single precision are dominated by round-off, but in double precision round-off remains unimportant down to time-steps as small as 10^{-5}. Errors here are in units of energy, $E^* = E/\varepsilon$.

Figure 4.4 compares the Verlet and Gear algorithms. The quantity plotted is the rms global error per step,

$$\langle \text{ge} \rangle = \sqrt{\frac{1}{M} \sum_{k=1}^{M} \left[E^*(0) - E^*(k \, \Delta t) \right]^2} \qquad (4.59)$$

At large time-steps the two algorithms exhibit about the same degree of energy conservation; however, the stability limit for the Verlet algorithm occurs at a higher Δt than that for Gear. For time-steps $\Delta t^* < 0.01$, the Gear algorithm provides better energy conservation; specifically, least-squares fits to the data give

$$\langle \text{ge} \rangle \propto (\Delta t)^{2.04} \quad \text{for Verlet} \qquad \langle \text{ge} \rangle \propto (\Delta t)^{2.97} \quad \text{for Gear} \qquad (4.60)$$

That is, Gear's algorithm is effectively one order higher in computing total energy than is the Verlet algorithm.

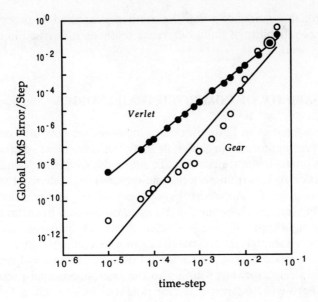

FIGURE 4.4 Based on energy conservation, Gear's algorithm is about one order higher in accuracy than Verlet's. Shown here is the rms global error in total energy per time-step computed through a one-dimensional collision between two Lennard-Jones atoms. Closed circles are from Verlet's method (4.36) and (4.38); the line is a least-squares fit and has slope 2.04. Open circles are from Gear's algorithm (Section 4.4.3); the line has slope 2.97 and is also from a least-squares fit. The units of the time-step are $\Delta t^* = \Delta t / \sigma (m / \varepsilon)^{1/2}$ and those of the energy are $E^* = E / \varepsilon$.

These tests are instructive, but a plot such as Figure 4.4 does not give an unambiguous determination of the accuracy of computed trajectories because total energy has contributions from both positions and velocities, quantities that may be computed to different orders. For example, the Verlet and Gear methods tested in Figure 4.4 both treat positions and velocities differently. Moreover, the same order of energy conservation can be obtained in several ways (see Exercise 4.12). The situation with predictor–correctors can be further complicated by a corrector whose order differs from that of the predictor.

We would prefer to test algorithms directly on individual positions and velocities, but analytic calculations are not possible for potential functions studied by molecular dynamics (else, why do a simulation?). Analytic determinations of position and velocity *are* possible for the ODHO, and that model has been used to test finite-difference methods commonly used in molecular dynamics [10]. Results of those tests are consistent with Figure 4.4: high-order Gear methods are more accurate than Verlet's method. But we may question whether the performance of a method applied to the ODHO provides reliable guidance when the method is applied to other potentials. The issue of reliability of finite-difference methods is pursued in more depth in the next section.

4.5   RELIABILITY OF COMPUTED TRAJECTORIES

From characteristics of finite-difference methods, we now turn our attention to the more important problem of how reliable those methods are in estimating phase-space trajectories. To make the discussion quantifiable, we use simulation results from periodic one-dimensional systems of soft spheres; the code is given in Appendix I. A schematic one-dimensional system is shown in Figure 4.5. One-dimensional systems offer an important advantage for testing algorithms: if we allow interactions only between nearest neighbors, then even though their dynamics cannot be computed analytically, their equilibrium properties *can* be computed directly in statistical mechanics. A particularly straightforward solution to the one-dimensional problem for an arbitrary, hard-core, nearest-neighbor potential $u(x)$ is given by Takahashi [11].

The quantities studied here are the configurational integral Δ, the density $\rho = N/\langle L \rangle$, the internal energy $\langle \mathscr{U} \rangle$, and the total energy E,

$$\Delta = \int_0^\infty \exp\left[\frac{-u(x) + Px}{kT} \right] dx \tag{4.61}$$

$$\langle L \rangle = \frac{N}{\Delta} \int_0^\infty x \exp\left[-\frac{u(x) + Px}{kT} \right] dx \tag{4.62}$$

$$\langle \mathscr{U} \rangle = \frac{N}{\Delta} \int_0^\infty u(x) \exp\left[-\frac{u(x) + Px}{kT} \right] dx \tag{4.63}$$

$$\frac{E}{N\varepsilon} = \frac{kT}{2\varepsilon} + \frac{\langle \mathscr{U} \rangle}{N\varepsilon} \tag{4.64}$$

The integrals in (4.61)–(4.63) can be computed numerically using Simpson's rule. The only aggravation in using these expressions is that they are derived in the *NPT* ensemble, so the pressure P and temperature T must be specified[†]; in contrast, the simulations are done on an isolated system for which the density and total energy are fixed. Otherwise, one-dimensional systems are easily programmed, are inexpensive to simulate, and have analytic results with which to compare.

[†]In three dimensions pressure is force per unit area; however, in one dimension there is no area and the quantity P is merely a force. Nevertheless, we will continue to refer to P as a pressure.

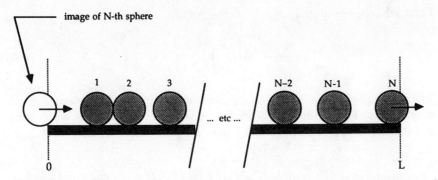

FIGURE 4.5 Schematic representation of a periodic one-dimensional system of "spheres." The system density is N/L, where L is the length of the available space. In this drawing the Nth sphere is leaving the system and its periodic image is entering at the opposite end. Because the interactions are purely repulsive at small separations, sphere positions relative to one another are maintained throughout a simulation. All the one-dimensional calculations considered in this book allow interactions only between nearest neighbors.

4.5.1 Ergodicity

In Chapter 2 we found that to obtain meaningful values for properties, the dynamics of an N-body system must be at least ergodic and preferably a K-flow. That is, the system must evolve through nearly all available phase space and time averages must be computed over representative portions of the space. For isolated systems the available phase space is the hypersurface of constant total energy; on that surface, interactions among particles cause the phase-space trajectory to be chaotic. However, chaotic motion may not develop if the degrees of freedom are too restricted or if interactions are too pathological. For example, one-dimensional hard-sphere simulations are not ergodic because in one dimension hard spheres merely exchange velocities on collision. In such systems, whatever velocity distribution initially assigned is maintained throughout the simulation, Maxwell's velocity distribution does not develop, and the full constant-energy surface is not sampled.

Is ergodicity achieved in one-dimensional N-body dynamics if hard interactions are replaced by soft? Or will it be achieved only if the velocity change on collision is greater than some threshold value (on average)? To answer these questions, we consider two one-dimensional periodic systems: (a) purely repulsive soft spheres with nearest-neighbor interactions given by

$$\frac{u(x)}{\varepsilon} = \left(\frac{x}{\sigma}\right)^{-12} \tag{4.65}$$

and (b) soft spheres with nearest-neighbor interactions given by the

Lennard-Jones potential

$$\frac{u(x)}{\varepsilon} = 4\left[\left(\frac{x}{\sigma}\right)^{-12} - \left(\frac{x}{\sigma}\right)^{-6}\right] \tag{4.66}$$

In these equations, ε and σ are the natural units of energy and distance, respectively. To monitor development of velocity distributions, we compute instantaneous values of the kinetic part of Boltzmann's H-function [12, 13],

$$H(t) = \int_{-\infty}^{\infty} f(v)\ln f(v)\, dv \tag{2.57}$$

where $f(v)$ is the velocity distribution at time t. If the distribution is

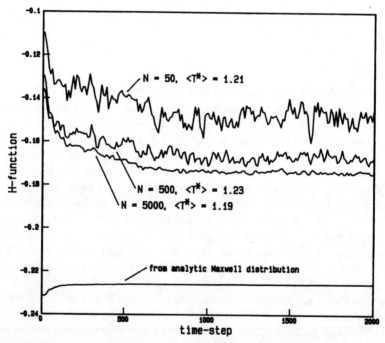

FIGURE 4.6 Tests for ergodicity and approach to equilibrium in one-dimensional simulations of soft spheres interacting with the repulsive x^{-12} potential at density $\rho\sigma = 0.65$ and average temperatures $\langle T^* \rangle = \langle kT/\varepsilon \rangle \approx 1.20$. Instantaneous values of the Boltzmann H-function (2.57) are plotted here at intervals of 10 time-steps, where $\Delta t/\sigma(m/\varepsilon)^{1/2} = 0.004$. Each simulation was started from uniform distributions of sphere positions and velocities. The bottom line is the H function computed from the Maxwell velocity distribution at the running average temperature of the $N = 500$ simulation.

Maxwellian, then $f(v)$ is given by (2.54):

$$f(v) \equiv \frac{N(v)}{N} = \sqrt{\frac{m}{2\pi kT}} \exp\left(-\frac{mv^2}{2kT}\right)$$

and H has a particular value that depends only on temperature.

Figure 4.6 shows the instantaneous H-function computed for the one-dimensional x^{-12} substance at $\rho\sigma = 0.65$, $\langle kT/\varepsilon \rangle = 1.2$. Results are shown for three system sizes: 50, 500, and 5000 soft spheres. Each run was started from uniform distributions of sphere positions and velocities, and the figure clearly shows the initially assigned velocities relaxing to a stationary distribution. In each simulation, the stationary distribution required about 900 integration time-steps to develop [here the time steps are $\Delta t^* = \Delta t/\sigma(m/\varepsilon)^{1/2} = 0.004$]. However, also shown in the figure is the H-function

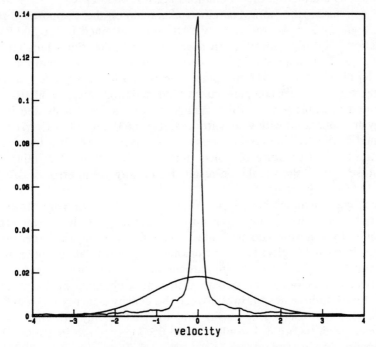

FIGURE 4.7 At some state conditions simulations of the one-dimensional x^{-12} substance are not ergodic. Shown here is the average velocity distribution obtained from a 500-particle simulation at $\rho\sigma = 0.65$ and $\langle kT/\varepsilon \rangle = 1.20$. The average was obtained over 10,000 time-steps after discarding 2000 equilibration steps. The lower line is the analytic result for the Maxwell velocity distribution at $\langle kT/\varepsilon \rangle = 1.20$. The velocities are plotted in units of $v^* = v(m/\varepsilon)^{1/2}$.

TABLE 4.2   Comparison of Simulation and Theoretical Results for Equilibrium Properties of One-Dimensional Soft-Sphere Systems

Substance	Method	kT/ε	$P\sigma/\varepsilon$	$\rho\sigma$	$U_c/N\varepsilon$
x^{-12}	Simulation	1.201	2.081	0.65	0.167
	Theory	1.201	2.081	0.627	0.177
x^{-12}	Simulation	1.168	18.548	1.0	1.448
	Theory	1.168	18.548	0.995	1.456
Lennard-Jones	Simulation	1.214	1.716	0.65	-0.463
	Theory	1.214	1.716	0.647	-0.463

Each simulation used 500 spheres and averages were accumulated over 10,000 time steps. Theoretical values of density and internal energy were computed from (4.61)–(4.63) using the temperature and pressure obtained from the simulations.

for the Maxwell distribution (2.54) at temperatures very near the running average temperatures of the simulations. It is evident that the velocity distributions attained in the simulations are not Maxwellian.

To emphasize the non-Maxwellian character of the computed distributions, Figure 4.7 shows the velocity distribution attained in the 500-particle simulation. The distribution is an average over 10,000 time-steps after discarding the 2000 steps shown in Figure 4.6. Also shown in Figure 4.7 is the Maxwell distribution (2.54): the simulation is clearly not ergodic, and consequently, time averages computed from the simulations are erroneous. This is illustrated in Table 4.2, in which simulation values for density and internal energy are compared with values from (4.61)–(4.63). Since (4.61)–(4.63) were obtained in the *NPT* ensemble, those equations were solved using values of temperature and pressure obtained from the simulations. The table shows that at $\rho\sigma = 0.65$, the simulation results for density are in error by 3.7% and those for U are in error by 5.7%.

Thus, we conclude that at $\rho\sigma = 0.65$, $kT/\varepsilon = 1.2$, changing from hard spheres to soft repulsive spheres is not sufficient to make the motion ergodic. Evidently, to achieve ergodicity, collisions should provide still larger possible changes in particle velocities. To promote more mixing of velocities, we can contemplate further changes in particle interactions or changes in state condition. For example, the velocity distribution obtained in the simulation more closely approaches Maxwell's when the density is increased from 0.65 to 1.0. Simulation results for density and internal energy are then in good agreement with theoretical results (within 0.5%), as shown in Table 4.2.

Likewise, if, instead of changing state condition, we change the x^{-12} potential (4.65) to the Lennard-Jones model (4.66), then for $N \geq 500$ the computed H-function closely approaches the Maxwell H-function. Moreover, the Lennard-Jones substance achieves equilibrium more quickly than the x^{-12} substance: thus, according to Figure 4.8, the Lennard-Jones H-function stabilizes in about 400 time-steps, while the x^{-12} substance needed about 900 steps at the same state condition. In Figure 4.9 the average velocity distribu-

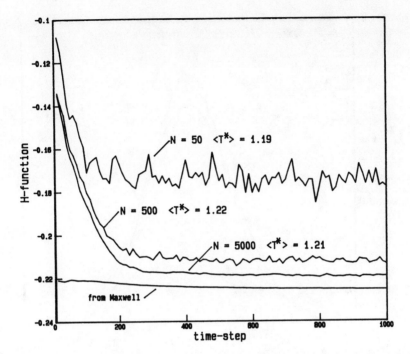

FIGURE 4.8 Tests for ergodicity and approach to equilibrium in one-dimensional simulations of soft spheres interacting with the Lennard-Jones (12,6) potential at density $\rho\sigma = 0.65$ and average temperatures $\langle T^* \rangle = \langle kT/\varepsilon \rangle \approx 1.20$. Quantities plotted are the same as Figure 4.6, but now the simulations become nearly ergodic as the number of particles increases.

tion attained in the 500-particle Lennard-Jones simulation is compared with the Maxwell distribution; the agreement is good (note the finer scale on the ordinate compared to that in Figure 4.7) but not perfect. Simulation values for density and internal energy are also in good agreement with theoretical values (see Table 4.2).

From these simple one-dimensional simulations, we draw the following conclusions:

(a) Monitoring the *H*-function by itself tracks only the development of a stationary velocity distribution; that distribution may not be Maxwellian.

(b) Access to the full constant-energy surface requires the ability to exchange a spectrum of energies among degrees of freedom. A particular simulation may fall below the threshold of sufficient energy exchange because of restrictions in the number of degrees of freedom,

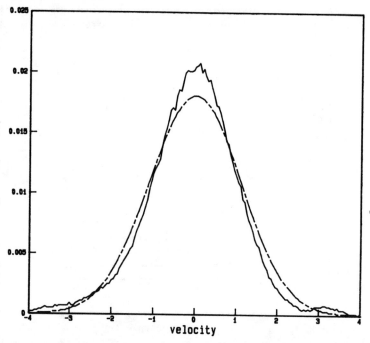

FIGURE 4.9 Simulations of the one-dimensional Lennard-Jones $(12,6)$ substance are nearly ergodic. The solid line is the average velocity distribution obtained from a 500-particle simulation at $\rho\sigma = 0.65$ and $\langle kT/\varepsilon \rangle = 1.22$ The broken curve is the analytically computed Maxwell distribution at $\langle kT/\varepsilon \rangle = 1.22$. Calculational details and units are as in Figure 4.7.

because of state condition, or because of the nature of interparticle interactions.

(c) The degrees of freedom may be restricted by the number of particles or by system geometry or both. Simulations of materials in severely confined spaces, for example, may pose inherent ergodicity problems.

(d) Adjusting one of the variables in item (b), say state condition, in attempts to exceed the ergodicity threshold may *not* be sufficient to compensate for constraints in the other variables—degrees of freedom and interactions.

4.5.2 Difficulties in Assessing Algorithms

Now that Section 4.5.1 has discussed ergodicity problems that may arise in one-dimensional simulations, we use the one-dimensional Lennard-Jones substance to illustrate difficulties in assessing the performance of finite-difference algorithms. The algorithms studied here are the Verlet and Gear

TABLE 4.3   Simulation Results Obtained from Verlet and Gear Algorithms for 500 Lennard-Jones Spheres in One Dimension[a]

Algorithm	$\rho\sigma$	$\langle kT/\varepsilon \rangle$	$\langle P\sigma/\varepsilon \rangle$	$\langle \mathscr{U}/N\varepsilon \rangle$	$E/N\varepsilon$
Gear	0.5	1.490	1.229	−0.306	0.4392
Verlet	0.5	1.482	1.218	−0.303	0.4378
Deviations		0.5%	0.9%	1.%	—
Gear	0.65	1.341	1.939	−0.451	0.2197
Verlet	0.65	1.334	1.926	−0.449	0.2186
Deviations		0.5%	0.7%	0.4%	—
Gear	0.8	0.930	2.711	−0.642	−0.1767
Verlet	0.8	0.928	2.698	−0.642	−0.1783
Deviations		0.2%	0.5%	< 0.1%	—
Gear[b]	1.0	1.122	31.949	0.488	1.049
Verlet[b]	1.0	1.112	31.891	0.485	1.041
Deviations		0.9%	0.2%	0.6%	—

[a]All simulations used time-step $\Delta t^* = 0.004$ and averages were accumulated over 20,000 time-steps, except where noted otherwise.
[b]These runs used $\Delta t^* = 0.002$ and 20,000 time-steps.

methods presented in Section 4.4, but the particular methods are secondary to the primary lessons to be learned.

Table 4.3 compares equilibrium properties obtained from simulations using the Verlet and Gear algorithms applied to one-dimensional Lennard-Jones systems. All runs involved 500 particles and used only nearest-neighbor interactions. At each state condition, the Verlet and Gear runs were started from exactly the same particle positions and velocities; thus, each pair of runs was done at the same density and started from the same point in phase space. The table contains temperatures, pressures, and internal energies averaged over 20,000 time-steps; the total energies $E/N\varepsilon$ are instantaneous values at the end of each run. From these final energies, it is evident that the two algorithms do not follow the same trajectories, or even the same surfaces, in phase space. Consequently, averages for T, P, and $\mathscr{U}$ from the two algorithms are not identical; their values deviate by as much as 1%. This agreement is good but not perfect. Can we explain these discrepancies? Is one algorithm—Verlet or Gear—more accurate?

Perhaps we can identify the better algorithm by comparing the simulation results from Table 4.3 with theoretically exact values from (4.61)–(4.63). Such comparisons are given in Table 4.4. Since the theory (4.61)–(4.63) requires temperature and pressure as input, we solved those equations at each state condition using average values of T and P from pairs of Verlet and Gear runs cited in Table 4.3. Thus, the properties to be compared in Table 4.4 are density and internal energy. The table shows that the simulation results for density are in good agreement with the theory; the deviations in ρ are < 0.6% and can be attributed to the small differences in T and P and to the

TABLE 4.4 Comparison of Simulation Results with Theoretical Values (4.61)–(4.64) for Equilibrium Properties of One-Dimensional Lennard-Jones Substances[a]

Method	$\rho\sigma$	Deviation in $\rho\sigma$ (%)	$\langle kT/\varepsilon \rangle$	$\langle P\sigma/\varepsilon \rangle$	$\langle \mathscr{U}/N\varepsilon \rangle$	Deviation in $\langle \mathscr{U}/N\varepsilon \rangle$ (%)	$E/N\varepsilon$
Gear	0.5	0.6	1.490	1.229	−0.306	5.5	0.4392
Verlet	0.5	0.6	1.482	1.218	−0.303	6.5	0.4378
Theory	0.497	—	1.486	1.223	−0.324	—	0.4194
Gear	0.65	0.3	1.341	1.939	−0.451	<0.1	0.2197
Verlet	0.65	0.3	1.334	1.926	−0.449	0.4	0.2186
Theory	0.648	—	1.338	1.932	−0.451	—	0.2177
Gear[b]	0.65	0.2	2.188	3.311	−0.376	1.9	0.7186
Verlet[b]	0.65	0.2	2.193	3.327	−0.378	2.4	0.7186
Theory	0.646	—	2.190	3.320	−0.369	—	0.7258
Gear	0.8	0.6	0.930	2.711	−0.642	1.4	−0.1767
Verlet	0.8	0.6	0.928	2.698	−0.642	1.4	−0.1783
Theory	0.795	—	0.929	2.704	−0.633	—	−0.1685
Gear[a]	1.0	0.2	1.122	31.949	0.488	0.8	1.049
Verlet[c]	1.0	0.2	1.112	31.891	0.485	1.4	1.041
Theory	0.998	—	1.117	31.920	0.492	—	1.051

[a]Computational details given in Table 4.3.
[b]These runs used $\Delta t^* = 0.002$ and 40,000 time-steps.
[c]These runs used $\Delta t^* = 0.002$ and 20,000 time-steps.

N-dependence of the finite (500-particle) simulations. The simulation values for the internal energy deviate by roughly 2% from the exact values; these deviations are largely due to the N-dependence of $\mathscr{U}$. The results from both algorithms deviate to about the same degree from the exact results. Results for $\mathscr{U}$ and E obtained with the Gear algorithm are in marginally better agreement with the exact values than are those obtained with Verlet; however, the Gear algorithm is only apparently better. Within their combined uncertainties, the Gear and Verlet results for $\langle \mathscr{U} \rangle$ and E are essentially the same.

So which algorithm is better? More precisely, what causes the differences in results given in Table 4.3? To compare algorithms means to compare truncation errors, but by calculating properties from trajectories, as done for Tables 4.3 and 4.4, we invariably introduce round-off errors. In those tables the differences in results are not caused primarily by differences in the algorithms but rather by differences in the way the algorithms are implemented—round-off (Section 4.2). We now try to assess the effects of round-off error by using the same code in different computational environments; that is, we change the compiler and then change the computer.

For the tests described in what follows, calculations were done on workstations running the UNIX operating system. Most versions of UNIX provide

Fortran compilers that contain optimizing features to enhance execution speed of Fortran code. Optimization can be selected at one of three levels (1, 2, or 3), with level 3 being the most drastic; that is, level 3 should execute fastest. Here "optimized code" will always refer to optimization at level 3. Optimization may change the order in which some lines of code are executed, but optimization should not change the logical sense of the code. Therefore, optimized and nonoptimized codes constitute two implementations of the same source program and may provide insight into round-off error.

To compare results from different implementations, we executed optimized and nonoptimized compilations of *one* molecular dynamics program applied to exactly the same conditions: same system of 100 Lennard-Jones spheres in one dimension, same density $\rho\sigma = 0.65$, same initial positions and velocities and therefore the same initial value of total energy $E/N\varepsilon = 0.2086$, same time-step $\Delta t^* = 0.004$, same double-precision arithmetic, and same computer. As a measure of differences in the two calculations, we monitored the instantaneous temperature (kT/ε); temperature differences between the optimized and nonoptimized runs are shown in Figure 4.10. For the first 3500 time-steps the temperatures of the two runs were identical to four significant figures, but after 4000 time-steps, the two temperatures differed by as much as 0.2 (15% of the average T). At the end of 10,000 time-steps the average temperatures of the two runs differed by 0.3%, as shown in Table 4.5. Also shown in Table 4.5 are average values for other properties obtained from these two runs.

To demonstrate conclusively that these deviations are caused by round-off error, we halved the time-step (from 0.004 to 0.002) and repeated both the optimized and nonoptimized runs. These simulations used the Gear algorithm, and recall from Figure 4.4 that if we halve the time-step, truncation errors in the Gear algorithm should *decrease* substantially (the size of the decrease depends on the particular property monitored). Runs with the smaller time-step were extended to 20,000 steps, so the total lengths of trajectory obtained from both sets of runs were the same:

$$M\Delta t^* = 0.004 \times 10,000 = 0.002 \times 20,000 \qquad (4.67)$$

With the smaller time-step, the deviations in average temperature, pressure, and internal energy all *increase* by factors of 4 or more, as shown in Table 4.5. It is evident that the deviations appearing in Figure 4.10 and Table 4.5 result from round-off error, not truncation error. The Verlet algorithm behaves in the same fashion; although as shown in Table 4.5, for the same size time-step, the Verlet algorithm exhibits larger round-off than does Gear. However, we emphasize that this reflects differences in implementation, not differences in the algorithms.

Finally, we show in Table 4.5 that the round-off errors found here are not restricted to particular hardware. Most calculations were done on a SUN-SPARC station 1, which has a central processing unit (CPU) based on a

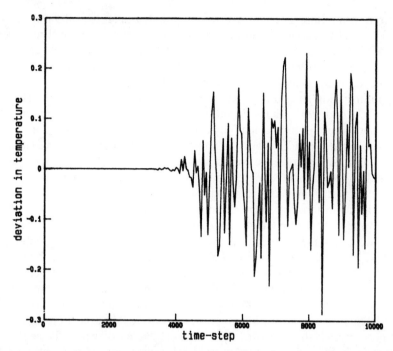

FIGURE 4.10 Different compilations of the same source code can produce different phase-space trajectories. Shown here are deviations in instantaneous temperature (kT/ε) caused by different compilations of the same molecular dynamics source program. The two compilations differed only in that one was optimized while the other was not. Each run applied the Gear algorithm to 100 Lennard-Jones spheres in one dimension at $\rho\sigma = 0.65$ and total energy $E/N\varepsilon = 0.2086$. Both runs were performed on the same SUNSPARC station 1 running under the SunOS version of UNIX.

$r$educed $i$nstruction $s$et $c$hip (RISC). Some calculations were done on a MASSCOMP 5500, which is based on the Motorola 68010 chip. Both machines run under the UNIX operating system. Both machines showed similar round-off behavior: deviations between optimized and nonoptimized runs were about the same on both machines.

As we might expect with round-off error, deviations such as appear in Table 4.5 are affected by the values assigned to run parameters such as the number of particles, time step, and state condition. Just as it is possible to find combinations of these parameters that produce measurable deviations, so too is it possible to find other combinations that produce no apparent deviations in trajectories over, say, 10,000 time-steps. Such calculations do not imply that round-off is negligible, but rather that, in those cases, round-off occurs in the same or in compensating ways.

TABLE 4.5   Effects of Hardware and Compiler on Round-off Error in One-Dimensional Lennard-Jones Simulations[a]

Machine	Compiler	Algorithm	$\langle kT/\varepsilon \rangle$	$\langle P\sigma/\varepsilon \rangle$	$\langle \mathcal{U}/N\varepsilon \rangle$	$E/N\varepsilon$
SUN	No optimizer	Gear	1.311	1.879	−0.447	0.2086
	Optimizer	Gear	1.307	1.883	−0.445	0.2085
		Deviations	0.3%	0.2%	0.4%	—
SUN	No optimizer	Gear[b]	1.320	1.907	−0.452	0.2083
	Optimizer	Gear[b]	1.304	1.876	−0.444	0.2083
		Deviations	1.2%	1.9%	1.8%	—
MASSCOMP	No optimizer	Gear	1.312	1.900	−0.448	0.2085
	Optimizer	Gear	1.317	1.873	−0.450	0.2090
		Deviations	0.4%	1.4%	0.4%	—
SUN	No Optimizer	Verlet	1.322	1.899	−0.453	0.2082
	Optimizer	Verlet	1.305	1.853	−0.444	0.2083
		Deviations	1.3%	2.4%	2.0%	—

[a]All these runs are for 100 particles at $\rho\sigma = 0.65$. Each run was for 10,000 time-steps of size $\Delta t^* = 0.004$, except where noted otherwise. All calculations used double-precision (64-bit-word) arithmetic.
[b]These runs used $\Delta t^* = 0.002$ and 20,000 time-steps.

We conclude from this discussion that when two finite-difference algorithms produce average properties and total energy within about 1% of one another (Table 4.3), we are unable to judge the relative accuracy of the algorithms. Uncertainties of about 1% also occur because of accumulated round-off error (Table 4.5); these errors arise from the finite number of digits carried in a calculation, and they are aggravated by the way in which the logic of a code is organized, by the compiler, and by the hardware. A kind of uncertainty principle is at work here in that at some level of accuracy we cannot decouple truncation error from round-off.

4.6 SUMMARY

Finite-difference methods serve as the foundation for soft-body molecular dynamics. In general, the methods are reliable and easily implemented; nevertheless, as with any numerical method, they introduce two kinds of errors: truncation error, which pertains to the method, and round-off error, which pertains to how the method is implemented. Truncation error is controlled by decreasing the step size Δt while round-off is controlled primarily by using high-precision arithmetic. The magnitude of those errors reflects accuracy, while the propagation of those errors reflects the stability of the algorithms.

Although finite-difference methods are commonly characterized by their order based on local truncation error, it is not sufficient to merely choose an algorithm that is high order. We should be more concerned with total global error: how large is it and how does it propagate? The answers to these kinds of questions are not easily resolved; they depend not only on the choice of algorithm but also on the differential equations being solved, on how the equations are formulated, on how the algorithm is implemented on a machine, on how the hardware stores floating-point numbers, on whether the hardware truncates or rounds values to a finite number of digits, and so on. Truncation error refers to the accuracy with which an algorithm computes positions and velocities, but with no exact results for particle trajectories we must use indirect means for assessing algorithms.

However, at some level of accuracy indirect tests fail to discriminate between finite-difference algorithms: judgments of accuracy become confounded by ever-present round-off errors. Thus, for computing particle trajectories using small time-steps, the Gear algorithm is more accurate than Verlet's (Figure 4.4); however, these two algorithms incur similar uncertainties in computing time-average values for properties. The Verlet algorithm is stable for larger time-steps than Gear, but this observation must be tempered by loss of accuracy if time-steps near the stability limit are used (Figure 4.2). For simulations of soft spheres, there is little to choose between the two methods, and the decision is largely one of personal and historical preference. In simulations of molecules, rotational and internal degrees of freedom typically demand a smaller time-step than translational motion, and algorithms having accuracy higher than the simple Verlet, (4.36) and (4.38), are preferred.

But how accurately are individual particle trajectories computed? For short durations, a computed trajectory appears to be accurate, the degree of accuracy depending on the local truncation error of the particular algorithm used. But for long elapsed times, accumulated round-off error disrupts the dynamic connection between points on a computed trajectory. We can distinguish between these long and short durations by monitoring trajectories through a velocity-reversal calculation [12, 13] (see Exercise 4.16).

Even though lengthy trajectories may be adversely affected by round-off, time averages for many equilibrium properties—those having short relaxation times—are little affected. Some error can be tolerated in individual particle trajectories so long as the N-particle system remains near the constant-energy surface. Thus, we have a somewhat paradoxical situation in that modestly reliable particle trajectories can combine to produce reliable time averages. This is not so paradoxical as it may sound: time averages do not really depend on how samples are drawn from a population but on whether the population sampled is representative of the system under investigation. In other words, time averages do not depend on how consecutive samples may be related—by Newtonian differential equations or by approximate finite-difference equations or by stochastic processes—instead, averages depend on the samples being tolerably close to the constant-energy surface for the state condition simulated.

We may venture further by conjecturing that round-off errors could be realistic and perhaps even desirable consequences of using finite-difference methods. Round-off error might be interpreted as simulating uncontrolled fluctuations that occur in real systems. No truly isolated system can be studied experimentally, and in experiments small uncontrolled perturbations (a voltage surge or a mechanical vibration for example) induce local gradients that usually dissipate quickly and have little impact on measured time averages. In molecular simulations it appears that a similar role is played by round-off error.

REFERENCES

[1] P. Hagedorn, *Non-Linear Oscillations*, 2nd ed, Clarendon, Oxford, 1988.

[2] W. H. Press, B. P. Flannery, S. A. Teukolsky, and W. T. Vetterling, *Numerical Recipes*, Cambridge University Press, New York, 1986, Chapter 15.

[3] W. G. Hoover, *Molecular Dynamics*, Lecture Notes in Physics, Vol. 258, Springer-Verlag, Berlin, 1986.

[4] L. Verlet, "Computer 'Experiments' on Classical Fluids. I. Thermodynamical Properties of Lennard-Jones Molecules," *Phys. Rev.*, **159**, 98 (1967).

[5] D. Beeman, "Some Multistep Methods for Use in Molecular Dynamics Calculations," *J. Comput. Phys.*, **20**, 130 (1976).

[6] W. C. Swope, H. C. Andersen, P. H. Berens, and K. R. Wilson, "A Computer Simulation Method for the Calculation of Equilibrium Constants for the Formation of Physical Clusters of Molecules: Application to Small Water Clusters," *J. Chem. Phys.*, **76**, 637 (1982).

[7] H. J. C. Berendsen and W. F. van Gunsteren, "Practical Algorithms for Dynamic Simulations" in *Molecular – Dynamics Simulation of Statistical – Mechanical Systems,* G. Ciccotti and W. G. Hoover, Eds., North-Holland, Amsterdam, 1986.

[8] A. Rahman, "Correlations in the Motion of Atoms in Liquid Argon," *Phys. Rev.*, **136** (2A), 405 (1964).

[9] C. W. Gear, *Numerical Initial Value Problems in Ordinary Differential Equations*, Prentice-Hall, Englewood Cliffs, NJ, 1971, Chapter 9.

[10] G. D. Venneri and W. G. Hoover, "Simple Exact Test for Well-Known Molecular Dynamics Algorithms," *J. Comput. Phys.*, **73**, 468 (1987).

[11] H. Takahashi, "A Simple Method for Treating the Statistical Mechanics of One-Dimensional Substances," *Proc. Physico-Math. Soc. Jpn.*, **24**, 60 (1942); English translation in *Mathematical Physics in One Dimension*, E. H. Lieb and D. C. Mattis, Eds., Academic, New York, 1966, p. 25.

[12] J. Orban and A. Bellemans, "Velocity-Inversion and Irreversibility in a Dilute Gas of Hard Disks," *Phys. Lett.*, **24A**, 620 (1967).

[13] B. J. Alder and T. E. Wainwright, "Molecular Dynamics by Electronic Computers" in *Proceedings of the International Symposium on Transport Processes in Statistical Mechanics*, I. Prigogine, Ed., Interscience, New York, 1958, p. 97.

EXERCISES

4.1   A shotgun shell containing 100 pellets is fired at a large paper target. Discuss how the pattern of holes in the target can illustrate (a) low accuracy and low precision, (b) low accuracy and high precision, (c) high accuracy and low precision, and (d) high accuracy and high precision. (This problem is a corruption of an example used by W. E. Deming in *Some Theory of Sampling*, Wiley, New York, 1950; Dover reprint, New York, 1960, pp. 19–20.)

4.2   Explain how stable finite-difference methods can produce unstable trajectories in phase space.

4.3   Use a hand calculator to compute values for the Lennard-Jones potential from $x = 0.6$ to $x = 1.3$ at increments of $\Delta x = 0.1$. Use each of the two forms

$$u(x) = 4(x^{-12} - x^{-6}) \qquad u(x) = 4x^{-6}(x^{-6} - 1)$$

In both sets of calculations use all numbers, including intermediate values, rounded to three decimal places. Explain your results.

4.4   Write a program that applies Euler's method to the ODHO problem;
☒   that is, iteratively implement (4.7) and (4.8). For particular choices of the spring constants (say, $\gamma = 1$ and $m = 1$) and initial conditions [say, $x(0) = 1$ and $v(0) = 1$], use your program to generate the phase-plane trajectory for various time-steps Δt. Monitor the total energy throughout your calculations. Convince yourself that no matter how small you make Δt, the trajectory always diverges.

4.5   Write a program that applies the modified Euler method (4.29) and
☒   (4.30) to the ODHO.

   (a) Generate phase-plane trajectories, such as shown in Figure 4.2. Study how the trajectories differ and how well total energy is conserved for time steps close to and on either side of the stability limit given by (4.32).

   (b) Analytically determine the order of the method (4.27) and (4.28). Then computationally confirm your result by calculating, for various time-steps, $x(t)$ and $v(t)$ and comparing with the analytical solution to the ODHO (see Exercise 2.2).

4.6   Analytically determine the stability bound on the time-step Δt for the following second-order method applied to the ODHO:

$$x(t + \Delta t) = x(t) + \dot{x}(t)\,\Delta t + \tfrac{1}{2}\ddot{x}(t)(\Delta t)^2$$

$$v(t + \Delta t) = v(t) + \dot{v}(t)\,\Delta t + \tfrac{1}{2}\ddot{v}(t)(\Delta t)^2$$

4.7 Analytically determine the order and the stability bound on Δt for the method

$$x(t + \Delta t) = x(t) + \dot{x}(t + \Delta t)\,\Delta t - \tfrac{1}{2}\ddot{x}(t)(\Delta t)^2$$

$$v(t + \Delta t) = v(t) + \dot{v}(t)\,\Delta t + \tfrac{1}{2}\ddot{v}(t)(\Delta t)^2$$

applied to the ODHO.

4.8 Consider 50 Lennard-Jones spheres confined to a periodic one-dimensional system. The spheres interact only with nearest neighbors. Initially the 50 spheres are uniformly spaced on the line of length L and all spheres are given the same initial velocity. Describe the phase-space trajectory traced by this system.

4.9 Use the Taylor expansions in Section 4.4.1 to show that Verlet's central difference estimator for velocity (4.38) is first order.

4.10 Analytically compute the value of the H-function for one-dimensional motion in which the velocities of N particles are uniformly distributed between $-a$ and $+a$, where a is dimensionless. Recall that the distribution $f(v)$ is normalized so that

$$\int f(v)\,dv = 1$$

4.11 Consider one Lennard-Jones sphere constrained to move only in one
☒ dimension between two fixed Lennard-Jones spheres. The centers of the fixed spheres are separated by 5σ.

(a) Sketch the form of the potential $u(r)$ felt by the movable sphere. Sketch the form of its phase-plane trajectory.

(b) Express the equations of motion in Hamiltonian form. Generate solutions to the equations of motion by writing a program that uses the fourth-order Runge–Kutta algorithm:

$$r(t + \Delta t) = r(t) + \tfrac{1}{6}(s_1 + 2s_2 + 2s_3 + s_4)\,\Delta t$$

where the s_i are estimates of the slope at various points on $[t, t + \Delta t]$:

$$s_1 = \dot{r}(t)$$

$$s_2 = \dot{r}\left(t + \tfrac{1}{2}\Delta t, r(t) + \tfrac{1}{2}s_1\,\Delta t\right)$$

$$s_3 = \dot{r}\left(t + \tfrac{1}{2}\Delta t, r(t) + \tfrac{1}{2}s_2\,\Delta t\right)$$

$$s_4 = \dot{r}\left(t + \Delta t, r(t) + s_3\,\Delta t\right)$$

Analogous equations apply for the velocities. Using trial runs, locate the stability bound on the time-step. Then for a particular choice of initial conditions, study how the phase-plane trajectory and energy conservation respond to changes in the time-step. Is the behavior consistent with a fourth-order algorithm?

(c) Study how the computed phase-plane trajectories are affected by round-off error by doing parallel calculations in single and in double precision.

(d) Express the equations of motion in Newtonian form and write a program that generates solutions via the Verlet algorithm. Repeat (b) and (c) using Verlet.

(e) Repeat (b) and (c) using the Gear predictor–corrector algorithm. Then compare the three algorithms.

4.12   The purpose of this exercise is to relate local truncation errors in total energy E to those in particle positions r_i and velocities v_i. Write the true total energy E_T as a double first-order Taylor expansion about the computed total energy E_c,

$$E_T = E_c + \sum_i \frac{\partial E}{\partial v_i} e_{vi} + \sum_i \frac{\partial E}{\partial r_i} e_{ri}$$

where e_v and e_r are the local truncation errors in velocity and position, respectively. For finite-difference algorithms of order m in velocity and n in position,

$$e_v \propto (\Delta t)^{m+1} \qquad e_r \propto (\Delta t)^{n+1}$$

(a) Now show that the local truncation error in total energy can be written as

$$E_T - E_c = A(\Delta t)^{m+1} + B(\Delta t)^{n+1}$$

where A and B are independent of the time-step Δt. Thus, algorithms of different orders in position and velocity can produce errors of the same order in total energy.

(b) The Verlet algorithm (4.36) and (4.38) has $m = 1$ and $n = 3$. Show that in this case the global error in total energy per step varies as $(\Delta t)^2$, as appears in Figure 4.4.

4.13   Show that Verlet's algorithm (4.36) and (4.38) is invariant under time reversal. That is, if we decrement time rather than increment time $(\Delta t \rightarrow -\Delta t)$ (4.36) and (4.38) are unchanged.

4.14 In multistep methods, such as Verlet's algorithm, use of periodic boundary conditions in previous time-steps will distort subsequently calculated values for particle displacements, $[x(t) - x(t - \Delta t)]$. Write a segment of simulation code that circumvents this problem for Verlet's algorithm, (4.36) and (4.38).

4.15 (a) Consider the Gear algorithm with third-order predictor

$$\mathbf{r}_i(t + \Delta t) = \mathbf{r}_i(t) + \dot{\mathbf{r}}_i(t)\, \Delta t + \ddot{\mathbf{r}}_i(t)\frac{(\Delta t)^2}{2!} + \mathbf{r}_i^{(iii)}(t)\frac{(\Delta t)^3}{3!}$$

$$\dot{\mathbf{r}}_i(t + \Delta t) = \dot{\mathbf{r}}_i(t) + \ddot{\mathbf{r}}_i(t)\, \Delta t + \mathbf{r}_i^{(iii)}(t)\frac{(\Delta t)^2}{2!}$$

$$\ddot{\mathbf{r}}_i(t + \Delta t) = \ddot{\mathbf{r}}_i(t) + \mathbf{r}_i^{(iii)}(t)\, \Delta t$$

$$\mathbf{r}_i^{(iii)}(t + \Delta t) = \mathbf{r}_i^{(iii)}(t)$$

Show that this predictor is invariant under time reversal ($\Delta t \to -\Delta t$). By the same procedure, but involving more algebra, one may also show that the fifth-order predictor of Section 4.4.3 is also invariant.

(b) Now show that the Gear correctors (4.51)–(4.57) are invariant under time reversal; hence, the full Gear algorithms are invariant. (Note: Making the change $\Delta t \to -\Delta t$ does *not* mean changing the sequence of steps from *PEC* to *CEP*; reversing the steps in the algorithm constitutes a change in algorithm and teaches nothing about the temporal behavior of the original *PEC* method.)

4.16 Use the one-dimensional Lennard-Jones program in Appendix I to do ☒ the following:

(a) Perform calculations to determine the limiting value of the time-step Δt beyond which the Gear algorithm is unstable. Study how this stability limit varies with changes in density and temperature.

(b) Perform the velocity-reversal calculation suggested in Section 4.6: Assign and store on disk initial positions for N particles, carry the simulation forward M steps, reverse the sign of all odd-order derivatives of the positions, and carry the simulation back M steps. If there were no round-off error, the final positions would be the same as the initial values. Compute the rms error by

$$\overline{\Delta x} = \sqrt{\frac{1}{N}\sum_i^N \left[x_i^{\text{final}} - x_i^{\text{initial}} \right]^2}$$

Study how these errors vary with M, with Δt, and with state condition. For example, make log-log plots of the rms error versus M and versus Δt.

4.17 Consider monitoring the H-function through a velocity-reversal calculation, such as is described in Exercise 4.16(b). If the calculation is started from an arbitrary, nonequilibrium situation, during the first half of the calculation (M steps) the H-function decreases toward its equilibrium value. However, after the velocities are reversed, the H-function will *increase* as the system moves away from equilibrium. Now imagine that the situation at time-step M (with velocities reversed) was initially (and arbitrarily) assigned to start a new run and you monitor the H-function to track the approach to equilibrium. From an arbitrarily assigned initial condition, you expect the system to relax to equilibrium and so expect H to decrease, but in fact H increases. In other words, the system appears to be spontaneously moving to a state of lower entropy, in violation of the second law of thermodynamics. Explain. (*Hint:* What happens after M time-steps are executed?)

4.18 Modify the one-dimensional program in Appendix I to use the Verlet
☒ algorithm (4.36) and (4.38) and repeat Exercise 4.16. Note Exercise
◼ 4.14.

4.19 The following algorithm is one recommended by D. Beeman, *J. Comput. Phys.*, **20**, 130 (1976):

$$\mathbf{r}_i(t+\Delta t) = \mathbf{r}_i(t) + \dot{\mathbf{r}}_i(t)\,\Delta t + \tfrac{1}{6}\big[4\ddot{\mathbf{r}}_i(t) - \ddot{\mathbf{r}}_i(t-\Delta t)\big](\Delta t)^2$$

$$\dot{\mathbf{r}}_i(t+\Delta t)\,\Delta t = \mathbf{r}_i(t+\Delta t) - \mathbf{r}_i(t) + \tfrac{1}{6}\big[2\ddot{\mathbf{r}}_i(t+\Delta t) + \ddot{\mathbf{r}}_i(t)\big](\Delta t)^2$$

Show that this method is in fact the Verlet algorithm (4.36) for positions combined with a second-order estimator for velocities.

4.20 Write a program that performs the test, described in Section 4.4.4, for
☒ energy conservation through a collision of two Lennard-Jones atoms. First implement the program using the Verlet algorithm (4.36) and (4.38). Then repeat the test for the following algorithms and generate a plot such as Figure 4.4:

(a) The Beeman algorithm given in the previous exercise.

(b) The Swope et al. method [W. C. Swope, H. C. Andersen, P. H. Berens, and K. R. Wilson, *J. Chem. Phys.*, **76**, 637 (1982)]:

$$x(t+\Delta t) = x(t) + \dot{x}(t)\,\Delta t + \tfrac{1}{2}\ddot{x}(t)(\Delta t)^2$$

$$v(t+\Delta t) = v(t) + \tfrac{1}{2}(\Delta t)\big[\dot{v}(t) + \dot{v}(t+\Delta t)\big]$$

4.21 Modify the one-dimensional Lennard-Jones code in Appendix I so that
☒   evaluation and correction are done twice at each time-step; that is, use
$P(EC)^2$. Perform simulations and compare results and execution speed
of $P(EC)^2$ with those using PEC. Does $P(EC)^2$ offer any advantage
in the size of the time-step?

4.22 Modify the one-dimensional Lennard-Jones code in Appendix I to use
☒   the fourth-order Runge–Kutta (RK) algorithm (Exercise 4.11). Perform simulations and compare RK results and execution speed with
those using PEC. Does RK offer any advantage in the size of the
time-step?

4.23 Lyapunov's methods of stability analysis are an outgrowth of an observation originally attributable to Lagrange: In a conservative system, if a
rest position (all momenta equal to zero) corresponds to a minimum in
the potential energy, then the position is a stable equilibrium point.
Conversely, if a rest position is not at a minimum in the potential, then
the point is unstable (F. Brauer and J. A. Nohel, *The Qualitative
Theory of Ordinary Differential Equations*, Benjamin, New York, 1969;
Dover reprint, New York, 1989, Chapter 5.) A stable equilibrium point
is a point in phase space that has a neighborhood such that trajectories
started in the neighborhood are bounded and remain in the neighborhood of the equilibrium point.

(a) Using the phase-space trajectory for the ODHO as a guide, prove
Lagrange's stability criterion.

(b) Consider a particle in one dimension moving under the influence
of an oscillatory potential, such as that in Figure 2.6(a). Pick
several points on the potential curve and identify them as stable or
unstable according to Lagrange's criterion.

4.24 Lagrange's stability criterion (Exercise 4.23) provides a sufficient condition for stability, but the condition is not necessary.

(a) Does Lagrange's criterion teach us anything about the stability of
the motion of ideal-gas molecules, which have no potential energy?

(b) What about motions under the influence of purely repulsive forces,
such as hard spheres or the models described in Section 2.5?

4.25 To apply Lagrange's stability criterion (Exercise 4.23) to a conservative
system of N particles interacting under a potential $\mathscr{U}(\mathbf{r}^N)$, we must
seek points in phase space that satisfy

$$d\mathscr{U}(\mathbf{r}^N) = 0$$

that is,

$$\sum_i^N \frac{\partial \mathscr{U}}{\partial \mathbf{r}_i} d\mathbf{r}_i = -\sum_i^N F_i \, d\mathbf{r}_i = 0$$

Since the $d\mathbf{r}_i$ are all independent, this could be satisfied only if the force on each atom were zero. Could a rest position (all momenta zero) for the Lennard-Jones substance, at liquid densities, be found that would satisfy Lagrange's criterion? What about at low densities?

4.26   It is proposed that round-off error could be studied directly in a one-dimensional Lennard-Jones simulation (nearest-neighbor interactions only) by starting a run from the following special set of initial conditions: each atom would be assigned zero initial velocity. All nearest-neighbor pairs would be initially separated by the distance $\sigma 2^{1/6}$. All higher-order derivatives of the positions would be initially set to zero. Keeping in mind the principle of sensitivity to initial conditions and Lagrange's stability criterion (Exercise 4.23), would you expect atomic motion to develop during the run? If so, would repeating the simulation using different finite-difference algorithms provide any insight into the magnitudes and propagation of round-off errors?

4.27   Lagrange's stability criterion does *not* apply to nonconservative systems, that is, situations in which equations of motion are not derived
⊠   from some conserved Hamiltonian. For example, consider a fluid (e.g., the earth's atmosphere) of uniform depth and having a constant overall temperature difference from top to bottom. The problem is to model convective flow in the fluid column. By making a suitable truncation and simplification of the Navier–Stokes equation, Lorenz obtained the following model for convection (E. N. Lorenz, *J. Atmos. Sci.*, **20**, 130 (1963); reprinted in Hao Bai-Lin, *Chaos*, World Scientific, Singapore, 1984):

$$\frac{dx}{dt} = \text{Pr}(y - z) \qquad \frac{dy}{dt} = \mathcal{R}x - xz - y \qquad \frac{dz}{dt} = xy - \mathcal{B}z$$

Here x is a measure of the intensity of the convective motion, y measures the difference in temperature between rising and falling regions of fluid, and z measures the degree to which the temperature profile is distorted from a linear one. These three quantities, as well as the time t, have been made dimensionless. The parameters in these three differential equations are Pr, $\mathcal{R}$, and $\mathcal{B}$: Pr is the Prandtl number, $\mathcal{R}$ is the ratio of the Rayleigh number to its critical value, and $\mathcal{B}$ is related to the critical value of the Rayleigh number.

(a)   Write a program that uses the fourth-order Runge–Kutta algorithm (Exercise 4.11) to generate solutions for this set of equations.

(b)   When the ratio $\mathcal{R}$ is below a critical value, the equations have a steady state solution, the temperature profile varies linearly with depth of fluid, and no convective flow develops. Test this assertion by running your program using $\text{Pr} = 10$, $\mathscr{B} = \frac{8}{3}$, and values of $\mathscr{R}$ less than the critical value of $470/19$. Time steps of the order of $\Delta t = 0.05$ are appropriate. Test initial conditions near the steady state solution $\{0,0,0\}$ as well as at points away from $\{0,0,0\}$.

(c) Conversely, when $\mathscr{R}$ exceeds its critical value, the computed trajectory is unstable and convective motion develops. Test this by running your program using the same parameter values as in (b), but use $\mathscr{R} = 28$ ($>470/19$). Initial conditions near $\{0,1,0\}$ are sufficient to establish the convective motion. Plot $x(t)$, $y(t)$, and $z(t)$ versus t, and plot $x(t)$ versus $z(t)$.

(d) Perform two runs as in (c), but change the initial conditions slightly (say, from $\{0,1,0\}$ to $\{0,1.1,0\}$), form the difference in one variable $[x_1(t) - x_2(t)]$, and plot versus time. What do you conclude about sensitivity to initial conditions and about the stability of these differential equations?

5

SOFT SPHERES

As illustrated in Figure 5.1, dynamic modeling divides into two tasks: developing a model and using the model in a simulation. The first task, model development, includes choosing a form for the intermolecular potential and then deriving appropriate equations of motion. In this chapter we tackle simulation of substances modeled as soft spheres, substances for which the prototypical potential is that proposed by Lennard-Jones. In Section 5.1 we describe the Lennard-Jones model, how it is often modified for use in molecular dynamics, and how those modifications help simplify the simulation.

After we have chosen the model potential, we must formulate equations of motion. Since in this book we consider only isolated systems, the equations of motion are simply obtained from Newton's second law. Those equations are discussed in Chapter 2 and methods for solving them are described in Chapter 4, so equations of motion are ignored here.

Again as in Figure 5.1, the second modeling task—the simulation—itself divides into two parts: generating phase-space trajectories and analyzing the trajectories for properties. Trajectory generation claims most of our attention in this chapter; in particular, trajectory generation decomposes into initialization (Section 5.2), equilibration (Section 5.3), and production (Section 5.4). Trajectory analysis is discussed in detail in Chapters 6 and 7; however, some analysis must be done during each run to check that the run is proceeding correctly, to identify the onset of equilibrium, and to help decide when the run can be stopped. This ongoing analysis is described in Section 5.5. Although trajectory generation is discussed here in terms of the Lennard-Jones model, in fact, the discussions in this chapter generally apply to any continuous pair potential acting between spherical objects.

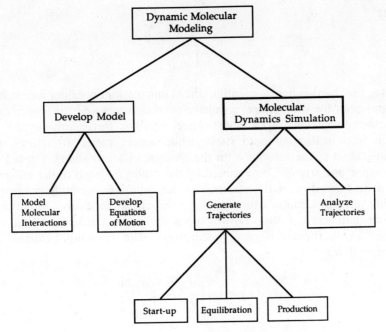

FIGURE 5.1 Steps in dynamic modeling of matter.

5.1 INTERMOLECULAR POTENTIAL MODELS

As the first step in performing a simulation we must choose a functional form for the intermolecular potential $\mathcal{U}(\mathbf{r}^N)$. Nearly always, we take the potential to be pairwise additive; that is, we assume the interaction energy among N atoms is a sum of isolated two-body contributions. For spheres, the pairwise additive form is

$$\mathcal{U}(\mathbf{r}^N) = \sum_{i<j}\sum u(r_{ij}) \tag{5.1}$$

Recall $\mathbf{r}^N$ is a shorthand for the set of sphere position vectors (Figure 2.1), $\mathbf{r}^N = \{\mathbf{r}_1, \mathbf{r}_2, \mathbf{r}_3, \ldots, \mathbf{r}_N\}$, u is a two-body potential, and r_{ij} is the scalar distance between the centers of atoms i and j, $r_{ij} = |\mathbf{r}_i - \mathbf{r}_j|$. The form (5.1) neglects simultaneous multibody interactions which, if included, would drastically increase the time required for a simulation.

In 1924 J. E. Lennard-Jones introduced [1] a useful model for the soft-sphere pair potential

$$u(r) = k\varepsilon\left[\left(\frac{\sigma}{r}\right)^n - \left(\frac{\sigma}{r}\right)^m\right] \tag{5.2}$$

where

$$k = \frac{n}{n-m}\left(\frac{n}{m}\right)^{m/(n-m)} \tag{5.3}$$

Unlike the hard-sphere potential, the Lennard-Jones model attempts to account both for short-range, repulsive overlap forces and for longer range, attractive dispersion forces. Short-range repulsive forces prevent the substance from collapsing onto itself, while longer range attractions deter disintegration of the substance (in the absence of a container). These forces have range and strength determined by the values assigned to the integers n and m ($n > m$). For m the common choice is $m = 6$, primarily because the leading term in London's theory [2] for dispersion varies as $1/r^6$. Popular wisdom then sets $n = 2m = 12$, which has the merit of a kind of logic while lacking any particular physical justification. The resulting Lennard-Jones (12, 6) model is

$$u(r) = 4\varepsilon\left[\left(\frac{\sigma}{r}\right)^{12} - \left(\frac{\sigma}{r}\right)^{6}\right] \tag{5.4}$$

The remaining parameters in (5.4) are σ, the distance to the zero in $u(r)$, and ε, the energy at the minimum in $u(r)$; see Figure 5.2. Since intermolecular forces are necessarily conservative, the force that results from the potential (5.4) is

$$F(r) = -\frac{du(r)}{dr} = 24\frac{\varepsilon}{\sigma}\left[2\left(\frac{\sigma}{r}\right)^{13} - \left(\frac{\sigma}{r}\right)^{7}\right] \tag{5.5}$$

By convention, repulsive forces are positive while attractive forces are negative.

5.1.1 Truncated Potential

In a system of N atoms, the double sum in (5.1) accumulates $\frac{1}{2}N(N-1)$ unique pair interactions. Thus, if all pair interactions are sampled during a simulation, the number of such samples increases with the square of the number of atoms. Further, if the allowed range of interaction between atoms is increased, say from r to $r + \Delta r$, then the number of sampled interactions increases as r^2; that is,

$$N(r, \Delta r) \approx \rho V(r, \Delta r) \approx 4\pi\rho r^2 \Delta r \tag{5.6}$$

Here ρ is the number density, $N(r, \Delta r)$ is the number of atoms in a spherical shell of radius r and thickness Δr, and $V(r, \Delta r)$ is the volume of the shell.

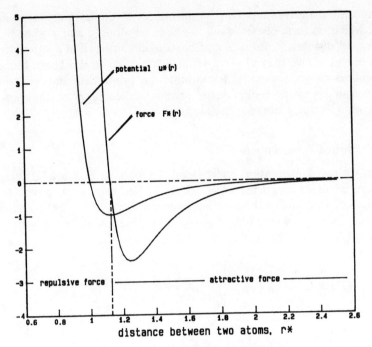

FIGURE 5.2   Lennard-Jones $(12, 6)$ pair potential (5.4) and pair force (5.5). The units here are $r^* = r/\sigma$, $u^* = u/\varepsilon$, and $F^* = F\sigma/\varepsilon$.

From Figure 5.2 we see that the Lennard-Jones potential extends over a modest range of pair separations; as a result, we can achieve a considerable savings in computer time by neglecting pair interactions beyond some distance r_c. Therefore the Lennard-Jones potential actually used in simulations is a truncated version of (5.4),

$$
u(r) = \begin{cases} 4\varepsilon\left[\left(\dfrac{\sigma}{r}\right)^{12} - \left(\dfrac{\sigma}{r}\right)^{6}\right] & r \le r_c \\ 0 & r > r_c \end{cases} \tag{5.7}
$$

A common choice for r_c is 2.5σ, at which $u = -0.0163\varepsilon$ and the force $F = -0.039\varepsilon/\sigma$. Thus, when $r_{ij} = r_c$, atom j makes only a small contribution to the force on atom i.

Because of the truncation at r_c, a simulation can provide only a portion of those properties, such as the internal energy and pressure, that are directly related to the potential. Simulation results for such properties must be corrected for long-range interactions ($r > r_c$) that are neglected during a run. To estimate these long-range corrections, formulas are given in Chapter 6.

Truncating the potential at r_c introduces a similar truncation into the force which, in turn, causes small impulses on atoms i and j whenever their separation distance r_{ij} crosses r_c. Consequently, instead of a strictly constant total energy E, we may observe small fluctuations in E. These fluctuations little affect values computed for equilibrium properties, and of course, the effects can be made negligible by simply increasing r_c at the expense of increased computer time for the simulation.

5.1.2   Shifted-Force Potential

When we test a simulation code, small fluctuations in the total energy can obscure small errors, frustrating the test. For testing purposes, the step change in $u(r)$ and $F(r)$ can be removed by shifting $F(r)$ vertically so that the force goes smoothly to zero at r_c. Hence, define a shifted force $F_s(r)$ by [3]

$$F_s(r) = \begin{cases} -\dfrac{du}{dr} + \Delta F & r \le r_c \\ 0 & r > r_c \end{cases} \tag{5.8}$$

where ΔF is the magnitude of the shift,

$$\Delta F = -F(r_c) = \left(\frac{du}{dr}\right)_{r_c} \tag{5.9}$$

The shifted-force potential $u_s(r)$ corresponding to $F_s(r)$ can be derived from

$$F_s(r) = -\frac{du_s(r)}{dr} \tag{5.10}$$

or

$$\int_0^{u_s} du_s = -\int_\infty^r F_s(r)\, dr \tag{5.11}$$

Substituting (5.8) into (5.11) and integrating gives

$$u_s(r) = \begin{cases} u(r) - u(r_c) - [r - r_c]\left(\dfrac{du}{dr}\right)_{r_c} & r \le r_c \\ 0 & r > r_c \end{cases} \tag{5.12}$$

Figure 5.3 compares this shifted-force potential with the full Lennard-Jones model.

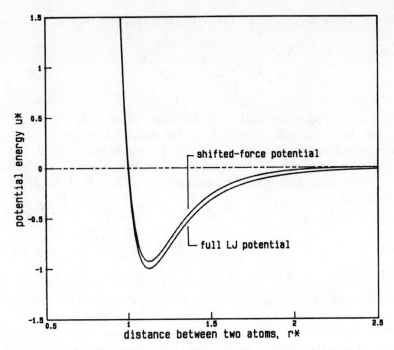

FIGURE 5.3 Comparison of the full Lennard-Jones (LJ) potential (5.4) and the shifted-force potential (5.12), with the shift applied at $r_c = 2.5\sigma$. Here $u^* = u/\varepsilon$ and $r^* = r/\sigma$.

Using (5.8) and (5.12) in place of (5.7) removes energy fluctuations that occur because of the truncation of u and F. However, if the shifted-force potential is used to evaluate properties, then corrections must be included to account for both the long-range interactions (as discussed in Section 5.1.1) and the shift [3]. The shifted-force corrections are approximate and cumbersome to compute, and so we recommend that the shifted-force model be used only as an aid in verifying codes: it should not be used to determine properties in production runs.

5.1.3 Minimum-Image Criterion

In a system whose N atoms interact via a pairwise additive potential, the force vector exerted by atom j on atom i is given, at any instant, by

$$\mathbf{F}_i = -\frac{\partial u(\mathbf{r}_{ij})}{\partial \mathbf{r}_i} \tag{5.13}$$

However, in periodic systems forces can be exerted by image atoms, as well as by real atoms (Section 2.9), and so (5.13) is replaced by

$$\mathbf{F}_i = -\sum_\alpha \frac{\partial u(\mathbf{r}_{ij} - \boldsymbol{\alpha} L)}{\partial \mathbf{r}_i} \tag{5.14}$$

Here $\mathbf{r}_{ij} - \boldsymbol{\alpha} L = \mathbf{r}_i - (\mathbf{r}_j + \boldsymbol{\alpha} L)$, L is the length of one edge of the primary cell, and $\boldsymbol{\alpha}$ is the cell translation vector discussed in Section 2.9. In a D-dimensional system the sum over $\boldsymbol{\alpha}$ in (5.14) is a D-fold sum, as in (2.106).

When the forces are short ranged compared to L, we need consider only those image cells that adjoin the primary cell. Then in three dimensions (5.14) becomes

$$\mathbf{F}_i = -\sum_{\alpha_x=-1}^{+1} \sum_{\alpha_y=-1}^{+1} \sum_{\alpha_z=-1}^{+1} \frac{\partial u(\mathbf{r}_{ij} - \boldsymbol{\alpha} L)}{\partial \mathbf{r}_i} \tag{5.15}$$

The triple sum in (5.15) accumulates forces from atom j and each of its 26 images; however, of the 27 terms in (5.15), only one has a separation $|\mathbf{r}_{ij} - \boldsymbol{\alpha} L|$ less than $\frac{1}{2}L$. Therefore, if the pair potential $u(r_{ij})$ is truncated at $r_c \leq \frac{1}{2}L$, then either atom j or, at most, only *one* of its images can exert a force on atom i. The interaction distance is necessarily the smallest of the 27 possibilities: this identifies the *minimum image criterion* for computing forces. Thus, in periodic systems, when the interaction is truncated at $r_c \leq \frac{1}{2}L$, the minimum image criterion selects from the triple sum in (5.15) the one nonzero term—that for the smallest of the 27 distances $\{|\mathbf{r}_{ij} - \boldsymbol{\alpha} L|, (\alpha_x = -1, 1), (\alpha_y = -1, 1), (\alpha_z = -1, 1)\}$.

To develop a procedure for finding the minimum image distance, we can consider a one-dimensional system with periodic boundaries, as in Figure 5.4. In the figure the primary "cell" has width L and contains two particles, $i(0)$ and $j(0)$. Because the system is one dimensional, the primary cell has only two adjacent image cells, one identified by $\alpha_x = +1$, the other by $\alpha_x = -1$. In one dimension the force, analogous to (5.15), is

$$F_i = -\sum_{\alpha_x=-1}^{+1} \frac{\partial u(x_{ij} - \alpha_x L)}{\partial x_i} \tag{5.16}$$

To obtain the force, we must consider the three values of α_x, any one of which might give the smallest distance $|x_{ij} - \alpha_x L|$:

Situation A in the figure ($\alpha_x = 0$) has $x_{ij} < \frac{1}{2}L$. In this case the distances between i and either j image are greater than $\frac{1}{2}L$ and hence greater than r_c. The images do not interact with i, and therefore the

Situation A: $x_{ij} < L/2$

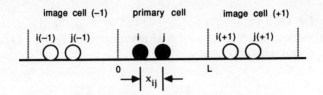

Situation B: $x_{ij} < -L/2$ $x_{ij} \rightarrow x'_{ij} = x_{ij} + L$

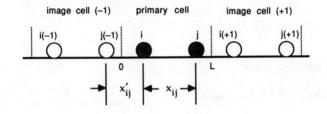

Situation C: $x_{ij} > L/2$ $x_{ij} \rightarrow x'_{ij} = x_{ij} - L$

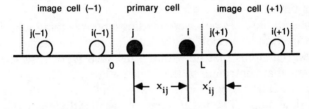

FIGURE 5.4 Three possible situations that can occur when applying the minimum image criterion for determining force between atoms i and j in a one-dimensional periodic system.

distance between i and j in the primary cell is the appropriate distance to use in (5.16) for evaluating the force.

Situation B in Figure 5.4 ($\alpha_x = -1$) has $x_{ij} < -\frac{1}{2}L$. In this case $|x_{ij}| > r_c$ and j does not interact with i via the truncated potential. However, the image $j(-1)$ is within r_c of i and does interact; so the distance x_{ij} becomes

$$x_{ij} \rightarrow x_i - x_{j(-1)} = x_i - \left[x_{j(0)} - L \right] = x_{ij} + L \qquad (5.17)$$

The other image, $j(+1)$ lies more than $\frac{1}{2}L$ from i and therefore does not interact.

Situation C in Figure 5.4 ($\alpha_x = +1$) has $x_{ij} > \frac{1}{2}L$. In this case $x_{ij} > r_c$ and j does not interact with i. However, the image $j(+1)$ is within r_c of i and does interact; so

$$x_{ij} \to x_i - x_{j(+1)} = x_i - \left[x_{j(0)} + L \right] = x_{ij} - L \qquad (5.18)$$

For a cubic container in three dimensions, the minimum image criterion applies separately to each Cartesian component of the pair separation vector $\mathbf{r}_{ij}$; thus, for the x component, we use $x_{ij} \to x_{ij} - \alpha_x L$ where $\alpha_x = 0$ if $-\frac{1}{2}L \le x_{ij} \le \frac{1}{2}L$, $\alpha_x = -1$ if $x_{ij} < -\frac{1}{2}L$, and $\alpha_x = +1$ if $x_{ij} > \frac{1}{2}L$. Similarly for y_{ij} and z_{ij}.

5.1.4 Neighbor Lists

The most time-consuming part of a molecular dynamics simulation is calculation of the forces on the atoms. For a fluid of N atoms, the forces are computed by sampling, at each time step of the simulation, $\frac{1}{2}N(N-1)$ unique r_{ij} distances. But when a truncated potential is used, the force equals zero for any $r_{ij} > r_c$ (where r_c is the potential cutoff distance described in Section 5.1.1); consequently, evaluating distances r_{ij} that are greater than r_c wastes computer time.

This wasted time can be saved by using a bookkeeping scheme such as that originally devised by Verlet [4]. For each atom i, the method maintains a list of neighboring atoms that lie within a distance r_L of i; so the list identities those atoms that contribute to the force on atom i. The same neighbor list is used over several consecutive time steps, and it is updated periodically, say every 10 time steps. The list distance r_L is slightly larger than r_c so that j atoms can cross r_c and still be properly considered in evaluating the force on i. Typically, $r_L = r_c + 0.3\sigma$.

For a three-dimensional Lennard-Jones fluid at density $\rho\sigma^3 = 0.8$, each atom has about 75 neighbors lying within a radius $r = 2.8\sigma$ (which is the appropriate list distance when $r_c = 2.5\sigma$). However, in the neighbor list for atom i, we need store only the identities of those neighbors j having $j > i$, because for $j < i$ atom i appears in the list as a neighbor of j. On average, about half the neighbors of i have $j > i$, so we need, for the neighbor list, about $\frac{75}{2} \approx 40$ storage locations per atom. Therefore to store the neighbor list in a simulation of 256 atoms, we use an array, call it LIST, containing about 11,000 elements. We then use the elements of another array, NPOINT, to locate in LIST the neighbors of a particular atom. The relation between NPOINT and LIST is illustrated in Figure 5.5. By using two one-dimensional arrays rather than one two-dimensional array, we avoid the computational

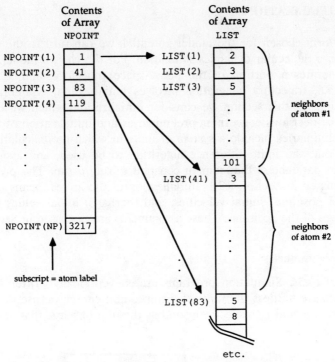

FIGURE 5.5 Use of two one-dimensional arrays to store a neighbor list for each atom during a simulation of a system containing N atoms.

overhead involved in dealing with double subscripted variables. A neighbor list can provide substantial gains in execution speed, as shown in Table 5.1.

The scheme just described is but one way by which a neighbor list can be implemented. It has the disadvantage that it does not vectorize to run on parallel machines; however, modifications can be made to overcome this handicap [5–9].

TABLE 5.1   Speed of Execution of Lennard-Jones Program with and without a Neighbor List, $\langle kT / \varepsilon \rangle = 1.095$

		Average Number of Neighbors Stored	Relative Number of Time Steps Executed	
$\rho\sigma^3$	N	per Atom	No List	With List[a]
0.6	108	27.2	6.7	9.7
	256	27.5	1	3.3
0.85	108	28.0	5.3	8.1
	256	38.1	1	2.6

[a]$r_c = 2.5\sigma$, $r_L = 2.8\sigma$, list up-dated every 10 time-steps.

5.2   INITIALIZATION

With a form chosen for the model potential, we can turn to the use of the model in a molecular dynamics simulation. For the chosen model, a simulation generates a portion of the phase-space trajectory, and as shown in Figure 5.6, trajectory generation involves initialization, equilibration, and production. In this section we consider initialization, which we divide into two parts: decisions concerning preliminaries and initialization of the atoms. The preliminaries include a system of units in which the calculation will be carried out, the finite-difference algorithm to be used, and assignment of values to parameters that remain constant during a run. The preliminaries are discussed in Section 5.2.1. Initialization of the atoms means assignment of initial positions, initial velocities, and (perhaps) initial values for higher derivatives of the positions. These assignments are discussed in Section 5.2.2.

5.2.1   Preliminaries

System of Units. Simulation programs are conventionally written so that all quantities are unitless. As units of distance and energy we use the potential parameters σ and ε, respectively, and as the unit of mass, that of one atom.

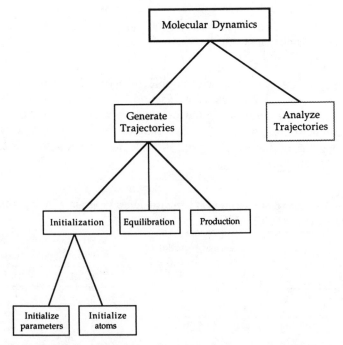

FIGURE 5.6 Trajectory generation divides into initialization, equilibration, and production.

TABLE 5.2   System of Units[a] Used in Soft-Sphere Molecular Dynamics Programs

Fundamental Quantities

Mass	m = mass of one atom
Length	σ
Energy	ε
Time	$\sigma\sqrt{m/\varepsilon}$

Derived Quantities

Adiabatic compressibility	$\kappa_s^* = \kappa_s \varepsilon / \sigma^3$
Configurational internal energy	$U_c^* = U_c / N\varepsilon = \langle \mathscr{U}^* \rangle = \langle \mathscr{U} / N\varepsilon \rangle$
Density	$\rho^* = N\sigma^3 / V$
Force	$F^* = F\sigma / \varepsilon$
Heat capacity	$C_v^* = C_v / Nk$
Radial position	$r^* = r / \sigma$
Pressure	$P^* = P\sigma^3 / \varepsilon$
Temperature	$T^* = kT / \varepsilon$
Thermal pressure coefficient	$\gamma_v^* = \gamma_v \sigma^3 / k$
Total energy	$E^* = E / N\varepsilon$
Velocity	$v^* = v\sqrt{m/\varepsilon}$

[a]W. Thomson, Lecture XVI, p. 158: "The distinguishing feature of an engineer is the quickness with which he can reduce from square feet to acres, and so on. If his brain were free from that, he might do more elsewhere, and have more time to find out about the properties of matter."

The unitless (or reduced) forms for quantities are indicated by an asterisk; for example, (5.4) for the Lennard-Jones potential becomes

$$u^*(r^*) = 4\left[\left(\frac{1}{r^*}\right)^{12} - \left(\frac{1}{r^*}\right)^6\right] \tag{5.19}$$

where $u^* = u/\varepsilon$ and $r^* = r/\sigma$. The reduced forms for other derived quantities are given in Table 5.2.

State Condition. For equilibrium molecular dynamics performed on isolated systems, the independent thermodynamic properties are the number of molecules N, system volume V, and total energy E. Then for a one-phase system, specifying values for the number density $\rho^* = N\sigma^3/V$ and the energy per atom $E^* = E/N\varepsilon$ determines the thermodynamic state. Usually, $100 < N < 1000$, with the particular value of N taken to be large enough to properly capture the phenomena of interest yet small enough to prevent the simulation from being prohibitively expensive. Having picked N, the required value of V is obtained from the assigned density ρ^*.

Assigning a value to E^* may be accomplished by scaling the atomic velocities. Let subscripts D and A designate "desired" and "actual" values for properties. For example, based on the initially assigned atomic positions and velocities, the actual total energy is

$$E_A^* = E_{kA}^* + \mathcal{U}^* \tag{5.20}$$

Using the desired set point for the total energy E_D^* and the computed potential energy $\mathcal{U}^*$, we can compute a required value for the kinetic energy

$$E_{kD}^* = E_D^* - \mathcal{U}^* \tag{5.21}$$

Then we scale each initially assigned component of a velocity vector according to

$$v_{ix}^{*\,\text{new}} = v_{ix}^* \sqrt{\frac{E_{kD}^*}{E_{kA}^*}} = v_{ix}^* \sqrt{\frac{E_D^* - \mathcal{U}^*}{E_A^* - \mathcal{U}^*}} \tag{5.22}$$

The same scaling is also applied to the y- and z-components. By substituting these new velocities into

$$E^* = \frac{1}{2N} \sum_i \mathbf{v}_i^* \cdot \mathbf{v}_i^* + \mathcal{U}^*(r^N) \tag{5.23}$$

we can verify that the set-point energy E_D^* is indeed obtained. Here, as in all that follows, we follow the convention in Table 5.2 that sets the atomic mass to unity.

Usually, instead of density and energy (ρ^*, E^*), we would like to establish the thermodynamic state using density and temperature (ρ^*, T^*). This may be achieved by scaling the initially assigned velocities to obtain a desired kinetic energy rather than a desired total energy. The procedure takes advantage of the relation (2.43) between absolute temperature and time-average kinetic energy,

$$T^* = \frac{1}{3N} \left\langle \sum_i \mathbf{v}_i^* \cdot \mathbf{v}_i^* \right\rangle \tag{5.24}$$

Thus, rather than apply the scaling (5.22), we scale the initial velocities $\mathbf{v}_i(0)$ so that initially we have the desired temperature T_D^*

$$v_{ix}^{*\,\text{new}} = v_{ix}^* \sqrt{\frac{T_D^*}{T_A^*}} \tag{5.25}$$

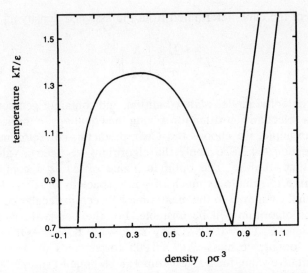

FIGURE 5.7 Temperature–density phase diagram for the pure Lennard-Jones (12, 6) substance. The vapor–liquid saturation curve was computed using the Nicolas et al. [3] equation of state. The melting lines were taken from Hansen and McDonald [10].

where T_A^* is the actual temperature computed from (5.24) using the initially assigned velocities. A similar scaling is applied to the y- and z-components of $v_i^*(0)$.

Because kinetic energy is not a constant of the motion, the scaling (5.25) is less effective in achieving a set-point temperature than is the scaling (5.22) in achieving a set-point energy. During the initial phase of a run, as the system relaxes from its initial condition, energy will be exchanged between the potential and kinetic modes; consequently, the temperature will drift from its set point. To compensate, the scaling (5.25) is normally continued over the full equilibration phase. After scaling for a few hundred time-steps, the state condition will be close but not exactly equal to the desired one. To facilitate the choice of state conditions, we give in Figure 5.7 the temperature–density phase diagram for the pure Lennard-Jones (12, 6) substance.

Boundary Conditions, Algorithm, and Time-Step. For simulations of bulk fluids, unwanted surface effects are removed by using periodic boundary conditions (pbc). Periodic boundary conditions are discussed in Section 2.9, and so here we merely note that to apply pbc, the volume holding the atoms must be space filling. In three dimensions the simplest space-filling shape is the cube: the cube provides the simplest pbc transformations (2.114). Therefore most bulk substance simulations are done on cubic containers, with the length L^*

of one edge of the cube determined from the assigned density ρ^*

$$L^* = \frac{L}{\sigma} = \left(\frac{N}{\rho^*} \right)^{1/3} \tag{5.26}$$

With the potential model, state condition, and container geometry chosen, we must now select an algorithm for solving the equations of motion. To have a concrete example, we choose the Gear predictor–corrector method discussed in Section 4.4.3. To apply the algorithm, we need a value for the integration time-step Δt. The optimum value of Δt is a compromise. It should be large to sample as much of phase space as possible; however, it should be small compared to the mean time between molecular collisions, for otherwise the algorithm will be unstable. For the Lennard-Jones fluid, an acceptable value is $\Delta t^* = 0.005$. Using $\varepsilon / k = 120$ K, $\sigma = 3.405$ Å, and the argon value for atomic mass, $\Delta t^* = 0.005$ corresponds to $\Delta t \approx 10^{-14}$ sec. With this tentative value for Δt, test runs can be done to study how the code performs with changes in time-step.

5.2.2   Initial Conditions for Atoms

To start the finite-difference algorithm, atomic positions $\mathbf{r}_i$ and their time derivatives must be assigned at time $t = 0$. The positions $\mathbf{r}_i$ are measured with respect to some space-fixed frame, such as in Figure 5.8, and their initial values $\mathbf{r}_i(0)$ may be assigned according to some lattice structure or taken from a previous simulation. They should not be assigned randomly because random assignments often create artificially large overlaps of adjacent atoms; such overlaps produce an unphysically large repulsive force that can cause numerical failure of the finite-difference algorithm. When positions are assigned to a lattice, a face-centered-cubic (fcc) structure is usually used because argon crystallizes into an fcc structure. The construction of an fcc lattice is discussed in Section 3.4 and need not be repeated here.

Initial velocities $\mathbf{v}_i(0)$ may be randomly assigned or they may be taken from a previous simulation. For randomly assigned velocities let ξ_x, ξ_y, and ξ_z represent three random numbers each uniformly distributed over the interval $[-1, +1]$. Then the Cartesian components of $\mathbf{v}_i(0)$ are assigned by

$$v_{ix}^*(0) = \frac{\xi_x}{\xi} \qquad v_{iy}^*(0) = \frac{\xi_y}{\xi} \qquad v_{iz}^*(0) = \frac{\xi_z}{\xi} \tag{5.27}$$

where $\xi \equiv \sqrt{\xi_x^2 + \xi_y^2 + \xi_z^2}$. These initial velocities are scaled either according to (5.22) to produce the set-point total energy or according to (5.25) to produce the set-point temperature.

Since no external forces act on the system, the total linear momentum should be conserved (Section 2.10). Moreover, the substance is static; there-

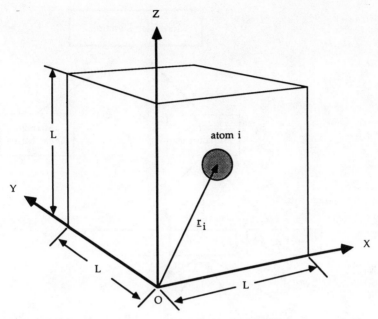

FIGURE 5.8 Space-fixed reference frame located with origin at one vertex of the cubic container having edges of length L. Sphere location vectors $\mathbf{r}_i$ are measured in this frame.

fore, we adjust the scaled initial velocities so that the total linear momentum is initially zero. The adjustment takes the form

$$v_{ix}^{*\text{new}}(0) = v_{ix}^*(0) - \frac{1}{N} \sum_j v_{jx}^*(0) \qquad (5.28)$$

with similar adjustments for $v_{iy}(0)$ and $v_{iz}(0)$.

Values for the initial accelerations $\mathbf{a}_i(0)$ are determined from the positions $\mathbf{r}_i(0)$ by computing the force on each atom, via (5.5), and applying Newton's second law. Thus, at this point we have values for $\mathbf{r}_i(0)$, $\mathbf{v}_i(0)$, and $\mathbf{a}_i(0)$. Besides these, the Gear algorithm uses the third, fourth, and fifth derivatives of the positions. To obtain initial values for these higher derivatives of $\mathbf{r}_i(0)$, we may evaluate derivatives of the force (5.5), but the calculations are tedious. Alternatively, initial values can be taken from the end of a previous simulation. A third scheme is to set the higher derivatives initially to zero,

$$\mathbf{r}_i^{(\text{iii})}(0) = \mathbf{r}_i^{(\text{iv})}(0) = \mathbf{r}_i^{(\text{v})}(0) = 0 \qquad (5.29)$$

This means that for the first few time-steps of a run the fifth-order predictor is not being used; however, within about 50 time-steps, values of the higher

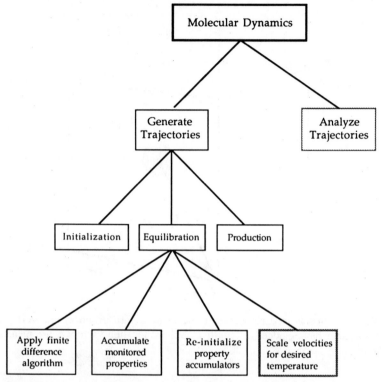

FIGURE 5.9 Equilibration involves applying a finite-difference method to generate the trajectory and accumulating selected properties to monitor the system's relaxation from initial conditions. The velocity scaling is optional. At the end of equilibration, property accumulators are reset to zero so that properties are computed only along the equilibrium trajectory.

derivatives are accumulated and thereafter the full predictor–corrector algorithm is operational.

5.3 EQUILIBRATION

With initialization accomplished, the run itself can begin. Over the first few hundred time-steps the system relaxes from the arbitrarily assigned initial conditions and approaches equilibrium; this relaxation phase is called *equilibration*. As suggested in Figure 5.9, equilibration involves at least three, possibly four, distinct activities. The first is use of the finite-difference algorithm to generate the phase-space trajectory. The second is our monitoring of the trajectory by following selected properties. The third activity is optional: during equilibration we may continually scale the atomic velocities to keep the system temperature close to a desired value. The last activity occurs at the end of equilibration when property accumulators are reinitialized to zero, so that properties are computed only over the equilibrium phase of the run. The duration of equilibration varies: it depends on how far removed the initial conditions are from the equilibrium state. The goal of equilibration is to ensure that values computed for equilibrium properties are not affected by how the simulation was started.

5.3.1 Approach to Equilibrium

During equilibration, running averages for most properties are not stable; rather, averages will be evolving toward their equilibrium values. However, because the system is isolated, the total energy should have the same value at each time-step even when the system is not yet at equilibrium. Conservation of total energy is sensitive to many small errors that can inadvertently creep into a simulation code; so, during equilibration without velocity scaling, energy conservation provides an important and direct indicator of errors in the program.

The evolution to equilibrium necessarily involves equilibration in both configuration space and momentum space, so we need quantities that monitor both. Just as for hard spheres, we use the positional order parameter λ to monitor the disintegration of the initial fcc lattice and use the instantaneous (kinetic) Boltzmann H-function to monitor the development of the equilibrium distribution of atomic velocities. Recall from Section 3.5.1 that the order parameter is defined by

$$\lambda = \tfrac{1}{3}\left[\lambda_x + \lambda_y + \lambda_z\right] \qquad (3.41)$$

where

$$\lambda_x = \frac{1}{N}\sum_i^N \cos\left(\frac{4\pi x_i}{a}\right) \qquad (3.42)$$

and a is the length of one edge of the fcc unit cell. Also in Section 3.5.2, we took one component of the H-function to be

$$H_x = \sum_{\Delta v_x} f(v_x)\ln f(v_x)\,\Delta v_x \qquad (3.45)$$

where $f(v_x)$ is the distribution of the x-component of atomic velocities,

$$f(v_x)\,\Delta v_x = \frac{1}{N}\sum_i^N \delta(v_x - v_{xi})\,\Delta v_x \qquad (3.44)$$

The behaviors of λ and H for soft spheres should be the same as for hard spheres: equilibration is not complete until λ is fluctuating about zero and the H-function is near a value that is consistent with the Maxwell velocity distribution. Consult Section 3.5 for detailed discussions of these quantities.

5.3.2 Identification of Equilibrium

For an isolated system, a necessary and sufficient condition for identifying equilibrium is that the system entropy be a maximum. Unfortunately, entropy is not a measurable property and cannot be readily evaluated from the time average of some mechanical quantity (see Section 6.3). We are therefore left with the problem of identifying equilibrium by accumulating a number of necessary conditions and claiming that these are also sufficient.

Thermodynamic equilibrium encompasses thermal equilibrium, mechanical equilibrium, and chemical equilibrium. Thermal equilibrium means the absence of any driving forces for heat transfer, mechanical equilibrium means the absence of driving forces that would deform the size or shape of the system, and chemical equilibrium means the absence of driving forces that promote chemical reactions, phase transitions, and diffusional mass transfer. These driving forces must be absent not only across system boundaries but also across boundaries between any arbitrarily identified, macroscopic parts of the system.

Equilibrium eludes simple identification in molecular dynamics, first, because it is a macroscopic concept: it applies over a finite duration to a system of macroscopic size. Sampling the system at one instant is not sufficient to identify an equilibrium state. Second, equilibrium pertains to a finite duration that is important to the observer [11]. For example, the carbon-steel blade of a knife appears, over a few hours, to be in equilibrium with its surroundings; apparently, no driving forces are present to disturb its thermal, mechanical, or chemical equilibrium. However, after a few days of exposure to humid air, the blade has rusted. Is the knife in equilibrium with its surroundings or not? The answer depends on what time scale is of interest to the observer.

Here are some necessary conditions that an isolated system at equilibrium should satisfy:

1. The total number of molecules N and total energy E should be constants, independent of time. Since E is constant, fluctuations in the kinetic energy and potential energy must be equal in magnitude but out of phase with one another.

2. Each Cartesian component of the velocities should, on a time average, describe a Maxwell distribution (2.54),

$$f(v_x^*)\, \Delta v_x^* = \frac{1}{\sqrt{2\pi T^*}} \exp \frac{-v_x^{*2}}{2T^*} \Delta v_x^* \qquad (5.30)$$

where $f(v_x^*)\Delta v_x^*$ is the probability of finding a molecule whose x-component of velocity lies in $v_x^* \pm \frac{1}{2}\Delta v_x^*$. Recall, in our system of units (Table 5.2) the molecular mass is unity. As described in Section 3.5.2, the probability density $f(v_x^*)$ can be obtained from a molecular dynamics simulation by counting, at one instant, the number of molecules having x-velocities in $v_x^* \pm \frac{1}{2}\Delta v_x^*$, and averaging that number over time,

$$f(v_x^*)\Delta v_x^* = \frac{1}{N}\left\langle \sum_i^N \delta[v_x^* - v_{ix}^*]\Delta v_x^* \right\rangle \qquad (5.31)$$

The y- and z-components of velocity should also have the distribution (5.30); moreover, the distribution should be independent of the orientation of the reference frame that defines the directions x, y, and z. A consequence of the Maxwell distribution of velocities is equipartition of kinetic energy (2.55). That is, each Cartesian component of the molecular velocities should, on a time average, give the same kinetic energy and hence the same temperature,

$$\frac{1}{N}\left\langle \sum_i v_{ix}^{*2} \right\rangle = \frac{1}{N}\left\langle \sum_i v_{iy}^{*2} \right\rangle = \frac{1}{N}\left\langle \sum_i v_{iz}^{*2} \right\rangle = T^* \qquad (5.32)$$

Again, this equality should be satisfied regardless of the directions of x, y, and z.

3. Thermodynamic properties, such as the temperature, configurational internal energy, and pressure, should be fluctuating about stable average values. These averages should be independent of how the equilibrium state was attained. Thus, at a specified state condition, averages should be reproduced when a run is repeated from different assignments of initial positions and initial velocities. The magnitudes of the fluctuations depend on the system size, that is, on the number of atoms N; in particular, the fluctuations decrease as $N^{1/2}$. Thus, when we increase the number of atoms from 108 to 500, the magnitudes of fluctuations are roughly halved, as illustrated in Figure 5.10.

4. Property averages should be stable to small perturbations. If the state is disturbed, for example by momentarily adding then removing a small amount of heat, thermodynamic properties should recover their equilibrium values. Such small perturbations often can disrupt metastable states, which otherwise may exhibit some features of equilibrium.

To illustrate, Figure 5.11 shows the consequences of artificially increasing the instantaneous temperature in a simulation. The simulation used 108 atoms interacting via the Lennard-Jones shifted-force potential (5.12). At 5000 time-steps after equilibration, the velocities were scaled so as to increase the temperature by a factor of 1.5. This scaling was continued for 20 time-steps (0.17 psec), after which the velocities were rescaled for another 20 time-steps to regain the original value of the total energy ($E/N\varepsilon = -1.48$).

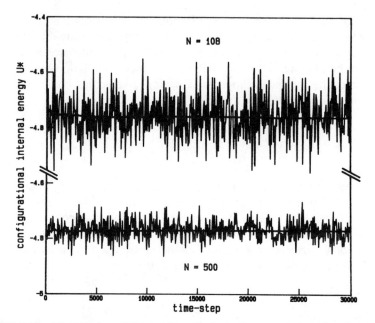

FIGURE 5.10 Magnitudes of fluctuations in instantaneous properties decrease as the number of particles is made larger. Here the configurational internal energy $\mathcal{U}^* = \mathcal{U}/N\varepsilon$ is shown from two simulations of the Lennard-Jones fluid, both at $\rho\sigma^3 = 0.7$ and $\langle kT/\varepsilon \rangle = 1.19$. The upper panel is for $N = 108$ atoms; the lower for $N = 500$ atoms. In each plot, the nearly horizontal line is the running average of the fluctuating values. Values are plotted here at intervals of 50 time-steps, where the time-step is $\Delta t^* = \Delta t/\sigma(m/\varepsilon)^{1/2} = 0.004$.

The figure indicates that both the instantaneous internal energy and the instantaneous pressure were unaffected by the perturbation.

5. If the system is divided into macroscopic parts, time averages for each property should be the same in each part. For example, imagine that we arbitrarily divide the system in half. When averaged over a finite duration, the number of molecules, temperature, total energy, and pressure in the two halves should each be the same.

None of these five conditions, individually or collectively, is sufficient to prove the existence of an equilibrium state. However, the more of these we can demonstrate, the stronger the case for claiming equilibrium, at least over the time-scale of the runs. These tests need not be applied to every run; rather, they are typically implemented in pilot runs before extensive production runs are done. However, they should be applied periodically during a series of production runs, especially after substantial changes have been made in state condition or potential model.

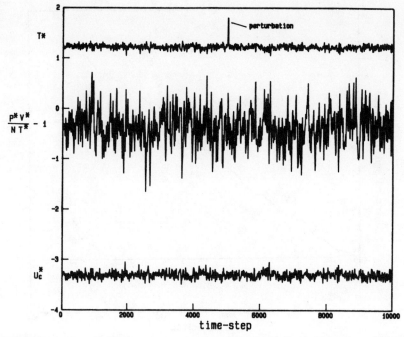

FIGURE 5.11 At equilibrium a system is stable to small perturbations. Here the instantaneous pressure and configurational internal energy are unaffected when, at time-step 5000, the temperature is momentarily increased by a factor of 1.5. These results are from a run of 108 atoms at $\rho\sigma^3 = 0.6$ and $\langle kT/\varepsilon \rangle = 1.22$. The potential used was the Lennard-Jones shifted-force model (5.12), with the shift at 2.5σ. The top line is temperature kT/ε, the middle line is pressure $P/\rho\langle T \rangle - 1$, and the bottom line is the potential energy $\mathcal{U}^* = \mathcal{U}/N\varepsilon$.

5.3.3 Fluid or Solid?

When simulations are performed at state points near the melting curve or whenever a simulation is started from a lattice structure, there is concern as to whether the system is solid or fluid throughout the simulation. We mention here three quantities that can be used to distinguish fluid from solid-like behavior. First is the positional order parameter, already discussed in Sections 5.3.1 and 3.5.1. A second indicator is the running mean-square displacement Δr^{*2}, defined by

$$\Delta r^{*2}(t) = \frac{1}{N} \sum_i \left[\mathbf{r}_i^*(t) - \mathbf{r}_i^*(0) \right]^2 \tag{5.33}$$

For a solid, $\Delta r^2(t)$ remains nearly constant, while for a fluid it increases almost linearly with time. (When $\Delta r^2(t)$ is averaged over many time origins,

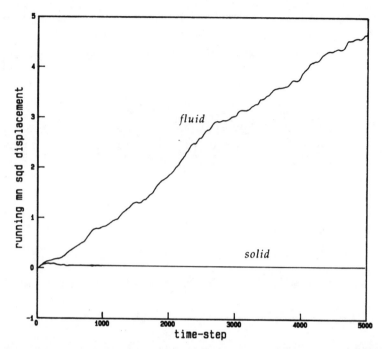

FIGURE 5.12 The running mean-square displacement (msd) can often be used to distinguish a fluid from a solid. Here the running msd, $\Delta r^{*2}(t)$ from (5.33), of Lennard-Jones atoms in a fluid state (upper line) is compared with that in a solid state (lower line). The simulations were each done using 256 atoms at $\rho\sigma^3 = 0.9$. The fluid state was at $\langle kT/\varepsilon \rangle = 1.087$, while the metastable solid was at $\langle kT/\varepsilon \rangle = 0.80$. See Figure 5.7 for the location of these states on the Lennard-Jones phase diagram.

it should be exactly linear; see Chapter 7.) In other words, a fluid has a self-diffusion coefficient that is several orders of magnitude larger than that for a solid. The fluid and solid behaviors of $\Delta r^2(t)$ are compared in Figure 5.12.

A third indicator is the radial distribution function $g(r)$. For partially crystallized fluids or partially melted solids, $g(r)$ may exhibit remnants of the underlying lattice structure; such remnants will be absent in the distribution functions of completely fluid phases. This use of the radial distribution function is discussed in Section 6.4.

5.4 PRODUCTION

Once equilibration is complete and we are satisfied that the system is at equilibrium, the production phase of the run can begin. Production introduces nothing new—the two basic activities used in equilibration are also used in production: the code repeatedly applies the finite-difference algorithm to follow the phase-space trajectory and it accumulates contributions to properties. At intervals during production, we may save to disk or tape certain quantities that can best be analyzed after the run is finished. Examples include quantities that lead to fluctuation properties (Chapter 6), to time correlation functions (Chapter 7), and to estimates of statistical uncertainties (Section 5.5).

The only remaining issue, then, is how long should the production run be? Just as for hard spheres, the purpose of extending the length of a run is to reduce statistical uncertainties (see Section 3.8.2). Recall that statistical uncertainties decrease as the root of the run length. We determine the length of run by first doing a short pilot run in which we compute properties and their statistical uncertainties. We then identify the magnitude of statistical uncertainty we are willing to tolerate, and finally, to obtain the needed duration of the production run, we scale the pilot run length by the square of the ratio of pilot uncertainty to desired uncertainty.

5.5 ASSESSING RELIABILITY OF RESULTS

At the end of a production run we are faced with the problem of assessing the reliability of the results. The problem is the same as for hard spheres, so if you managed to get this far without having read Section 3.8, you should go back now and read it. This section, like Gaul, is divided into three parts: Section 5.5.1 lists checks that can be used to test the code itself, Section 5.5.2 discusses statistical error, and Section 5.5.3 considers systematic error.

5.5.1 Checking Reliability of the Code

Here we list several tests that can be used to check a soft-sphere simulation code. The list parallels that given in Section 3.8.1 for hard spheres, and as there, the intention is merely to provide a listing for easy reference. The tests cited here are only necessary conditions; they are not sufficient to prove that a code is correct.

Equilibration

1. During equilibration, if no velocity scaling is being done, the total energy should be essentially constant.
2. By the end of equilibration:
   (a) The positional order parameter should be fluctuating about zero, with fluctuations having magnitudes $\leq \sqrt{N}/N$ (Section 3.5.1).
   (b) The H-function should be near a value consistent with the Maxwell velocity distribution (Sections 2.7.1 and 3.5.2).

Conservation Principles (Section 2.10)

3. The number of spheres in the primary cell should be constant with time.
4. Each Cartesian component of the total linear momentum should be zero.
5. The total energy should be constant with time.

Values of Properties

6. The kinetic energy should be equally partitioned among its three Cartesian components (Section 2.7.1).
7. Instantaneous property values should be stable, fluctuating about constant values.
8. Averages should be reproducible, within statistical uncertainties, when runs are repeated at the same state condition but from different initial conditions.

5.5.2 Statistical Error

Control of statistical error is linked to the duration of the production run (Section 5.4) and to the particular way we choose to sample the computed trajectory. Our first concern is to avoid serial correlations between instantaneous property values that are close to one another on the phase-space trajectory. In the discussion of statistical sampling in Section 2.8.4, we pointed out that a common strategy for avoiding serial correlations is to divide a computed phase-space trajectory into segments whose durations are longer than the relaxation time for the property of interest. For the Lennard-Jones fluid, we found (Figure 2.12) that relaxation times for simple properties, such as the pressure, are about one mean collision interval. Using time-steps of size $\Delta t^* = 0.004$, a collision interval is typically 40–50 time-steps; the exact duration depends on temperature and density. In Section 2.8.4 we suggested three sampling procedures that can help avoid serial correlations: stratified systematic sampling, stratified random sampling, and coarse graining. Any of the three can be used for soft spheres, and in fact, if we save to disk both instantaneous and running average values for properties, then we can apply both coarse graining and stratified systematic sampling and use the one that gives the smaller statistical uncertainty. Since we have discussed coarse graining in Section 3.8.2, we consider here stratified sampling.

In stratified systematic sampling for a property x, we simply draw instantaneous values $x(t)$ at a regular interval whose duration is greater than the relaxation time for serial correlations. Typically, every 50 time-steps is sufficient for simple thermodynamic properties. For example, from a production run of 50,000 time-steps, we would have a stratified systematic sample composed of $M = 1000$ values of $x(t)$ whose arithmetic average $\langle x \rangle$ approximates the time average. Further, since serial correlations have been disrupted, the sample is effectively a random one with the standard deviation of the average given by (2.96) combined with (2.77); that is,

$$\sigma_{\langle x \rangle} = \frac{1}{\sqrt{M}} \sqrt{\frac{1}{M-1} \sum_{k}^{M} [\langle x \rangle - x(k\Delta\tau)]^2} \qquad (5.34)$$

Here, $\Delta\tau$ is the interval at which the values $x(t)$ are drawn (in our example, $\Delta\tau = 50$ time-steps).

Note that the equations used to compute the average $\langle x \rangle$ and its standard deviation in stratified systematic sampling also apply in stratified random sampling; however, the sampling procedure differs in that a value of $x(t)$ is drawn randomly from each trajectory segment of duration $\Delta\tau$. Stratified random sampling incurs a practical disadvantage: we must either store all instantaneous values of $x(t)$ computed during a run and subsequently sample the segments randomly or else contrive to sample $x(t)$ randomly during the run. The latter option is a cumbersome addition to the code.

5.5.3   Systematic Error

Provided we use statistically unbiased estimates for trajectory averages (see Section 2.8.2), many kinds of systematic errors reveal themselves by causing drift in the total energy or drift in running averages for properties. Examples include blunders in coding and truncation error caused by using too large a time-step Δt in the finite-difference method. This is why, during a run, we monitor running averages for selected properties. To have a concrete example, lay a straight edge on the lines representing the running averages in Figure 5.10. You should find that once the system is stabilized (at about time step 2000), there is no consistent drift, at least within the scale of the plots. Because of the inherent instability of the phase-space trajectory, systematic errors may, as the run proceeds, grow exponentially (see Sections 2.4 and 2.5). For this reason, some errors that we overlook or neglect during the first 10,000 time-steps become glaringly important after 40,000 time-steps. We are then faced with a job of detective work in tracking down the source of the error.

Those errors that cause drift in properties we might call *friendly* systematic errors because they signal their presence. More insidious are errors that cause a constant nongrowing displacement of averages from their true values. These we might call *hidden* systematic errors. As an example, there are ways to miscode the implementation of periodic boundary conditions without affecting conservation of total energy or affecting the consistency of computed running averages: the values provided by the run appear stable and internally consistent, but nevertheless, they are wrong. Unlike friendly errors, hidden errors evade easy identification and simple analysis: a plot, such as Figure 5.10, does not offer any evidence for the presence or absence of hidden errors. Much of the problem is that hidden errors produce consistent behavior and the human psyche is tuned to accept consistent behavior as normal: we are then tempted to accept normal behavior as correct. It is difficult to offer many guidelines that will help reveal hidden errors; a few generalities are suggested here.

Checking for a Normal Distribution. Some systematic errors not only displace the computed average from the true average, but they also bias the distribution of values about their average. Thus, one test for systematic error is whether values exhibit the expected distribution. For example, simple thermodynamic properties are expected to define a Gaussian distribution about their average. An illustration using the temperature computed during a Lennard-Jones simulation is shown in Figure 2.11.

To compare a computed distribution to a known distribution, we may compute *chi squared* (see Section 3.8.3),

$$\chi^2 = \frac{1}{n_b - 3} \sum_{k=1}^{n_b} \frac{[No_k - Ne_k]^2}{Ne_k} \qquad (3.62)$$

The calculation separates the range of observed x-values into n_b bins; Ne_k is the number of x-values we expect to find in bin k, while No_k is the number actually found. When the expected distribution is a Gaussian, the values of Ne_k can be obtained from (3.63). Details for computing χ^2 are given in Section 3.8.3. Recall that if χ^2 is zero, then the observed distribution agrees perfectly with the expected distribution. Conversely, if χ^2 is large, then the observed distribution is not the expected one. We are satisfied if χ^2 is of order unity.

Thermodynamic Limit. When simulation results are intended to approximate properties for a system of macroscopic size (one that contains of the order of Avogadro's number of molecules), systematic error will be present because of the small size of the simulation system. To negate that error, we extrapolate the simulation results to the thermodynamic limit (3.64),

$$\lim_{\substack{N \to \text{large} \\ V \to \text{large}}} (\text{simulation results})_{N/V \text{ fixed}} \approx \text{macroscopic results}$$

To evaluate this limit from simulations of soft-body potentials, we need to specify values for two intensive variables. Often we prefer to use density and temperature, but in simulations of isolated systems temperature is difficult to reproduce from one run to another. It is easier to evaluate the limit at fixed density and fixed total energy per molecule ($E/N\varepsilon$). A target value for the total energy can readily be attained by scaling velocities during equilibration, as described in Section 5.2.1. If you fix density and total energy during the limiting process but wish to report the limiting state in terms of ρ and T, then the thermodynamic limit for the temperature must also be evaluated.

Just as for hard-sphere simulations, there are two possible contributors to finite size effects: (a) simple N-dependence, which is known to be expressible as a power series in $1/N$, and (b) anomalous N-dependence, which is caused by periodic boundary conditions. The relative importance of these two effects depends on the property under investigation, on state condition, on the intermolecular force law, and on the environmental variables used to fix the thermodynamic state (i.e., the ensemble).

The constant (E, ρ)-limit is illustrated in Figure 5.13. In that figure, the limit has been evaluated for the temperature and configurational internal energy of the Lennard-Jones fluid at $E/N\varepsilon = -2.980$ and $\rho\sigma^3 = 0.7$. For

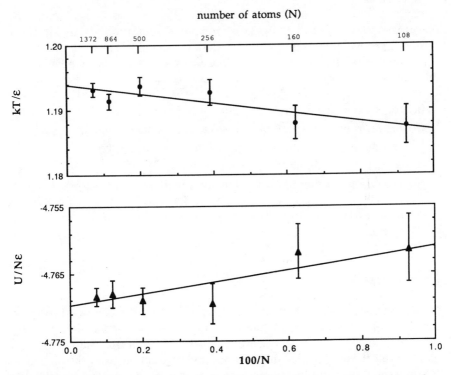

FIGURE 5.13   The thermodynamic limit evaluated for the temperature and configurational internal energy of the Lennard-Jones fluid at constant density $\rho\sigma^3 = 0.7$ and constant total energy $E/N\varepsilon = -2.980$. Error bars are statistical uncertainties based on the standard deviation of the means, shown at the 95% confidence level. Straight lines are weighted least-squares fits, with weights proportional to the statistical precision.

$N > 108$, each point in the figure was obtained from a simulation of 50,000 time-steps, with averages computed by stratified systematic sampling applied at 50-time-step intervals. For $N = 108$ the points are from two simulations whose results were linearly interpolated to the set-point energy $E/N\varepsilon = -2.980$. The error bars in the figure represent statistical uncertainties at the 95% confidence level. Note that for runs of the same length, the uncertainties decrease (within rounding) as the root of N. The straight lines in the figure are weighted least-squares fits, with the weights proportional to $1/$uncertainties. Note also that for a series of runs at the same total energy, the variations with N of T and U_c must compensate so that

$$\tfrac{3}{2}\langle kT/\varepsilon \rangle + \langle \mathcal{U}/N\varepsilon \rangle = \text{const} \tag{5.35}$$

From the fits, the limiting values are estimated to be $\langle kT/\varepsilon \rangle = 1.194 \pm 0.001$ and $\langle \mathcal{U}/N\varepsilon \rangle = -4.770 \pm 0.002$. The procedure is given in Appendix E for estimating uncertainties in least-squares parameters from uncertainties in the original data.

Independent Tests. Recall that simply repeating runs probes only statistical errors; reproducibility implies nothing about systematic errors. To probe for systematic errors beyond the suggestions just offered, we can repeat runs in different computing environments. For example, we may

(a) repeat runs using the same code on different computers,
(b) repeat runs using the same code but different compilers,
(c) repeat runs using the same algorithm but different code,
(d) repeat runs using different algorithms, or
(e) repeat runs using different simulation methods (e.g., we may compare results from molecular dynamics applied in different ensembles or compare results from molecular dynamics with those from Monte Carlo).

5.6   SUMMARY

If this book were a novel, then Chapter 5 would be the climax—everything that has come before (even Chapter 3) prepares you for Chapter 5, everything that comes after is, in a sense, a consequence. In this chapter we have emphasized those activities that demand most attention from the simulator: (a) preparing for a run, (b) monitoring a run's early stages, and (c) analyzing the results. Typically, those new to simulation will spend much effort in preparation, modest effort in monitoring, and little effort in analysis. But experienced simulators take a more balanced view. After all, analysis provides the payoff: the numbers painstakingly extracted from the analysis must justify all earlier effort.

The soft-sphere codes provided in Appendices I and L should relieve you of some of the burden of preparation, but they do not relieve you of the responsibility of knowing what you are doing. Codes should not be used as black boxes: they should be studied, exercised, tinkered with. A most instructive tinkering is akin to Exercise 5.34—how many ways can you find to introduce hidden errors into a code? If you can achieve the insight that hidden errors are easily introduced and easily overlooked, and simultaneously if you feel challenged by the possibility that all hidden errors can be eliminated, then you are on your way to becoming a successful simulator.

REFERENCES

[1] J. E. Lennard-Jones, "The Determination of Molecular Fields. I. From the Variation of the Viscosity of a Gas with Temperature," *Proc. Roy. Soc. (Lond.)*, **106A**, 441 (1924); "The Determination of Molecular Fields. II. From the Equation of State of a Gas," *Proc. Roy. Soc. (Lond.)*, **106A**, 463 (1924).

[2] F. London, "Properties and Applications of Molecular Forces," *Zeit. Physik. Chem. B*, **11**, 222 (1930).

[3] J. J. Nicolas, K. E. Gubbins, W. B. Streett, and D. J. Tildesley, "Equation of State for the Lennard-Jones Fluid," *Mol. Phys.*, **37**, 1429 (1979).

[4] L. Verlet, "Computer 'Experiments' on Classical Fluids. I. Thermodynamical Properties of Lennard-Jones Molecules," *Phys. Rev.*, **159**, 98 (1967).

[5] S. G. Lambrakos, J. P. Boris, I. Chandrasekhar, and B. Gaber, "A Vectorized Near-Neighbors Algorithm of Order N for Molecular Dynamics Simulations," *Ann. NY Acad. Sci.*, **482**, 85 (1986).

[6] S. Gupta, "Vectorization of Molecular Dynamics Simulation for Fluids of Nonspherical Molecules," *Comp. Phys. Commun.*, **48**, 197 (1988).

[7] A. A. Chialvo and P. G. Debenedetti, "On the Use of the Verlet Neighbor List in Molecular Dynamics," *Comp. Phys. Commun.*, **60**, 215 (1990).

[8] A. A. Chialvo and P. G. Debenedetti, "On the Performance of an Automated Verlet Neighbor List Algorithm for Large Systems on a Vector Processor," *Comp. Phys. Commun.*, **64** 15–18 (1991).

[9] J. Boris, "A Vectorized Near Neighbors Algorithm of Order N Using a Monotonic Logical Grid," *J. Comput. Phys.*, **66**, 1–20 (1986).

[10] J. P. Hansen and I. R. McDonald, *Theory of Simple Liquids*, 1st ed., Academic, London, 1976, p. 351.

[11] S. K. Ma, *Statistical Mechanics*, World Scientific, Philadelphia, PA, 1985, pp. 2–4.

EXERCISES

5.1 Using only what you now know and without referring to any resource, estimate the diameter (in ångstroms) of one water molecule, assuming water molecules are spheres.

5.2 Compute the pair separation at which the Lennard-Jones potential is a minimum. Express your answer in units of σ.

5.3 Consider two isolated argon atoms whose intermolecular potential energy function approximately obeys the Lennard-Jones model (5.4) with $\varepsilon/k = 120$ K, $\sigma = 3.4$ Å, and k is Boltzmann's constant. The two atoms are initially separated by the distance r_m: the point at which the intermolecular force (*not* the potential) is zero.

  (a) Compute the work required to increase the separation from r_m to ∞.

  (b) Compute the work required to decrease the separation from r_m to 0.9σ.

  (c) Why do the answers to both (a) and (b) have the same sign?

5.4 Set $m = 9$ and $n = 18$ in the general Lennard-Jones potential model (5.2) and then evaluate the functional forms for the potential and force. Graphically compare your results with those given in Figure 5.2 for the (12, 6) model.

5.5 To overcome small impulses caused by using a truncated potential, some early simulators tried shifting the potential rather than the force. That is, they used

$$u(r) = \begin{cases} u_{\mathrm{LJ}}(r) - u_{\mathrm{LJ}}(r_c) & r \le r_c \\ 0 & r > r_c \end{cases}$$

where u_{LJ} is the Lennard-Jones model (5.4) and $u_{\mathrm{LJ}}(r_c)$ is the potential evaluated at the truncation point. Show that the force resulting from this shifted potential still contains a discontinuity at r_c.

5.6 Starting from the fact that the total energy is conserved in an isolated system, prove that the force on one atom is equal to the negative gradient of its potential energy function.

5.7 The second virial coefficient is related to the intermolecular pair ☒ potential by

$$B^* = \frac{B}{2\pi N_A \sigma^3} = \int_0^\infty \left[1 - \exp\left(-\frac{u^*(r^*)}{T^*} \right) \right] r^{*2} \, dr^*$$

where N_A is Avogadro's number, $r^* = r/\sigma$, $u^* = u/\varepsilon$, and $T^* = kT/\varepsilon$. Write a program that computes B^* for the Lennard-Jones potential. Use the program to test how well the values of the Lennard-Jones parameters for argon ($\varepsilon/k = 120$ K, $\sigma = 3.405$ Å) reproduce experimental data for B. For argon, selected values of the second virial coefficient (in cm^3/gmol) are

$T(K)$	85	100	150	200	300	500	700	1000
B	-251	-183.5	-86.2	-47.4	-15.5	$+7.0$	$+15$	$+22$

These values are from J. H. Dymond and E. B. Smith, *The Virial Coefficients of Pure Gases and Mixtures*, Oxford University Press, Oxford, 1980.

5.8 **(a)** Use Newton's second law to show that in terms of the potential parameters ε and σ, the unit of time is $\sigma(m/\varepsilon)^{1/2}$, where m is the mass of one atom.

(b) Using the Lennard-Jones values of ε and σ for argon, convert the duration of one integration time-step from $\Delta t^* = 0.004$ to picoseconds.

5.9 In a preliminary test run, 108 Lennard-Jones atoms were followed over 1000 time-steps. The configurational internal energy and its statistical uncertainty were computed to be $\langle \mathcal{U}/N\varepsilon \rangle = -4.76 \pm 0.02$. The target for the production run is to reduce the uncertainty in $\langle \mathcal{U}/N\varepsilon \rangle$ to 0.001 using 500 atoms. If the same sampling procedure is to be used in production as was used in the test, how many time-steps should be computed in the production run?

5.10 From a molecular dynamics run on the Lennard-Jones fluid in an isolated system, a colleague reports an average temperature $\langle kT/\varepsilon \rangle = 1.24 \pm 0.03$ and an average configurational internal energy $\langle \mathcal{U}/N\varepsilon \rangle = -4.86 \pm 0.01$. The quoted uncertainties are said to be wholly statistical. Do you believe it?

5.11 Do a simulation of the Lennard-Jones substance at $\rho\sigma^3 = 0.6$ and $\boxtimes$ $kT/\varepsilon = 1.4$ using 256 atoms and 10,000 time-steps. Save to disk the instantaneous values of temperature, internal energy, and pressure. Now study how averages and their statistical uncertainties are affected by the sampling method used. In particular, compare (i) stratified systematic sampling, (ii) stratified random sampling, and (iii) coarse graining.

5.12 Write a program that reads a set of instantaneous property values from $\boxtimes$ disk, computes the distribution about their average, and compares their distribution to a Gaussian via χ^2(3.62). Apply your program to simulation results for the temperature, internal energy, and pressure of the Lennard-Jones substance at $\rho\sigma^3 = 0.6$ and $kT/\varepsilon = 1.4$. Study how χ^2 is affected by the number of values used and the interval at which they are sampled from the simulation. Explore the effect on χ^2 of changing the number of atoms, the density, and the temperature.

5.13   A simulation is performed in a cube of edge L using periodic boundary conditions and a pair potential truncated at $r_c < \frac{1}{2}L$. Consider the computation of the force exerted on atom 1 by atom 2. Prove that there is at most *one* image of atom 2 that lies within r_c of atom 1 and that could possibly contribute to the force on atom 1.

5.14   Devise a scheme for storing the neighbor list in one double-subscripted array rather than in the two single-subscripted arrays described in Section 5.1.4.

5.15   List at least five errors that could reside in the Lennard-Jones molecular dynamics code, yet the code would still display conservation of total energy.

5.16   One possible and subtle source of systematic error is error in the intrinsic functions that are provided by high-level programming languages: SQRT, EXP, and ABS in Fortran, for example. Devise and implement a program that exercises all the intrinsic functions used in the Lennard-Jones program given in Appendix L. Your program should test (randomly, perhaps?) the full range of argument values that might be encountered during a simulation.

5.17   (a) Verify that the scaling (5.22) applied to initially assigned velocities does indeed provide the desired value of the total energy.

    (b) Verify that the adjustment of velocities (5.28) produces zero linear momentum.

5.18   Is it possible to apply the velocity scaling (5.22) so that a Lennard-Jones simulation is performed at zero total energy? If so, what is the physical significance, if any, of having $E = 0$?

5.19   Pick four or five state conditions (see Figure 5.7 for ranges of densities and temperatures) and do simulations of the Lennard-Jones fluid at each using, say, 2000 time-steps per run. Compare your results for the internal energy and pressure with those computed from the Nicolas et al. equation of state [3].

5.20   (a) At $\rho\sigma^3 = 0.6$ and $kT/\varepsilon = 1.4$, use the code in Appendix L to do a series of runs for increasing numbers of atoms ($N = 32, 108, 256, 500, 864$). Study how changing the number of atoms affects the (i) speed of execution, (ii) energy conservation, and (iii) averages for temperature, pressure, and internal energy. Be quantitative in reporting your conclusions.

    (b) At the same state condition as in (a), with 108 atoms, do a series of runs for various values of the integration time-step, from $\Delta t^* = 0.0005$ to $\Delta t^* = 0.05$. Can you identify the stability bound for the predictor–corrector algorithm? How is energy conservation affected by changing the step size?

5.21 Study the accumulation of round-off error by doing a sequence of
☒ simulations in single and double precision. Study how property values
and energy conservation differ and how those differences are affected
by changes in time-step and state condition.

5.22 How do you expect a simulation would behave if it were started in the
unstable region under the vapor–liquid dome, say, at $\rho\sigma^3 = 0.4$ and
$kT/\varepsilon = 1.1$? Be detailed in your description. Now try it.

5.23 How would values of the reduced time-step, temperature, internal
energy, and pressure be affected if, in the Lennard-Jones code, the
masses of the atoms were all changed from 1 to 10?

5.24 Start a Lennard-Jones simulation with 108 atoms at $\rho\sigma^3 = 0.95$ and
☒ $kT/\varepsilon = 1$ and monitor how the H-function and order parameter
behave. Explain.

5.25 Perform a Lennard-Jones simulation in the metastable region of the
☒ phase diagram, say at $\rho\sigma^3 = 0.7$, $kT/\varepsilon = 0.9$. Monitor how the H-
function, order parameter, and other properties behave. Do you expect
the behavior would be affected by increasing the number of atoms?

5.26 Do a simulation of the Lennard-Jones substance at $\rho\sigma^3 = 0.6$ and
☒ $kT/\varepsilon = 1.4$ using 108 atoms and 5000 time-steps.

    (a) At 1000 time-step intervals print the velocity distribution and
compare it with the Maxwell distribution at the same temperature.
About how many time-steps are needed for the Maxwell distribution to fully develop?

    (b) Test equipartition of kinetic energy (5.32) by modifying the code to
accumulate separately the average temperatures in the x-, y-, and
z-directions. Would your results change if you reoriented the
coordinate axes?

5.27 Use the Lennard-Jones code to perform the stability test described in
☒ Section 5.3.2 in connection with Figure 5.11. Note that to obtain an
unambiguous test, the system must be brought back to the same state
(density and total energy) as it had before the disturbance was introduced. This typically requires some finesse. Can you contrive other
disturbances, in addition to temperature, to introduce?

5.28 Contrive diagnostics that would test, during a simulation of the
Lennard-Jones substance, each of the following:

    (a) Conservation of mass in the primary cell.

    (b) Force and total energy are properly computed for all N atoms.

    (c) Expression used to evaluate the force is truly the negative gradient
of the potential energy function.

    (d) Periodic boundary conditions are properly implemented.

(e) Neighbor list is properly constructed, used, and up-dated.

(f) All atoms are actually changing positions at each time-step.

5.29 A simulation of the Lennard-Jones substance is performed using the code MDSS from Appendix L. The run used 256 atoms, the potential truncated at $r_c = 2.5\sigma$, a time-step $\Delta t = 0.004 \ \sigma(m/\varepsilon)^{1/2}$, and density $\rho\sigma^3 = 0.7$. During production, the following values for running average temperature, configurational internal energy, and pressure were generated:

Time-step	$\langle kT/\varepsilon \rangle$	$\langle \mathcal{U}/N\varepsilon \rangle$	$\langle P\sigma^3/\varepsilon \rangle$
50,000	1.997	-4.298	3.038
100,000	2.009	-4.292	3.067
150,000	2.022	-4.286	3.100
200,000	2.035	-4.279	3.136
250,000	2.049	-4.272	3.173

Are you satisfied the program was running properly? If not, list possible problems and diagnostic tests for each that would enable you to identify the problem.

5.30 (a) Test the one-dimensional Lennard-Jones code in Appendix I for
☒   systematic errors by executing the code, under the same conditions (number of atoms, time-step, density, temperature, and seed to random-number generator), on different compilers and then on different machines.

(b) Repeat for the three-dimensional code given in Appendix L.

5.31 Contrive pathological initial conditions, such as those described in
☒   Section 3.5 for hard spheres, and apply them to the Lennard-Jones substance. Monitor how the order parameter and H-function behave and determine whether equilibrium values for properties are affected by initial conditions.

5.32 A molecular dynamics simulation code for the Lennard-Jones substance is being tested using 108 atoms at fixed N, V, and E. In test runs the code exhibited the following behavior (one item per run). List possible causes for each. If a particular symptom has more than one possible cause, what would you do to isolate the problem?

(a) The temperature and total energy increase exponentially during equilibration, causing a numerical overflow before production starts.

(b) The run appears normal during equilibration (using velocity scaling to attain a set-point temperature), but at the end of equilibration (when velocity scaling terminates) the temperature and total energy increase exponentially.

(c) During equilibration, the *H*-function decreases to a small steady value, but the positional order parameter stabilizes near $\lambda = 0.8$.

(d) During equilibration, the order parameter decreases to and fluctuates about zero but the *H*-function does not decrease.

5.33 Devise and implement procedures that test separately the principal subroutines used in the Lennard-Jones code MDSS given in Appendix L.

5.34 Execute the Lennard-Jones code MDSS in Appendix L for 1000 time-steps and 108 atoms to produce a set of baseline data with which to compare results from the following. Then introduce the following "errors" into the simulation code (one error per run except where instructed otherwise), execute for 1000 time-steps, and compare with the baseline data. After each run describe the nature of the error introduced and explain its consequences.

(a) In subroutine SSPDCT, line 13, replace X5 with Y5.

(b) In subroutine SSPDCT, line 13, replace X3 with Y3.

(c) In subroutine SSINLP, line 31, replace 4 with 2.

(d) In subroutine SSINLP, make the change (c) and also, in line 32, replace 48 with 24.

(e) In subroutine SSINLP, line 19, place a C in column 1.

(f) In subroutine PBC, line 16, place a C in column 1.

(g) In subroutine SSCORR, line 21, replace ALFA3 with ALFA1.

(h) Contrive two more errors of your own and test their consequences.

5.35 What problems arise in performing molecular dynamics on low-density gases?

5.36 How would the code MDSS in Appendix L need to be modified to simulate binary mixtures of Lennard-Jones atoms?

5.37 The intermolecular potential functions considered in this chapter are limited to pairwise additive forms. In spherically symmetric molecules, the first nonadditive interaction is the Axilrod–Teller triple dipole potential [B. M. Axilrod and E. Teller, *J. Chem. Phys.*, **11**, 299 (1943)]. For atoms 1, 2, and 3 forming a triangle of sides r_{12}, r_{13}, r_{23} and interior angles $\theta_{12}, \theta_{13}, \theta_{23}$, the Axilrod–Teller model is

$$u_{AT} = \frac{\alpha(3\cos\theta_{12}\cos\theta_{13}\cos\theta_{23} + 1)}{(r_{12}r_{13}r_{23})^3}$$

where α is the triple dipole constant. How would the Lennard-Jones code in Appendix L have to be modified to include the Axilrod–Teller interaction? How would the speed of execution of the simulation be affected by including this three-body potential?

6

STATIC PROPERTIES

In previous chapters we have focused attention on the simulation itself, that is, on computing the phase-space trajectory for a system of model molecules. In this chapter we turn our attention to the analysis of computed trajectories. The analysis involves evaluating macroscopic properties—properties that pertain, not to individual molecules, but to an entire system of molecules.[†] The properties we consider in this chapter are thermodynamic quantities and static structure.

We divide thermodynamic properties into three classes: (1) simple functions of the Hamiltonian, (2) the thermodynamic response functions, which are derivatives of the simple functions, and (3) the entropy and free energies. Generally, as we move from class 1 to class 3, these properties become more difficult to evaluate accurately. We discuss the simple functions in Section 6.1, the response functions in Section 6.2, and the entropic properties in Section 6.3.

Static structure of matter can be measured by the radial distribution function $g(r)$, which describes the spatial organization of molecules about a central molecule. The function $g(r)$ plays a central role in the pair distribution function theory of dense fluids and provides a signature for identifying the lattice structure of crystalline solids. The radial distribution function can be simply evaluated from simulation data using the method developed in Section 6.4.

[†]W. Thomson, Lecture XVII, p. 167: "As to the physical properties of matter, which are more properly subjects of interest and the subjects that we occupy ourselves with...."

Throughout this chapter the following notation is used. For a system of N particles, the phase-space trajectory is represented by the set of numbers

$$\left\{\left[\mathbf{r}^N(k\,\Delta t),\mathbf{p}^N(k\,\Delta t)\right],\, k=1,\ldots,M\right\}$$

Here $\mathbf{r}^N$ is the set of N position vectors and $\mathbf{p}^N$ is the set of N momentum vectors. The trajectory has been obtained from a simulation performed over M discrete times using the time-step Δt. From this trajectory the time average $\langle A\rangle$ of some function of the trajectory can be estimated by the sum

$$\langle A\rangle = \frac{1}{M}\sum_{k=1}^{M} A\left[\mathbf{r}^N(k\,\Delta t),\mathbf{p}^N(k\,\Delta t)\right] \tag{6.1}$$

As discussed in Section 2.6, the finite interval average (6.1) is an approximation to the infinite time average (2.41),

$$\langle A\rangle = \lim_{t\to\infty}\frac{1}{t}\int_{t_0}^{t_0+t} A(\mathbf{r}^N,\mathbf{p}^N)\,d\tau$$

Dimensionless quantities are identified with an asterisk, as in Table 5.2.

6.1 SIMPLE THERMODYNAMIC PROPERTIES

The class of simple thermodynamic functions contains those properties that are obtained from averages of either the Hamiltonian or its spatial and momentum derivatives. The class contains several properties; however, in this section we discuss only temperature, internal energy, pressure, and mean-square force on a molecule. Other members of the class are of less general importance and are not discussed here; an example is the surface tension in Fowler's model of the vapor–liquid interface [1].

6.1.1 Internal Energy and Temperature

For an isolated system the total internal energy E is just the Hamiltonian $\mathscr{H}$,

$$E = \mathscr{H}(\mathbf{r}^N,\mathbf{p}^N) = \text{const} \tag{6.2}$$

which divides into a kinetic part E_k and a configurational part $\mathscr{U}$,

$$E = E_k + \mathscr{U} \tag{6.3}$$

As shown in Appendix C, the average kinetic energy is proportional to the absolute temperature (2.43),

$$\langle E_k \rangle = \tfrac{3}{2} NkT = \frac{1}{2mM} \sum_{k=1}^{M} \sum_{i=1}^{N} \mathbf{p}_i(k \Delta t) \cdot \mathbf{p}_i(k \Delta t)$$

Here m is the mass of one particle. For soft spheres in a periodic system, the configurational internal energy is the average of the pair potential function $u(r)$,

$$U_c = \langle \mathcal{U} \rangle = \frac{1}{M} \sum_{k=1}^{M} \sum_{\alpha} \sum_{i<j} u\big[|\mathbf{r}_{ij}(k \Delta t) - \alpha L| \big] \qquad (6.4)$$

Recall u is a two-body potential, such as the Lennard-Jones model (5.4), $r_{ij} = |\mathbf{r}_i - \mathbf{r}_j|$ is the scalar distance between the centers of atoms i and j, L is the length of one edge of the primary cell, and α is the cell translation vector of Section 2.9.

Long-Range Corrections. Often in soft-sphere simulations we use a truncated pair potential (see Section 5.1.1), but we want property values for the full untruncated potential. In those situations, an expression such as (6.4) is incomplete because the simulation ignores interactions between atoms separated by distances greater than the cutoff distance r_c. Therefore, simulation results for $\langle \mathcal{U} \rangle$ must be corrected by adding an estimate for the long-range contribution U_{LR}. The estimation is done in the following way.

In the classical statistical mechanics of atomic, pairwise additive substances, the configurational internal energy U_c can be written [2, 3] as an integral over the pair potential $u(r)$ weighted with the radial distribution function $g(r)$,

$$\frac{U_c}{N} = 2\pi\rho \int_0^\infty u(r) g(r) r^2 \, dr \qquad (6.5)$$

where $\rho = N/V$ is the number density. The integral in (6.5) can be divided as

$$\frac{U_c}{N} = 2\pi\rho \int_0^{r_c} u(r) g(r) r^2 \, dr + 2\pi\rho \int_{r_c}^\infty u(r) g(r) r^2 \, dr \qquad (6.6)$$

The first term on the rhs of (6.6) is just $\langle \mathcal{U} \rangle / N$, which is evaluated from a molecular dynamics simulation via (6.4). The second term in (6.6) is the long-range correction. Therefore, the complete expression for the configura-

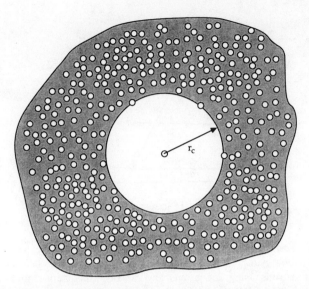

FIGURE 6.1 Simulation provides only a portion of most thermodynamic properties. During a simulation we obtain contributions to properties that arise from interactions between a central atom and others lying within a radius r_c. Contributions from beyond r_c are typically estimated using a mean-field approximation: we assume outlying atoms are uniformly distributed about the central atom and therefore present a small, constant contribution to thermodynamic properties.

tional internal energy is

$$\frac{U_c}{N} = \frac{\langle \mathcal{U} \rangle}{N} + U_{\mathrm{LR}} \tag{6.7}$$

In the absence of data for $g(r)$ for $r > r_c$, the long-range correction is typically estimated by

$$U_{\mathrm{LR}} = 2\pi\rho \int_{r_c}^{\infty} u(r) r^2 \, dr \tag{6.8}$$

That is, the substance is assumed to be uniform $[g(r) = 1]$ beyond the potential cutoff distance r_c. This amounts to a mean-field approximation for the long-range portion of the internal energy; see Figure 6.1. For the Lennard-Jones potential the integral in (6.8) can be performed analytically, resulting in

$$U_{\mathrm{LR}}^* = \frac{8\pi\rho^*}{3(r_c^*)^3} \left(\frac{1}{3(r_c^*)^6} - 1 \right) \approx \frac{-8\pi\rho^*}{3(r_c^*)^3} \tag{6.9}$$

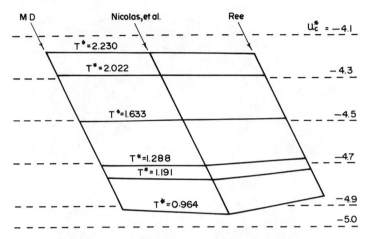

FIGURE 6.2   Tests of Ree and Nicolas et al. equations of state for the configurational internal energy U_c^* of the Lennard-Jones fluid at $\rho^* = 0.7$. The simulation results are from runs using 256 atoms and 500,000 time-steps and have statistical uncertainties of $\Delta U^* \leq 0.002$ at the 95% confidence level.

Thus, at a fixed number density, the long-range correction is merely a constant that is added to a molecular dynamics result for $\langle \mathscr{U}^* \rangle$.

If the shifted-force potential (5.12) is used in a simulation and if results are desired for the original unshifted potential, then corrections for the shift are needed. However, since the shifted-force potential is useful for testing code, not for making production runs, the expressions for the corrections are not given here (they may be found elsewhere [4]).

Sample Results. Figure 6.2 shows molecular dynamics results for U_c^* computed for the Lennard-Jones fluid at a density $\rho^* = 0.7$. The simulations were done using 256 atoms and the pair potential truncated at $r_c^* = 2.5$. Each run was started from an fcc lattice and each was continued for up to 500,000 time-steps. Runs at low temperatures used a time-step $\Delta t^* = 0.004$, while at high temperatures $\Delta t^* = 0.003$. Averages for properties were computed by stratified systematic sampling at intervals of 50 time-steps, and statistical uncertainties in U_c^* were computed to be $\Delta U^* \leq 0.002$ at the 95% confidence level. Over the temperature range $0.96 < T^* < 2.23$, the values of U_c^* at $\rho^* = 0.7$ are accurately represented by the cubic

$$U_c^* = -5.6597 + 0.8893T^* - 0.13588T^{*2} + 0.015593T^{*3} \qquad (6.10)$$

A quadratic in temperature will fit the U-data equally well, but this cubic provides more reliable values for the residual isometric heat capacity (see Section 6.2.1).

Figure 6.2 uses a two-way plot (see Appendix D) to compare these simulation results with U_c-values provided by empirical equations of state given by Nicolas et al. [4] and by Ree [5]. Both empirical equations are based on fits to collections of Lennard-Jones simulation data. At the density and temperatures studied here, both correlations reproduce U_c within 1% of nearly all our simulation values, though the Ree equation consistently provides less negative values of U_c than found in the simulation. The discrepancies are largest at the lowest temperatures. A careful inspection of Figure 6.2 leads us to anticipate that both correlations will reliably estimate the residual heat capacity $[C_v^R = (\partial U_c / \partial T)_v]$ at high temperatures, but at low temperatures the Ree equation will underestimate C_v^R, while the Nicolas et al. equation will overestimate C_v^R.

6.1.2   Pressure

The pressure P is related to molecular quantities through the virial equation of state, which is derived in Appendix B,

$$\frac{P}{\rho kT} = 1 - \frac{1}{3NkT} \left\langle \sum_{i<j} \sum r_{ij} \frac{du(r_{ij})}{dr_{ij}} \right\rangle \tag{6.11}$$

The first term on the rhs is the ideal-gas contribution, and the second term accounts for intermolecular forces, assuming a pairwise additive potential $u(r)$. In periodic systems, (6.11) can be evaluated from a phase space trajectory via

$$\frac{P_{md}}{\rho kT} = 1 - \frac{1}{3MNkT} \sum_{k=1}^{M} \sum_{\alpha} \sum_{i<j} \left| \mathbf{r}_{ij}(k\,\Delta t) - \boldsymbol{\alpha} L \right| \frac{du\left[\left| \mathbf{r}_{ij}(k\,\Delta t) - \boldsymbol{\alpha} L \right| \right]}{dr_{ij}} \tag{6.12}$$

Here L is the length of one edge of the primary cell and $\boldsymbol{\alpha}$ is the cell translation vector discussed in Section 2.9.

Just as for the internal energy, when a truncated potential is used, a long-range correction must be added to the molecular dynamics result for the virial. Thus the full expression for the pressure is

$$P = P_{md} + P_{LR} \tag{6.13}$$

The long-range correction can be estimated by a procedure analogous to that described above for the internal energy. The result is

$$\frac{P_{LR}}{\rho kT} = \frac{-2\pi\rho}{3kT} \int_{r_c}^{\infty} r \frac{du(r)}{dr} g(r) r^2 \, dr \tag{6.14}$$

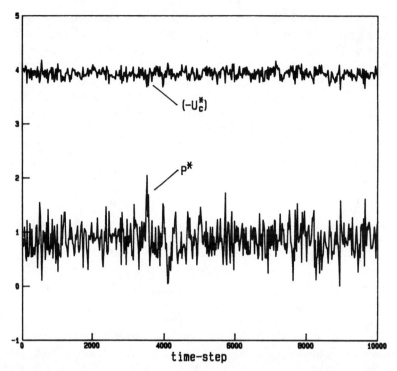

FIGURE 6.3 Fluctuations in the instantaneous pressure $P^* = P\sigma^3/\varepsilon$ are usually larger than those in the instantaneous configurational internal energy $\mathscr{U}^* = \mathscr{U}/N\varepsilon$. (Note that $-\mathscr{U}^*$ is plotted here.) These results are from a molecular dynamics simulation of 108 Lennard-Jones atoms at $\rho^* = 0.6$, $\langle T^* \rangle = 1.542$. Values are plotted here at intervals of 20 time-steps; the time-step was $\Delta t^* = 0.004$.

Assuming $g(r) = 1$ and applying (6.14) to the Lennard-Jones model, we find

$$\frac{P^*_{\text{LR}}}{\rho^* T^*} = \frac{-16\pi\rho^*}{3T^*(r^*_c)^3}\left(1 - \frac{2}{3(r^*_c)^6}\right) \approx \frac{-16\pi\rho^*}{3T^*(r^*_c)^3} \qquad (6.15)$$

which is again merely a constant to be added to the simulation result.

For soft spheres at liquid densities, the pressure exhibits larger fluctuations than does the internal energy; an example is provided in Figure 6.3. These large fluctuations occur because $r\,du/dr$ usually changes more quickly with r than does the potential $u(r)$. That is, for most pair separations

$$\left|\frac{d}{dr}\left(r\frac{du}{dr}\right)\right| > \left|\frac{du}{dr}\right| \qquad (6.16)$$

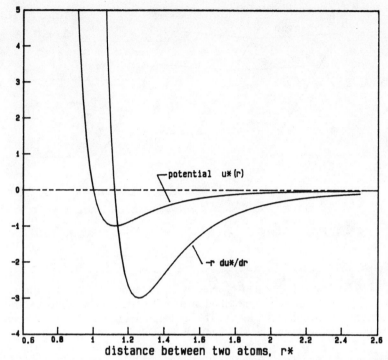

FIGURE 6.4 For the Lennard-Jones (12, 6) potential the magnitude of the slope of $\{-r^* \, du^*/dr^*\}$ is nearly everywhere greater than the magnitude of the slope of $u^*(r^*)$; therefore, fluctuations in the instantaneous pressure are larger than those in the internal energy, as shown in Figure 6.3.

For the Lennard-Jones (12, 6) potential, the inequality in (6.16) is violated only for pair separations near the minimum in $(-r \, du/dr)$; specifically, for

$$1.2445 \le r^* \le 1.2801 \tag{6.17}$$

Otherwise, Figure 6.4 shows that for any small change in r, the magnitude of the slope of $r \, du/dr$ is nearly everywhere greater than that of $u(r)$.

Molecular dynamics results for the residual pressure,

$$P_{res} = P - \rho kT \tag{6.18}$$

are shown in Figure 6.5 for the Lennard-Jones fluid at $\rho^* = 0.7$. These results are from the same simulations that gave the U_c results in Figure 6.2. The statistical uncertainties are $\Delta P^*_{res} \le 0.006$ at the 95% confidence level. Over the temperature range $0.96 < T^* < 2.23$, values of the residual pressure are

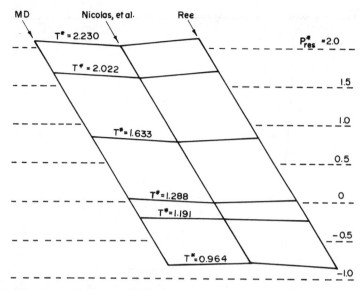

FIGURE 6.5 Tests of Ree and Nicolas et al. equations of state for the residual pressure P_{res}^* of the Lennard-Jones fluid at $\rho^* = 0.7$. Simulation results are from the runs described in Section 6.1.1 for U_c. The statistical uncertainties in these simulation results are $\Delta P_{res}^* \leq 0.006$ at the 95% confidence level.

accurately represented by

$$P_{res}^* = -4.1687 + 4.2808T^* - 0.9635T^{*2} + 0.1360T^{*3} \qquad (6.19)$$

Note that for the Lennard-Jones substance P_{res} and U_c are related by

$$P^* - \rho^*T^* = 2\rho^*\left[U_c^* + 4\langle r^{*-12}\rangle\right] \qquad (6.20)$$

where $\langle r^{*-12}\rangle$ is the time average per atom of the repulsive part of the potential.

Simulation results for the pressure of the Lennard-Jones substance have been fit by Nicolas et al. [4] and by Ree [5] to empirical equations in density and temperature. The 33-parameter Nicolas et al. equation applies to states in the region $\{0 \leq \rho^* \leq 1.2\}$ and $\{0.5 \leq T^* \leq 6\}$, while the 15-parameter Ree equation is restricted to states in $\{0.05 \leq \rho^* \leq 0.96\}$ and $\{0.76 \leq T^* \leq 2.698\}$. A third empirical equation has been devised by Adachi et al. [6]. With these equations any thermodynamic property may be computed by invoking the appropriate thermodynamic relation.

At the density shown in Figure 6.5 ($\rho^* = 0.7$), tests of the empirical equations indicate that the absolute deviations in the residual pressure are no more than 0.08 for the Nicolas et al. equation [4] and no more than 0.04

for the Ree equation [5]. Except at the lowest temperature, the Nicolas et al. equation consistently underestimates P_{res}. In contrast, at high temperatures the Ree equation overestimates P_{res} but at low temperatures it underestimates P_{res}.

6.1.3 Mean-Square Force

The mean-square force $\langle F^2 \rangle$ on an atom gives information about the shape of the repulsive part of the intermolecular pair potential. It is both the zero-time limit of the force autocorrelation function and the second frequency moment of the Fourier transform of the velocity autocorrelation function [7]. Experimentally, $\langle F^2 \rangle$ can be obtained from vapor–liquid isotopic separation factors [8–10].

Two methods are available for analyzing simulation data for $\langle F^2 \rangle$. The direct method is to form the time-average square of the force on, say, atom 1,

$$\langle F_1^2 \rangle = \left\langle \sum_{j \neq 1} \left(\nabla u(r_{1j}) \right)^2 \right\rangle \tag{6.21}$$

The symbol ∇ is the gradient operator, $\nabla \equiv \partial / \partial \mathbf{r}$. The statistical precision of the result can be improved by averaging over $\frac{1}{2}N$ unique atomic origins rather than just atom 1. Then interpreting the time average in (6.21) in the usual way, the expression for the mean-square force in periodic systems becomes

$$\langle F^2 \rangle = \frac{2}{MN} \sum_{k=1}^{M} \sum_{\alpha} \sum\sum_{i<j} \left(\nabla u \left[\left| \mathbf{r}_{ij}(k\,\Delta t) - \alpha L \right| \right] \right)^2 \tag{6.22}$$

The second method for evaluating $\langle F^2 \rangle$ is obtained from the identity [7, 11]

$$\left\langle W(\mathbf{r}^N) \nabla \mathcal{U}(\mathbf{r}^N) \right\rangle = kT \left\langle \nabla W(\mathbf{r}^N) \right\rangle \tag{6.23}$$

where W is any regular function of the atomic positions and $\mathcal{U}$ is the full potential energy function. For the mean-square force, we use $W = \nabla \mathcal{U}$, so

$$\langle F^2 \rangle = \left\langle \left(\nabla \mathcal{U}(\mathbf{r}^N) \right)^2 \right\rangle = kT \left\langle \nabla^2 \mathcal{U}(\mathbf{r}^N) \right\rangle \tag{6.24}$$

where $\nabla^2 \equiv \partial^2 / \partial \mathbf{r}^2$ is the Laplacian operator.[†] The working expression for

---

[†]W. Thomson, Lecture X, p. 98: "I took the liberty of asking Prof. Ball two days ago whether he had a name for this symbol ∇^2; and he has mentioned to me *nabla*, a humorous suggestion of Maxwell's. It is the name of an Egyptian harp which was of that shape. I do not know that it is a bad name for it. Laplacian I do not like for several reasons both historical and phonetical." Lecture XIII, p. 128: "I have another name from Prof. Ball for ∇, which is *atled*, or *delta* spelt backwards."

$\langle F^2 \rangle$ is now

$$\langle F^{*2} \rangle = \frac{2T^*}{MN} \sum_{k=1}^{M} \sum_{\alpha} \sum\sum_{i<j} \left(\nabla^2 u^* \left[\left| \mathbf{r}_{ij}^*(k\,\Delta t) - \alpha L^* \right| \right] \right) \qquad (6.25)$$

Even for the Lennard-Jones potential the Laplacian is tedious to evaluate, but after some effort we find [12]

$$\nabla^2 u^* = 24(22r^{*-14} - 5r^{*-8}) \qquad (6.26)$$

In using (6.22) we must not forget the long-range correction that applies to the simulation results for truncated potentials. If (6.25) is used, the long-range correction may be small and safely neglected. Thus for the Lennard-Jones model, the correction to (6.25) varies as r_c^{-5}, and for $r_c \geq 2.5\sigma$, the correction is negligible.

6.2   THERMODYNAMIC RESPONSE FUNCTIONS

The response functions reveal how simple thermodynamic quantities respond to changes in measurables, usually either the pressure or temperature. Hence they are derivative quantities. To cite a particular example, the constant-volume heat capacity measures how the internal energy responds to an isometric change in temperature,

$$C_v = \left(\frac{\partial E}{\partial T} \right)_V \qquad (6.27)$$

Two general methods are available for evaluating such properties. In the first method we use several simulations to determine values of the simple quantity (E in the case of C_v) as a function of the independent variable (T here). We then compute the response function separately from the simulations, either by numerical differentiation or by analytically differentiating an empirical fit to the simulation results for $E(T)$.

In the second method we evaluate the derivative analytically via statistical mechanics. The analytical form for the derivative always involves a fluctuation; for example, the isometric heat capacity is [2, 3]

$$C_v = \frac{1}{kT^2} \langle (\delta E)^2 \rangle \qquad (6.28)$$

where δE is the fluctuation of the internal energy about its average value

$$\delta E = E - \langle E \rangle \qquad (6.29)$$

and hence, $\langle (\delta E)^2 \rangle$ is the mean-square fluctuation

$$\langle (\delta E)^2 \rangle = \langle (E - \langle E \rangle)^2 \rangle \tag{6.30}$$

We then use simulation to accumulate this mean-square fluctuation and hence, by a relation such as (6.28), we obtain the response function.

Of these two methods, the first is often more accurate than the second, but the first has the disadvantage that several simulations must be done before the derivative can be estimated. The second method produces an estimate for the derivative from a single simulation; however, the derivative itself cannot be evaluated as the simulation proceeds. During a run the average $\langle E \rangle$ is accumulated, and instantaneous values $E(t)$ are stored on disk or tape. Only at the conclusion of the run can we compute the fluctuations needed in (6.30).

We can maneuver around this cumbersome realization of method 2 by expanding the quadratic in (6.30) to obtain the alternative form

$$\langle (\delta E)^2 \rangle = \langle E^2 \rangle - \langle E \rangle^2 \tag{6.31}$$

Note that since the lhs is necessarily positive, we must always have $\langle E^2 \rangle > \langle E \rangle^2$. The form (6.31) is appealing because it avoids the reanalysis of data required by (6.30): we merely accumulate both the average E and the average E^2 during the run. Unfortunately, (6.31) is often less accurate than (6.30) because of round-off errors—(6.31) suffers from the disease of small difference of large numbers. Aside from these computational details, the relaxation time for fluctuations is generally longer than that for the average quantity itself, and therefore, to obtain reliable results from method 2, long runs with large numbers of particles are needed.

Besides these practical difficulties in using fluctuation expressions, there is a theoretical problem: the relations between response functions and fluctuations depend on the choice of independent variables used to set the thermodynamic state. That is, fluctuation expressions depend on the statistical mechanical ensemble. Thus, the fluctuation expression (6.28) for C_v applies only in a system whose state is specified by fixed values of the number of atoms N, the volume V, and the temperature T; these quantities identify the canonical ensemble in statistical mechanics. In contrast, in the isolated system of molecular dynamics, the total internal energy E does not fluctuate and so C_v cannot be obtained from (6.28).

Most texts in statistical mechanics avoid developing fluctuation expressions in the microcanonical ensemble; instead, fluctuation expressions are developed in some other ensemble, and those results are converted, via Legendre transforms, to the microcanonical ensemble. This is the approach taken by Lebowitz et al. [13] and by Cheung [14]. However, as shown by Pearson et al. [15], one may also start directly in the microcanonical ensemble and develop all of thermodynamics without appealing to other ensembles. Since the microcanonical ensemble is central to isolated-system molecular dynamics and since the microcanonical ensemble typically receives meager attention in statistical mechanics texts, this direct approach is presented in Appendix C.

Unfortunately, the expressions for response functions obtained in the Legendre transform method differ somewhat from those obtained in the direct method. The two sets of response functions coincide as the number of atoms N is increased [15]; however, in simulations N may not be large. Apparently, the expressions obtained via Legendre transforms are strictly valid only when N is large, while the expressions obtained in the direct method apply for any N. This distinction is likely of little import and must usually be overshadowed by statistical uncertainties in the simulation results.

6.2.1 Isometric Heat Capacity

In Appendix C we show that the isometric heat capacity can be evaluated directly in the microcanonical ensemble, resulting in an expression containing the average reciprocal kinetic energy,

$$C_v^* \equiv \frac{C_v}{Nk} = \left[N - NT^* \left(\frac{3N}{2} - 1 \right) \left\langle E_k^{*-1} \right\rangle \right]^{-1} \tag{6.32}$$

Although (6.32) may not appear to involve a fluctuation, recall that T is proportional to the average kinetic energy, and so the rhs of (6.32) essentially involves the fluctuation of $[\langle E_k \rangle \langle E_k^{-1} \rangle]$ about unity.

Alternatively, Lebowitz et al. [13] show how fluctuations in the canonical situation are related to those in a molecular dynamics isolated system; for an isolated system containing spherical molecules, they find

$$C_v^* \equiv \frac{C_v}{Nk} = \frac{3}{2} \left[1 - \frac{2}{3NT^{*2}} \left\langle (\delta \mathscr{U}^*)^2 \right\rangle \right]^{-1} \tag{6.33}$$

where $\delta \mathscr{U}$ is the fluctuation in the total configurational internal energy. To maintain constant total energy in an isolated system, the mean-square fluctuations in the configurational energy must equal those of the kinetic energy; hence, (6.33) can also be written as

$$C_v^* = \frac{3}{2} \left[1 - \frac{2}{3NT^{*2}} \left\langle (\delta E_k^*)^2 \right\rangle \right]^{-1} \tag{6.34}$$

As N goes large, expressions (6.32) and (6.34) become identical (Exercise 6.17).

Usually the full heat capacity is not what we want; instead, we often need the residual part C_v^R, the part with the ideal-gas contribution removed. For spheres,

$$C_v^{*R} = C_v^* - \frac{3}{2} \tag{6.35}$$

Combining (6.32) and (6.35) produces

$$C_v^{*R} = \frac{-\left(\frac{3}{2}N - 1\right)\left[1 - \frac{3}{2}NT^*\langle E_k^{*-1}\rangle\right]}{N\left[1 - \left(\frac{3}{2}N - 1\right)T^*\langle E_k^{*-1}\rangle\right]} \tag{6.36}$$

while combining (6.33) and (6.35) produces

$$C_v^{*R} = \frac{\langle(\delta \mathcal{U}^*)^2\rangle}{NT^{*2} - \frac{2}{3}\langle(\delta \mathcal{U}^*)^2\rangle} \tag{6.37}$$

We have done calculational tests that compare values for C_v^R obtained from (6.36) and (6.37) and direct evaluation of the derivative (6.27). The tests were done on the Lennard-Jones fluid at $\rho^* = 0.7$. To compute the derivative (6.27), we used for U_c the simulation result given in (6.10); thus, differentiating (6.10) with respect to temperature gives

$$C_v^{*R} = 0.8893 - 0.27176T^* + 0.046779T^{*2} \tag{6.38}$$

which applies over the temperature range $0.96 < T^* < 2.23$. Values of C_v^{*R} computed from this equation are compared in Figure 6.6 with values obtained using the fluctuation expressions (6.36) and (6.37). In using these fluctuation expressions, intermediate averages such as $\langle U \rangle$ and $\langle E_k^{-1}\rangle$ *must* be computed to at least five significant figures. Figure 6.6 shows that the two fluctuation expressions are equally reliable in providing C_v; however, the simple quadratic (6.38) consistently underestimates C_v^R, though it is generally within the statistical uncertainties of the fluctuation results.

In Figure 6.7 the simulation results for C_v^R obtained from (6.37) are compared with values obtained from the Nicolas et al. [4] and Ree [5] correlations. The Ree correlation consistently underestimates C_v^R, while the Nicolas et al. correlation underestimates C_v^R at high temperatures but overestimates it at low temperatures.

6.2.2 Adiabatic Compressibility

The adiabatic compressibility κ_s measures how the system volume responds to a reversible adiabatic (hence, isentropic) change in pressure,

$$\kappa_s = \frac{-1}{V}\left(\frac{\partial V}{\partial P}\right)_s \tag{6.39}$$

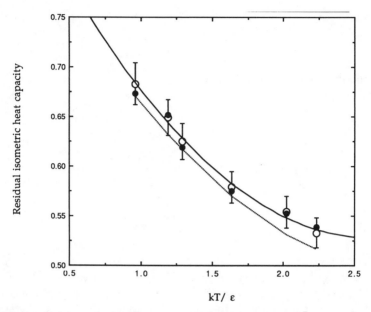

FIGURE 6.6 Comparison of various simulation results for C_v^{*R} for the Lennard-Jones fluid at $\rho\sigma^3 = 0.7$. Open circles were obtained from the fluctuation formula (6.37) involving the internal energy U_c. The error bars on these circles are ± 2 standard deviations for each average. The closed circles were obtained from the fluctuation formula (6.36) involving the reciprocal kinetic energy. The dotted line was obtained from (6.38), the derivative of the empirical fit to U_c. The solid line is a direct fit of a quadratic to the open circles. The simulations are described in Section 6.1.1.

There is no known way to specify a particular value of the entropy S in a molecular dynamics run, and so κ_s cannot be evaluated by numerically estimating the derivative $(\partial V / \partial P)_s$—we are forced to use a fluctuation expression. Fluctuation expressions for κ_s have been obtained directly in the microcanonical ensemble [15] and indirectly from the isothermal compressibility in the canonical ensemble [14]. The latter appears to be easier to implement in simulations, and so we consider only that expression here. Its general form involves pressure fluctuations $\langle (\delta P)^2 \rangle$, and for atomic substances it is

$$\kappa_s = \left[\frac{2}{3}P + \rho kT + \langle \theta \rangle - \frac{N}{\rho kT}\langle (\delta P)^2 \rangle \right]^{-1} \qquad (6.40)$$

where P is the full pressure, as in (6.13). For pairwise additive potentials

$$\theta = \frac{1}{9V}\sum_{i<j} r_{ij}^2 \frac{\partial^2 u(r_{ij})}{\partial r_{ij}^2} \qquad (6.41)$$

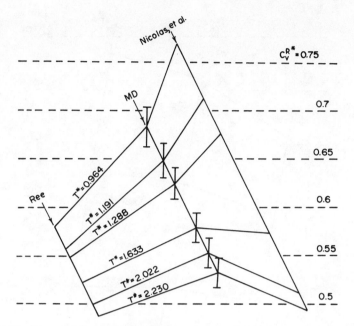

FIGURE 6.7 Tests of Ree and Nicolas et al. equations of state for the residual isometric heat capacity of the Lennard-Jones fluid at $\rho^* = 0.7$. The simulation results shown here are from (6.37), which appear as the open circles in Figure 6.6; the runs are described in Section 6.1.1. The error bars are statistical uncertainties taken to be ± 2 standard deviations of each average.

Moreover, when $u(r)$ is the Lennard-Jones potential, θ simplifies to [16]

$$\langle \theta_{LJ} \rangle = \frac{-8U_c}{V} + \tfrac{19}{3}(P - \rho kT) \tag{6.42}$$

Combining (6.40) and (6.42) gives

$$\kappa_s^* = \frac{\kappa_s \varepsilon}{\sigma^3} = \left[7P^* - \frac{16\rho^* T^*}{3} - 8\rho^* U_c^* - \frac{N}{\rho^* T^*} \langle (\delta P^*)^2 \rangle \right]^{-1} \tag{6.43}$$

Recall that U_c^* is intensive: it is the internal energy per atom.

Sample results obtained from (6.43) for the Lennard-Jones fluid are presented in Figure 6.8. For the states tested, the Nicolas et al. [4] equation gives κ_s within the uncertainties of the simulation results. The Ree [5] correlation is satisfactory at high temperatures but underestimates κ_s at low temperatures.

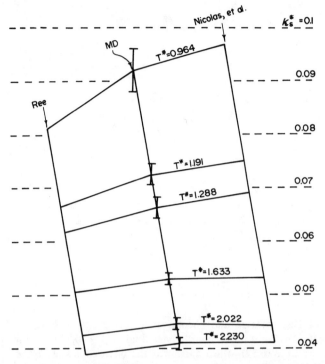

FIGURE 6.8 Tests of Ree and Nicolas et al. equations of state for the adiabatic compressibility κ_s^* of the Lennard-Jones fluid at $\rho^* = 0.7$. The simulation results are from the runs described in Section 6.1.1. The error bars are statistical uncertainties taken to be ± 2 standard deviations of each average.

6.2.3   Thermal Pressure Coefficient

The thermal pressure coefficient γ_v measures how the pressure responds to an isometric change in temperature,

$$\gamma_v = \left(\frac{\partial P}{\partial T} \right)_V \tag{6.44}$$

Like the isometric heat capacity, γ_v can be determined either directly from this definition or from fluctuation expressions [14, 15]. One such expression [14] involves cross-fluctuations in the pressure and kinetic energy $\langle \delta E_k \, \delta P \rangle$,

$$\gamma_v^* \equiv \frac{\gamma_v \sigma^3}{k} = \frac{2}{3} C_v^* \left[\rho^* - \frac{1}{T^{*2}} \langle \delta E_k^* \, \delta P^* \rangle \right] \tag{6.45}$$

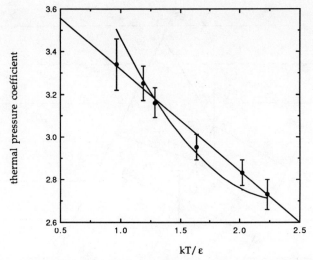

FIGURE 6.9 Comparison of various simulation results for the thermal pressure coefficient γ_v^* of the Lennard-Jones fluid at $\rho\sigma^3 = 0.7$. Points were obtained directly from simulation using the fluctuation formula (6.45). The error bars on these points are statistical uncertainties reported at ± 2 standard deviations of each average. The curved line is the quadratic (6.47), obtained from the derivative of the empirical fit (6.19) to the residual pressure. The straight line is a direct fit to the points. The simulations are described in Section 6.1.1.

where

$$\langle \delta E_k\, \delta P \rangle = \langle (E_k - \langle E_k \rangle)(P - \langle P \rangle)\rangle \qquad (6.46)$$

E_k is the extensive kinetic energy, and C_v is the full heat capacity, not just the residual part.

Simulation values of γ_v computed from the derivative (6.44) are compared in Figure 6.9 with values from the fluctuation expression (6.45). The derivative results were obtained by differentiating the simulation results (6.19) for the pressure; that temperature differentiation produces

$$\gamma_v^* = 4.9808 - 1.927 T^* + 0.408 T^{*2} \qquad (6.47)$$

which applies only at $\rho^* = 0.7$ and $0.96 < T^* < 2.23$. The γ_v values obtained from the empiricism (6.47) are generally within the uncertainties of the fluctuation results; however, the curvature of the quadratic (6.47) is clearly wrong at both low and high temperatures. A straight-line fit directly to the

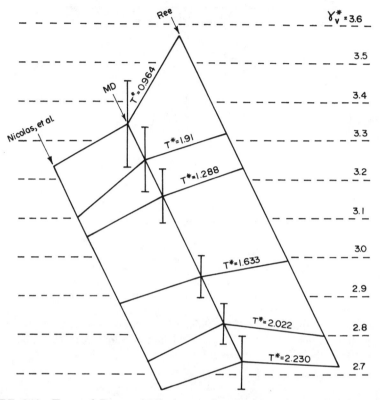

FIGURE 6.10 Tests of Ree and Nicolas et al. equations of state for the thermal pressure coefficient γ_v of the Lennard-Jones fluid at $\rho^* = 0.7$. The simulation results are from the runs described in Section 6.1.1. The error bars are statistical uncertainties taken to be ± 2 standard deviations of each average.

fluctuation results from (6.45) passes within the uncertainties of all the data in the figure.

In Figure 6.10 we use values of γ_v from the fluctuation expression (6.45) to test the Lennard-Jones equations of state [4, 5]. Except at low temperatures, both correlations provide γ_v within the uncertainties of the simulation results. The Nicolas et al. equation consistently underestimates γ_v, while the Ree equation underestimates γ_v at high temperatures and overestimates it at low temperatures.

6.2.4   Other Response Functions

With values for C_v, κ_s, and γ_v obtained from molecular dynamics, all remaining response functions can be evaluated from classical thermodynamic relations [17] without recourse to further analysis of simulation data. Thus,

$$\gamma = V\left(\frac{\partial P}{\partial E}\right)_V = \frac{V}{C_v}\gamma_v \quad \text{(Grüneisen parameter)} \tag{6.48}$$

$$\kappa_T^{-1} = \left[\frac{-1}{V}\left(\frac{\partial V}{\partial P}\right)_T\right]^{-1}$$

$$= \kappa_s^{-1} - \frac{T}{\rho}\frac{\gamma_v^2}{C_v/N} \quad \text{(isothermal compressibility)} \tag{6.49}$$

$$C_p = \left(\frac{\partial H}{\partial T}\right)_P = \frac{\kappa_T C_v}{\kappa_s} \quad \text{(isobaric heat capacity)} \tag{6.50}$$

$$\beta = \frac{1}{V}\left(\frac{\partial V}{\partial T}\right)_P = \kappa_T \gamma_v \quad \text{(volume expansivity)} \tag{6.51}$$

$$\mu = \left(\frac{\partial T}{\partial P}\right)_H = \frac{V(T\alpha - 1)}{C_p} \quad \text{(Joule–Thomson coefficient)} \tag{6.52}$$

$$w = \sqrt{\frac{1}{\rho m \kappa_s}} \quad \text{(sonic velocity)} \tag{6.53}$$

Here $H = E + PV$ is the enthalpy and m is the atomic mass.

6.3   ENTROPIC PROPERTIES

Of the many problems in molecular simulation, one of the most vexing is determination of entropic properties—the entropy, the Gibbs and Helmholtz free energies, and the chemical potential. The problem is addressed by an extensive literature and in several review papers [18–21]. The difficulty is that unlike the pressure and internal energy, entropic properties cannot be determined directly because they are not defined as time averages over a phase-space trajectory. Instead, they are related to the phase-space volume. Thus, in an isolated system in which N, V, and E are the independent variables, the entropy S is, from (C.5),

$$S = k \ln \Omega$$

In a system in which N, V, and T are independent, the Helmholtz free energy is

$$A = -kT \ln Q \tag{6.54}$$

and in a system in which N, P, and T are independent, the Gibbs free energy is

$$G = -kT \ln \Delta \tag{6.55}$$

In these equations Ω is the phase-space volume (aka the microcanonical partition function) available to the isolated system, Q is the phase-space volume available to the NVT system (the canonical partition function), and Δ is the phase-space volume available to the NPT system. Our first problem is to find an expression that gives one of S, G, or A in terms of some time average. We need evaluate only one of S, G, or A from simulation, for then the other two can be obtained from classical thermodynamics.

In an isolated system the entropy of a pure substance is related to other thermodynamic properties by the fundamental equation

$$dS = \frac{dE}{T} + \frac{P}{T} dV - \frac{\mu}{T} dN \tag{6.56}$$

where $\mu = G/N$ is the chemical potential. This equation suggests three routes to the entropy:

(a) manipulate the system energy E at fixed V and N,
(b) manipulate the system volume V at fixed E and N, or
(c) manipulate the number of molecules N at fixed E and V.

Just as in the experimental situation, each of these routes actually provides a *change* in entropy, not an absolute value. Although several schemes have been devised for realizing these routes via simulation, we discuss here only three. The first is *thermodynamic integration*, which may employ both routes (a) and (b). The other two schemes use route (c), in which we are to determine the response of S to a change in N: the *test particle method*, which evaluates the response in the form of a derivative, and the *coupling parameter method*, which evaluates the response in the form of an integral.

6.3.1 Thermodynamic Integration

The simplest methods for evaluating entropic properties are forms of thermodynamic integration. In these methods we determine the temperature, pressure, or density dependence of a simple thermodynamic property and then compute S, A, or G by integrating the appropriate thermodynamic relation. This strategy avoids introducing complicated time averages into a simulation program at the price of requiring several simulations to obtain one value of an entropic property.

As an example, we can use isolated-system simulations to compute the difference in entropy between two states at the same density (fixed N and V). In each of a series of simulations between the initial and final states (1 and 2), we would accumulate the average temperature and then, at the end of the series, compute ΔS via an integrated form of the fundamental equation (6.56):

$$S(E_2) - S(E_1) = \int_{E_1}^{E_2} \frac{dE}{T} \tag{6.57}$$

Alternatively, we can use temperature rather than total energy as the independent variable; thus, applying the chain rule, (6.57) becomes

$$S(T_2) - S(T_1) = \int_{T_1}^{T_2} \frac{1}{T}\left(\frac{\partial E}{\partial T}\right)_{NV} dT = \int_{T_1}^{T_2} \frac{C_v}{T} dT \tag{6.58}$$

where the integration must be done along a line of constant density (isochore). As an example, (6.38), which gives the temperature dependence of C_v for the Lennard-Jones fluid at $\rho^* = 0.7$, could be used in (6.58) to obtain ΔS.

Similarly, if we want the difference in entropy between two states at the same E and N (the states differ in volume and therefore in density), then the fundamental equation (6.56) gives the effect of a change in density as

$$S(\rho_2) - S(\rho_1) = -\int_{\rho_1}^{\rho_2} \frac{PN}{\rho^2 T} d\rho \quad (\text{fixed } E \text{ and } N) \tag{6.59}$$

To use (6.59), the total energy must be set to a prechosen value in each of a series of runs. This may be done by using (5.22) to scale the initial velocities.

Instead of (6.59), however, we more often want the difference in entropy between two states at the same T and N; then the fundamental equation (6.56) leads to

$$S(\rho_2) - S(\rho_1) = \frac{1}{T}[E(\rho_2) - E(\rho_1)] - \frac{N}{T}\int_{\rho_1}^{\rho_2} \frac{P}{\rho^2} d\rho \tag{6.60}$$

To use (6.60), we need simulation data for $E(\rho)$ and $P(\rho)$ along an isotherm; call it T_{set}. However, we cannot predetermine exactly the temperature in isolated-system simulations, although by scaling the atomic velocities during an equilibration phase, the final average temperature can be made close (generally within 5%) to the desired value. Call this final computed temperature T_{cal}. Then by using the heat capacity and thermal pressure coefficient, we can correct the simulation results from $E(\rho, T_{cal})$ and $P(\rho, T_{cal})$ to the needed values $E(\rho, T_{set})$ and $P(\rho, T_{set})$:

$$E(\rho, T_{set}) - E(\rho, T_{cal}) = \int_{T_{cal}}^{T_{set}} C_v \, dT \approx (T_{set} - T_{cal}) C_v(\rho, T_{cal}) \quad (6.61)$$

$$P(\rho, T_{set}) - P(\rho, T_{cal}) = \int_{T_{cal}}^{T_{set}} \gamma_v \, dT \approx (T_{set} - T_{cal}) \gamma_v(\rho, T_{cal}) \quad (6.62)$$

An alternative form of (6.60) is given in Exercise 6.31.

Thermodynamic integration is often the most reliable method for obtaining entropic properties because the integrands can usually be reduced to combinations of simple thermodynamic quantities and because integration tends to be forgiving of small uncertainties in individual contributions to integrands. In spite of its accuracy, thermodynamic integration is not often used because several simulations must be done to obtain one value of ΔS. In view of the ease and low cost of performing simulations, this attitude is largely outdated. As a general rule, we should start simply and leave sophisticated methods until we have experience with our problem. For example, (6.58) and (6.59) require nothing new: we need only run an isolated-system molecular dynamics program several times between the requisite initial and final states. If a simple method gives accurate results, why invest time implementing a sophisticated method? If you are simulating systems involving very large numbers of degrees of freedom, then multiple intermediate simulations may be overly expensive and more sophisticated methods must be considered. Otherwise, thermodynamic integration should be the method considered first.

6.3.2   Test Particle Method

Thermodynamic integration gives changes in entropy resulting from changes in either energy E or volume V; in contrast, the test particle method essentially estimates the change in S that occurs in response to a change in the number of molecules N. For a system of fixed E and V, such a change in S is proportional, according to the fundamental equation (6.56), to the chemical potential

$$\mu = -T \left(\frac{\partial S}{\partial N} \right)_{EV} \quad (6.63)$$

In particular, the chemical potential is related to the reversible work required to add molecules to the system. We now abandon thermodynamics for statistical mechanics and substitute (C.5) for S,

$$\mu = - kT \left(\frac{\partial \ln \Omega}{\partial N} \right)_{EV} \tag{6.64}$$

For large N this derivative is accurately approximated by the response on adding *one* molecule to the system; thus

$$\mu \approx - kT \left(\frac{\ln \Omega_{N+1} - \ln \Omega_N}{N + 1 - N} \right)_{EV} \approx - kT \left(\ln \frac{\Omega_{N+1}}{\Omega_N} \right)_{EV} \tag{6.65}$$

The problem now is to recast the ratio Ω_{N+1} / Ω_N into the form of a time or ensemble average, so that it may be computed in a simulation.

The reformulation of (6.65) and its subsequence use in simulation hinges on the special nature of the added particle [22]. This new molecule, the *test particle*, does not interact with the other N molecules in the system, it does not affect the N-particle dynamics, and hence, it does not disrupt the N-particle phase-space trajectory [23]. With this interpretation in mind, reformulation of (6.65) is straightforward in the canonical (NVT) ensemble [24, 25] (Exercise 6.27). It is more involved in the microcanonical (NVE) ensemble [20], and therefore the derivation is relegated to Appendix C. The result is

$$\mu = - kT \ln \left[\frac{1}{\rho \Lambda^3 \langle kT_{in} \rangle^{3/2}} \left\langle (kT_{in})^{3/2} \exp \left[- \frac{U_t}{kT_{in}} \right] \right\rangle \right] \tag{6.66}$$

where T is the time-average temperature, T_{in} is the instantaneous temperature at one step in the simulation, Λ is the thermal de Broglie wavelength, U_t is the test particle potential energy, and the angular brackets indicate a time average over the simulation. On subtracting from (6.66) the ideal-gas contribution at the same density and temperature (see Section C.4), we obtain the residual chemical potential

$$\mu_{res} = - kT \ln \left[\frac{1}{\langle kT_{in} \rangle^{3/2}} \left\langle (kT_{in})^{3/2} \exp \left[- \frac{U_t}{kT_{in}} \right] \right\rangle \right] \tag{6.67}$$

The quantity U_t is the potential energy of interaction between the test particle and the N real molecules in the system; it is the work that would be required should we add the test particle to the system, although it is not actually added. The value of U_t depends upon the position at which we propose to add the test particle (see Figure 6.11), and therefore the statistics of the time average in (6.67) are improved by proposing, at each sampled

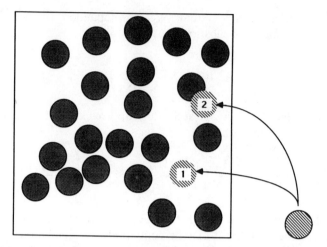

FIGURE 6.11 Determination of the energy U_t by proposing an insertion of a test particle into a system of N real atoms. The N atoms are shown here at one instant during a simulation. If the randomly selected point of insertion is at position 1, then U_t will make a contribution to the chemical potential, because $\mu \sim \exp(-U_t/kT)$. However, if the selected point of insertion is position 2, which is unrealistically near a real atom, then U_t is a large positive number and it makes no contribution: $\exp(-U_t/kT) = 0$.

time-step, multiple insertions of the test particle and computing U_t for each proposed insertion. Thus, for N_t proposed insertions, the time average in (6.67) is realized in a molecular dynamics simulation by

$$\left\langle (T_{\text{in}}^*)^{3/2} \exp\left[-\frac{U_{t,\text{MD}}^*}{T_{\text{in}}^*}\right]\right\rangle = \frac{1}{MN_t} \sum_{j}^{N_t} \sum_{k}^{M} (T_{\text{in}}^*)^{3/2} \exp\left[-\frac{1}{T_{\text{in}}^*} \sum_{i}^{N} u_{ij}^*(k\,\Delta t)\right]$$

$$(6.68)$$

The multiple insertions can be done for randomly selected positions in the N-molecule system or by laying down a regular three-dimensional lattice of test particle positions. Further, these two strategies can be combined; that is, use a regular lattice, but position it randomly in the system.

Long-Range Correction. The potential function u_{ij}^* appearing on the rhs of (6.68) is typically truncated at some distance r_{tc}. (This distance need not be the same as r_c, the distance at which the real-molecule interactions are truncated.) To account for interactions beyond r_{tc}, a correction is needed. Thus, for U_t in (6.67) we write

$$U_t = U_{t,\text{MD}} + U_{t,\text{LR}} \qquad (6.69)$$

where $U_{t,\text{MD}}$ is given by (6.68). The correction $U_{t,\text{LR}}$ is the "long-range" energy that must be overcome to add the test particle to the system. If we make the usual assumption for fluids, namely, that molecules are uniformly distributed for distances $r > r_{tc}$ (see Figure 6.1), then $U_{t,\text{LR}}$ is a constant mean-field contribution that is independent of the position chosen for the test particle. For pairwise additive potentials, U_t is *twice* the "shared" potential energy per particle [23]; hence, $U_{t,\text{LR}}$ is twice the long-range correction to the average configurational internal energy,

$$U_{t,\text{LR}} = 2U_{\text{LR}} \tag{6.70}$$

For the Lennard-Jones fluid, using (6.9) for U_{LR} in (6.70) gives

$$U_{t,\text{LR}}^* = -\frac{16\pi}{3}\frac{\rho^*}{r_{tc}^{*3}} \tag{6.71}$$

On putting (6.69) into (6.67), the residual chemical potential becomes

$$\mu_{\text{res}}^* = -T^* \ln\left[\frac{1}{\langle T_{\text{in}}^*\rangle^{3/2}}\left\langle (T_{\text{in}}^*)^{3/2} \exp\left[\frac{-U_{t,\text{MD}}^* - U_{t,\text{LR}}^*}{T_{\text{in}}^*}\right]\right\rangle\right] \tag{6.72}$$

where $\mu_{\text{res}}^* = \mu_{\text{res}}/\varepsilon$. Even though U_{LR} is a constant, such as in (6.71), it remains within the time average because it is weighted by the reciprocal instantaneous temperature. Equation (6.72) is the appropriate expression for evaluating the residual chemical potential in molecular dynamics performed on isolated systems. We emphasize that the truncation of U_t, which determines the value of U_{LR} in (6.71), occurs at r_{tc}. Since r_{tc} has no effect on the time needed for a simulation, we could make r_{tc} as large as possible, namely $\frac{1}{2}L$, where L is the length of one side of the primary cube.

Sample Results. Figure 6.12 shows the running average μ_{res} obtained from a simulation of 108 Lennard-Jones atoms at $\rho^* = 0.7$, $\langle T^*\rangle = 1.240$. The run was started from an fcc lattice and several hundred time-steps were executed before we started to accumulate property averages. The time-step was $\Delta t^* = 0.004$; the real-molecule and test particle potentials were both truncated at $r_c^* = r_{tc}^* = 2.5$. Test particle positions were assigned to a regular cubic lattice containing $8\times8\times8 = 512$ positions, and at every tenth time-step, this cubic lattice was randomly placed in the system to sample the test particle energy U_t.

We also show in Figure 6.12 values of the residual chemical potential provided by the Nicolas et al. [4] and Ree [5] equations of state. The values from those two equations are within 0.5% of one another. Similar agreement in μ_{res} was obtained from those two equations at all state conditions tested

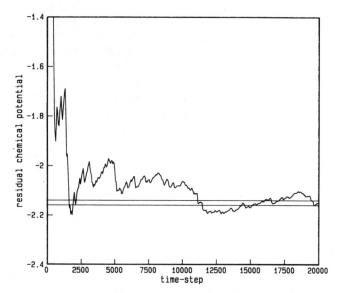

FIGURE 6.12 Running average residual chemical potential μ^{*}_{res} computed via the test particle method for 108 Lennard-Jones atoms at $\rho^{*} = 0.7$, $T^{*} = 1.240$. The test particle energy was obtained by inserting a cubic array of $(8 \times 8 \times 8 = 512)$ test particles into the system at every tenth time-step of the simulation. Both the real and test particle potentials were truncated at 2.5σ. The horizontal lines are values of μ_{res} obtained from empirical equations of state. The upper line is $\mu^{*}_{res} = -2.14$ from Ree [5]; the lower line is $\mu^{*}_{res} = -2.16$ from Nicolas et al. [4].

here, and so in what follows we refer only to values from the Nicolas et al. equation. Figure 6.12 shows that about 10,000 time-steps are needed for the simulation result to fluctuate within 5% of the Nicolas et al. value. Beyond time-step 11,000 the simulation value fluctuates within $\pm 2\%$ of the Nicolas et al. value. Fluctuations of that size must be within the uncertainty of the empirical fit used to obtain the Nicolas et al. equation.

Figures 6.13–6.16 are two-way plots (see Appendix D) that illustrate the magnitudes of various contributors to systematic error in μ_{res}. Test runs were generally done over 10,000 time-steps with the test particle lattice inserted randomly every tenth time-step. Occasionally runs were extended to 20,000 time-steps. Figure 6.13 shows how the number of atoms affects μ_{res} when using 1000 and 1331 test particles. In both sets of runs, 108 and 256 atoms give μ_{res} to within only 3–4% of the Nicolas et al. value, while with 500 atoms, μ_{res} is obtained to within 0.5%.

In Figure 6.14 we show how μ_{res} is affected by changing the number of test particles used in the lattice. For 108 real atoms (not shown in the figure), we find that 512 test particles are usually enough to give μ_{res} within 3% of the Nicolas et al. value; however, for 256 or more atoms, the test particle lattice should be increased to at least 1000 particles. If we were to approach the

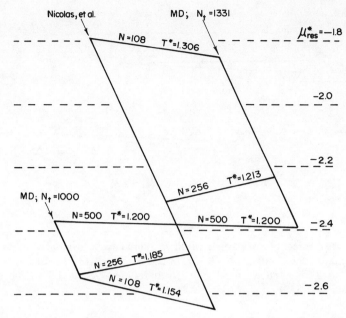

FIGURE 6.13 Effect of number of atoms N on molecular dynamics results for μ^*_{res} of the Lennard-Jones fluid at $\rho^* = 0.7$, $r^*_c = r^*_{tc} = 2.5$; N_t is the number of test particles used. At these state conditions runs with 108 and 256 atoms show measurable deviations from μ^*_{res} predicted by the Nicolas et al. equation of state; however, with 500 atoms agreement is satisfactory.

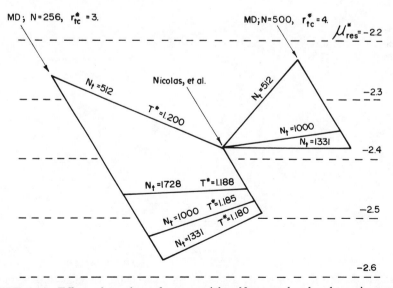

FIGURE 6.14 Effect of number of test particles N_t on molecular dynamics results for the residual chemical potential μ^*_{res} of the Lennard-Jones fluid at $\rho^* = 0.7$, $r^*_c = 2.5$. The simulation results using 500 atoms are all at $T^* = 1.200$.

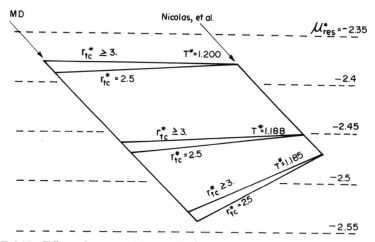

FIGURE 6.15 Effect of test particle cutoff r_c^* on molecular dynamics results for μ_{res} of the Lennard-Jones fluid at $\rho^* = 0.7$, $r_c^* = 2.5$. At $T^* = 1.185$ we used $N = 256$ and $N_t = 1000$, at $T^* = 1.188$ we used $N = 256$ and $N_t = 1728$, at $T^* = 1.200$ we used $N = 500$ and $N_t = 1331$.

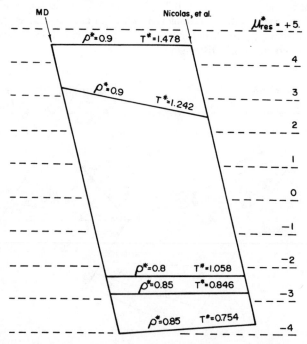

FIGURE 6.16 Effect of state condition on molecular dynamics results for the residual chemical potential μ_{res}^* of the Lennard-Jones fluid. All these results are from simulations that used $N = 256$, $N_t = 1331$, and $r_c^* = 2.5$.

thermodynamic limit over a series of simulations (i.e., over several runs, we increase both N and V but keep N/V fixed), then to maintain coverage of the system volume, we would also have to increase the number of test particles N_t.

The effects of the truncation point of the test particle potential are shown in Figure 6.15. In this series of runs, the truncation was systematically increased from 2.5 to 4.46σ, while maintaining the real-atom truncation at 2.5σ. With the long-range correction properly included, as in (6.72), Figure 6.15 shows that increasing r_{tc}^* from 2.5 to 3 produces no more than a 1% change in μ_{res}. Increasing r_{tc}^* beyond 3 has a negligible effect on μ_{res}.

Finally, in Figure 6.16 we show effects of state condition on test particle results for μ_{res}. We expect the test particle method to be less successful at high density, for there the real atoms are closely packed and little free volume is available in which to insert an additional atom. The figure shows that this effect becomes detrimental only at high densities and low temperatures. Regarding temperature effects, Figure 6.16 (as well as the isochoric plot in Figure 6.13) shows that the residual chemical potential becomes less negative as the temperature is increased at fixed density.

6.3.3   Coupling Parameter Method

The third method we describe for obtaining entropic properties is based on the concept of a coupling parameter [26]. This method may be interpreted either as a generalization of thermodynamic integration or as a kind of particle insertion. As in all simulation approaches to entropic properties, the immediate objective is to recast a statistical mechanical expression, such as (6.54) or (6.55), into an expression containing a time average. Here we seek such a relation for the entropy (C.5).

We begin by considering an isolated system of N atoms and letting

$$\mathscr{U} = \mathscr{U}(\mathbf{r}^N; \lambda) \tag{6.73}$$

be the potential energy function. Here λ is a scalar parameter that controls the strength of the interaction (or of some part of the interaction) among N atoms. It is continuous, differentiable, and bounded by λ_0 and λ_f: when $\lambda = \lambda_0$, the substance is in a reference condition, and when $\lambda = \lambda_f$, the potential takes on its full strength for the substance of interest. Thus, λ measures the degree of "coupling" through interactions among the atoms; in theories for electrolyte solutions, λ is called a "charging" parameter.

For the development that follows, we choose to interpret the entropy S in terms of the phase-space density of states ω rather than in terms of the phase-space volume Ω, as appears in (C.5) (cf. Appendix C and see ref. 15). Because $\mathscr{U}$ depends on λ, so too does ω, and analogous to (C.5), we have

$$S = k \ln \omega(\lambda) \tag{6.74}$$

Differentiating with respect to λ at fixed NVE and using (C.15) and (C.17), we obtain

$$\left(\frac{\partial S}{\partial \lambda}\right)_{NVE} = -k\left(\frac{3N}{2}-1\right)\left\langle E_k^{-1}\left(\frac{\partial \mathcal{U}}{\partial \lambda}\right)_{NVE}\right\rangle \qquad (6.75)$$

Here E_k is the instantaneous kinetic energy. Integrating (6.75) from the reference condition (λ_0) to the situation of interest (λ_f), we find

$$\Delta S = S(\lambda_f) - S(\lambda_0) = -k\left(\frac{3N}{2}-1\right)\int_{\lambda_0}^{\lambda_f}\left\langle E_k^{-1}\left(\frac{\partial \mathcal{U}}{\partial \lambda}\right)_{NVE}\right\rangle d\lambda \quad (6.76)$$

Note that in (6.76) both the reference condition and the substance of interest are at the same density ($\rho = N/V$) and same total energy (E). Nevertheless, by making suitable choices for λ_0, λ_f and $\mathcal{U}(\mathbf{r}^N; \lambda)$, many kinds of entropy differences can be computed from (6.76). For any such choice, it is the integrand in (6.76) that is computed as a time average in a simulation. With results in hand from several simulations, performed at discrete λ-values on $[\lambda_0, \lambda_f]$, we can evaluate the integral in (6.76) by a numerical quadrature. Since S is a state function, the path of integration over λ may be chosen for computational convenience.

Residual Entropy for E > 0. One common choice of reference condition is the ideal gas, and since here we are considering isolated systems, the ideal-gas reference is to be at the same density and total energy as the real substance. The strategy is to construct $\mathcal{U}(\mathbf{r}^N; \lambda)$ so that $\lambda_0 = 0$ produces the ideal gas, while $\lambda_f = 1$ produces the real substance. Then (6.76) gives the residual entropy

$$S_{\mathrm{res}}(\rho, E) = S_{\mathrm{real}}(\rho, E) - S_{\mathrm{ig}}(\rho, E) = S(\rho, E; \lambda_f) - S(\rho, E; \lambda_0) \quad (6.77)$$

$$S_{\mathrm{res}}(\rho, E) = -k\left(\frac{3N}{2}-1\right)\int_0^1\left\langle E_k^{-1}\left(\frac{\partial \mathcal{U}}{\partial \lambda}\right)_{NVE}\right\rangle d\lambda \qquad (6.78)$$

The angular brackets here mean a time average over the system whose atoms interact with potential $\mathcal{U}(\mathbf{r}^N; \lambda)$.

However, the residual property usually extracted from experiment is not (6.77), but rather that measured relative to an ideal gas at the same density and *temperature* as the real substance,

$$S_{\mathrm{res}}(\rho, T) = S_{\mathrm{real}}(\rho, T) - S_{\mathrm{ig}}(\rho, T) \qquad (6.79)$$

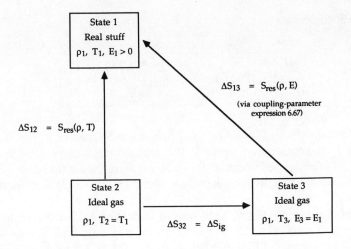

$$\Delta S_{12} = \Delta S_{32} + \Delta S_{13} \qquad \text{(which is 6.83)}$$

FIGURE 6.17 When the total energy $E_1 > 0$, a procedure can be devised for obtaining the residual entropy $S_{res}(\rho, T)$ by using (6.83), the coupling parameter expression (6.78), and isolated-system simulations.

Subtracting (6.77) from (6.79) allows us to compute $S_{res}(\rho, T)$ from a known value of $S_{res}(\rho, E)$,

$$S_{res}(\rho, T) = S_{res}(\rho, E) + \Delta S_{ig} \qquad (6.80)$$

where ΔS_{ig} is the response to changing the ideal-gas state from (ρ, T) to (ρ, E).

Equation (6.80) is illustrated schematically in Figure 6.17, where state 1 represents the real substance at the condition of interest $\{\rho_1, T_1, E_1\}$. State 2 is the ideal gas at the same density $(\rho_2 = \rho_1)$ and temperature $(T_2 = T_1)$ as state 1. State 3 is the ideal gas at the same density $(\rho_3 = \rho_1)$ and total energy $(E_3 = E_1)$ as the real substance. Since the total energy of an ideal gas is purely kinetic, we may immediately compute T_3. For spheres,

$$T_3 = \frac{2E_3}{3Nk} \qquad (6.81)$$

Thus, for an ideal gas composed of spheres, ΔS_{ig} in (6.80) is

$$\Delta S_{ig} = S_3(\rho, T_3) - S_2(\rho, T_2) = \frac{3Nk}{2} \ln \frac{T_3}{T_2} \qquad (6.82)$$

and (6.80) becomes

$$S_{res}(\rho, T) = S_{res}(\rho, E) + \frac{3Nk}{2} \ln \frac{T_3}{T_2} \tag{6.83}$$

Unfortunately, (6.83) can be used only for real systems in which the total energy E is greater than zero. If the real system energy is negative, then $S_{res}(\rho, E)$ is undefined because the ideal-gas reference state (state 3 in Figure 6.17) does not exist. In such cases, an alternative to (6.83) is needed.

Residual Entropy for $E \leq 0$. One way to avoid the undefined ideal-gas state when $E < 0$ is to refer the real system to a real state (call it state 4) having $E > 0$. The entropy difference between real states 1 and 4 could be found from simulation using thermodynamic integration. Thus, using (6.57), we have

$$\Delta S_{14} = S(\rho, E_1) - S(\rho, E_4) = \int_{E_4}^{E_1} \frac{dE}{T} \tag{6.84}$$

The entropy of state 4 could be obtained via the coupling parameter method, as in (6.83). The entire procedure is then

$$S_{res}(\rho, T) = S_{res}(\rho, E_4) + \Delta S_{ig} + \Delta S_{14} \tag{6.85}$$

This equation is interpreted schematically in Figure 6.18. Using (6.82) and (6.84) in (6.85) gives

$$S_{res}(\rho, T) = S_{res}(\rho, E_4) + \frac{3Nk}{2} \ln \frac{T_3}{T_2} + \int_{E_4}^{E_1} \frac{dE}{T} \tag{6.86}$$

where $S_{res}(\rho, E_4)$ is still obtained from (6.78).

Form for $\mathscr{U}(r^N; \lambda)$. To use (6.78) with simulation data to obtain $S_{res}(\rho, E)$, we must choose the λ-dependence of $\mathscr{U}$. A simple choice is a power law form [21, 27]

$$\mathscr{U}(\mathbf{r}^N; \lambda) = \lambda^m \mathscr{U}(\mathbf{r}^N) \tag{6.87}$$

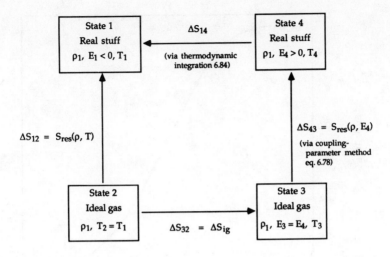

$$\Delta S_{12} = \Delta S_{32} + \Delta S_{43} + \Delta S_{14} \quad \text{(which is 6.85)}$$

FIGURE 6.18 When the total energy $E_1 < 0$, to evaluate the residual entropy $S_{res}(\rho, T)$, a scheme less direct than that in Figure 6.17 must be devised because ideal-gas states cannot have negative total energies. One possibility, (6.85), is to combine the coupling parameter method (6.78) with thermodynamic integration (6.84).

where m is a positive integer. Using (6.87) in (6.78), we obtain

$$S_{res}(\rho, E) = -mk\left(\frac{3N}{2} - 1\right)\int_0^1 \lambda^{m-1}\left\langle E_k^{-1}\mathscr{U}(\mathbf{r}^N)\right\rangle d\lambda \qquad (6.88)$$

We emphasize that the brackets indicate an average over an isolated-system phase-space trajectory for which the atoms interact with the potential $\mathscr{U}(\mathbf{r}^N; \lambda)$. The integration in (6.88) is along a path of constant ρ and E. In a simulation, to obtain a desired value of the total energy, the initially assigned velocities can be scaled according to the procedure described in Section 5.2.1.

The choice for the exponent m hinges on the following considerations [21, 27]. For realistic potential models, which are repulsive at short range, the time average in (6.88) will tend to diverge as $\lambda \to 0$. For r^{-n} repulsions, the limiting behavior of this divergence is $\lambda^{(md/n)-1}$ in a d-dimensional space. Moreover, if m is taken to be too large, then the high-λ portion dominates the integrand in (6.88). A workable compromise for the Lennard-Jones substance (which has $n = 12$) is $m = 4$. This is the value used in the following tests.

Sample Results for $E > 0$. We first consider a sample calculation using the coupling-parameter method with $E > 0$. The test used 108 Lennard-Jones atoms at $\rho^* = 0.7$, $E^* = E/N\varepsilon = 1$. Simulations were done to compute the

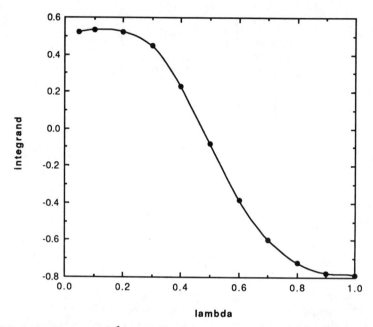

FIGURE 6.19   Integrand $\{\lambda^3 \langle \mathscr{U}/E_k \rangle\}$ of (6.88) for evaluating the residual entropy via the coupling parameter method. These results are from runs using 108 Lennard-Jones atoms at $\rho^* = 0.7$, $E^* = 1$.

integrand in (6.88) at eleven λ-values on $[0.05, 1]$. Each integrand was accumulated over only 500 time-steps; the values of the integrands are plotted in Figure 6.19. Extrapolating the integrand to $+0.52$ at $\lambda = 0$ and using Simpson's rule to compute the integral, (6.88) gives

$$\frac{S_{\text{res}}(\rho, E)}{Nk} = +0.578 \tag{6.89}$$

Putting this into (6.83), where $T_3^* = 2E^*/3 = 0.667$ is the temperature for $\lambda = 0$ and $T_2^* = T_1^* = 3.148$ is the temperature (from the simulation) for $\lambda = 1$, we obtain the coupling-parameter result for the residual entropy: $S_{\text{res}}/Nk = -1.75$.

This result may be compared with values obtained from empirical equations of state and from the test particle method. The test particle method produces μ_{res}, as described in Section 6.3.2. To compute S_{res}, we inserted μ_{res}, along with simulation results for U_c and P, into the integrated form of the fundamental equation (6.56),

$$\frac{TS_{\text{res}}}{N} = \frac{U_c}{N} + \frac{P_{\text{res}}}{\rho} - \mu_{\text{res}} \tag{6.90}$$

where the residual pressure is $P_{res} = P - \rho kT$. The comparison for $S_{res}(\rho, T)$ of the pure Lennard-Jones fluid at $\rho^* = 0.7$, $E^* = 1$, $T^* = 3.148$ is

$$\frac{S_{res}}{Nk} = -1.75 \quad \text{(coupling parameter)} \qquad \frac{S_{res}}{Nk} = -1.68 \quad \text{(test particle)}$$

$$\frac{S_{res}}{Nk} = -1.70 \quad \text{(Nicolas et al. [4])} \qquad \frac{S_{res}}{Nk} = -1.71 \quad \text{(Ree [5])}$$

These four values are within 4% of one another. The test particle result is from one simulation using 256 atoms and 7400 time-steps, so it actually required *more* computer time than did the eleven 108-atom, 500 time-step runs of the coupling-parameter method.

Sample Results for E < 0. Here we illustrate the procedure for situations in which the state of interest has $E < 0$. For this test we choose the pure Lennard-Jones substance at $\rho^* = 0.7$, $E^* = -1$ as the substance of interest and introduce $\rho_4^* = \rho_1^* = 0.7$, $E_4^* = +1$ as the intermediate state (state 4 in Figure 6.18). This intermediate state is a convenient choice here because in the previous section we found its residual entropy (6.89).

As illustrated in Figure 6.18, we now need a thermodynamic integration from state 4 to state 1. For this integration simulations were performed using 108 atoms at five states from $E_1^* = -1$ to $E_4^* = +1$ at increments of $\Delta E^* = 0.5$. The results for T and U_c are given in Table 6.1. Using these data in (6.84) and integrating by Simpson's rule, we find

$$\frac{\Delta S_{14}}{Nk} = \frac{1}{Nk} \int_{E_4}^{E_1} \frac{dE}{T} = -0.763 \tag{6.91}$$

Putting (6.89) and (6.91) into (6.86), we have, at this point,

$$\frac{S_{res}(\rho, T)}{Nk} = +0.578 - 0.763 + \frac{3}{2} \ln \frac{T_3^*}{T_2^*} \tag{6.92}$$

TABLE 6.1 Quantities Used in (6.91) to Obtain ΔS from Thermodynamic Integration

$E/N\varepsilon$	kT/ε	$\langle \mathcal{U}/N\varepsilon \rangle$
−1	2.127	−4.187
−0.5	2.387	−4.077
0	2.672	−4.003
0.5	2.922	−3.879
1	3.148	−3.715

Each simulation used 108 Lennard-Jones atoms at $\rho^* = 0.7$. Each average was accumulated over only 500 time steps after equilibration.

Referring to Figure 6.18, T_3^* is the temperature of the ideal gas at $E_3^* = E_4^* = +1$; thus, $T_3^* = 2E^*/3 = 0.667$. However, T_2^* is the temperature of the ideal gas at the conditions of the real substance in state 1; so, $T_2^* = T_1^*$, and from the data in Table 6.1, $T_1^* = 2.127$. Then for the Lennard-Jones fluid at $\rho^* = 0.7$, $T^* = 2.127$, $E^* = -1$, (6.92) gives

$$\frac{S_{res}}{Nk} = -1.93 \quad \text{(coupling parameter)}$$

The test particle method and empirical equations of state give

$$\frac{S_{res}}{Nk} = -1.95 \quad \text{(test particle)}$$

$$\frac{S_{res}}{Nk} = -1.89 \quad \text{(Nicolas et al. [4])}$$

$$\frac{S_{res}}{Nk} = -1.90 \quad \text{(Ree [5])}$$

These four values are within 3% of one another.

These coupling-parameter and thermodynamic integration tests were done using runs of only 500 time-steps after equilibration. The results counter claims that these methods require significantly more computer time than the test particle method. Recall from Figure 3.13 that the purpose of long runs is not to improve accuracy but to improve statistical precision. The precision of any of these short-run results is admittedly low; however, their accuracy was verified by comparing with independent results obtained from the test particle method. In the absence of such independent verification, coupling parameter and thermodynamic integration runs should typically be of several thousand time-steps. Thus, the total required computer time will be more than that needed for the test particle method, but not substantially more. The runs need not be overly long because only U_c, E, T, and perhaps the ratio U_c/T are needed for the integrands in (6.78) and (6.84). Unlike the test particle method, coupling parameter and thermodynamic integration methods do not fail at high density.

6.4 STATIC STRUCTURE

The radial distribution function $g(r)$ measures how atoms organize themselves around one another—"local structure." Specifically, it is proportional to the probability of finding two atoms separated by distance $r \pm \Delta r$. It plays a central role in statistical mechanical theories of dense substances, and for atomic substances, it can be extracted from x-ray and neutron diffraction experiments. Since molecular dynamics provides positions of individual atoms as functions of time, $g(r)$ can be readily computed from molecular dynamics trajectories. In Section 6.4.1 we derive an expression by which $g(r)$ can be obtained from simulation. The derivation is important not only for $g(r)$ but also because it serves as a model for obtaining expressions for more complex distribution functions, such as angular pair correlation functions in molecular substances and the space–time correlation functions discussed in Chapter 7. In Section 6.4.2 we summarize rudiments of the state dependence of $g(r)$.

6.4.1 Derivation of Simulation Expression

The radial distribution function is defined by

$$\rho g(\mathbf{r}) = \frac{1}{N}\left\langle \sum_i^N \sum_{j\neq i}^N \delta[\mathbf{r}-\mathbf{r}_{ij}] \right\rangle \tag{6.93}$$

Here N is the total number of atoms, $\rho = N/V$ is the number density, $\mathbf{r}_{ij}$ is the vector between centers of atoms i and j, and the angular brackets represent a time average. The mysteries of the δ-symbol are discussed in Appendix A. For homogeneous uniform substances, the structural arrangement of atoms depends only on the distance r between atoms and is independent of the orientation of the separation vector $\mathbf{r}$, so (6.93) reduces to

$$\rho g(r) = \frac{1}{N}\left\langle \sum_i^N \sum_{j\neq i}^N \delta[r - r_{ij}] \right\rangle \tag{6.94}$$

Now this double sum contains $N(N-1)$ terms; however, the distance r_{ij} is invariant under interchange of labels i and j, so only $\frac{1}{2}N(N-1)$ of those terms are unique. Therefore we can write (6.94) as

$$\rho g(r) = \frac{2}{N}\left\langle \sum_i^N \sum_{j<i}^N \delta[r - r_{ij}] \right\rangle \tag{6.95}$$

The normalization of $g(r)$ is obtained by integrating over all possible separations of two atoms,

$$\rho \int g(r)\, d\mathbf{r} = \frac{2}{N}\left\langle \sum_i^N \sum_{j<i}^N \int \delta[r - r_{ij}]\, d\mathbf{r} \right\rangle \tag{6.96}$$

On the rhs in (6.96) we have interchanged the order of time average, summation, and integration. The normalization condition of the δ-symbol, discussed in Appendix A [Equation (A.24)], is

$$\int \delta[r - r_{ij}] \, d\mathbf{r} = 1$$

So (6.96) reduces to

$$\rho \int g(r) \, d\mathbf{r} = N - 1 \approx N \qquad (6.97)$$

Equation (6.97) simply says that if we sit on one atom and count the atoms in the system, we find $N - 1$ other atoms. Equation (6.97) also serves as the basis for a probabilistic interpretation of $g(r)$,

$$\frac{\rho}{N-1} g(r) V(r, \Delta r) = \begin{array}{l} \text{probability that an atomic center lies in a} \\ \text{spherical shell of radius } r \text{ and thickness} \\ \Delta r \text{ with the shell centered on another atom} \end{array} \qquad (6.98)$$

Here $V(r, \Delta r) = \Delta \mathbf{r}$ is the volume of the spherical shell. The radial distribution function shows how the presence of one atom influences, on a time average, the positions of neighboring atoms. For separations less than about one atomic diameter, $g(r) = 0$. For large separations in fluids, one atom should have no influence on the position of another, the density will then be uniform, and $g(r) = 1$.

To obtain an expression for evaluating $g(r)$ from simulation data, we start by rewriting (6.96) using a small but finite shell thickness Δr,

$$\rho \sum_{\Delta r} g(r) V(r, \Delta r) = \frac{2}{N} \sum_{\Delta r} \left\langle \sum_{i}^{N} \sum_{j < i}^{N} \delta[r - r_{ij}] \Delta \mathbf{r} \right\rangle \qquad (6.99)$$

As discussed in Appendix A, the double sum on the rhs of (6.99) represents a counting operation; in particular, the sum is analogous to (A.9),

$$\sum_{i}^{N} \sum_{j < i}^{N} \delta[r - r_{ij}] \Delta \mathbf{r} = N(r, \Delta r) \qquad (6.100)$$

where $N(r, \Delta r)$ is the number of atoms found in a spherical shell of radius r and thickness Δr, with the shell centered on another atom. The equation that results from putting (6.100) into (6.99) must be satisfied term by term; that is, it should be obeyed for each spherical shell, so we find

$$g(r) = \frac{\langle N(r, \Delta r) \rangle}{\frac{1}{2} N \rho V(r, \Delta r)} \qquad (6.101)$$

Writing the time average explicitly over a total of M time-steps gives

$$g(r) = \frac{\sum\limits_{k=1}^{M} N_k(r, \Delta r)}{M(\frac{1}{2}N)\rho V(r, \Delta r)} \qquad (6.102)$$

where N_k is the result of the counting operation (6.100) at time t_k in the run. Physically, (6.102) can be interpreted as the ratio of a local density $\rho(r)$ (see Appendix A) to the system density ρ. The choice of a value for the shell thickness Δr is a compromise: it must be small enough to resolve important features of $g(r)$, but it must also be large enough to provide a sufficiently large sampling population for statistically reliable results. Values near $\Delta r = 0.025\sigma$ satisfactorily balance these competing factors.

Note that because the simulation is performed on a cubic container, we can conveniently determine $g(r)$ only to distances of at most $\frac{1}{2}L$, which is the radius of the largest sphere that can be inscribed in a cube of side L. For liquids whose molecules interact with short-range forces, $r < 3\sigma$ captures most of the interesting features in $g(r)$. However, if you need values of $g(r)$ for r beyond the size of the simulation container, numerical techniques are available for extending $g(r)$ to larger pair separations [28, 29].

6.4.2   Effect of Density on $g(r)$

The radial distribution function depends on density and temperature, and therefore, in computer simulation studies, $g(r)$ serves as a helpful indicator of the nature of the phase assumed by the simulated system. For atoms frozen onto the sites of regular crystal lattice structures—such as fcc, bcc, hcp— $g(r)$ takes the form of a sequence of delta symbols. This is illustrated in Figure 6.20, which compares $g(r)$ for close-packed fcc and simple cubic lattices. Differences in the $g(r)$ plots enable us to distinguish between these two structures; for example, the fcc lattice has fewer atomic pairs separated by $\sqrt{2}$ but more pairs separated by $\sqrt{3}$ than does the simple cubic lattice. Above each $g(r)$ plot we show one plane from the three-dimensional solid and indicate a few of the pair separations that occur in $g(r)$. Because regular crystalline solids are periodically repeating structures, the one-dimensional quantity $g(r)$ is sufficient to discriminate among various three-dimensional structures. If the atoms are vibrating about rather than fixed to the lattice sites, then the delta symbols in $g(r)$ resolve into Gaussians; but the positions and relative heights of those Gaussian distributions still allow determination of the crystalline structure.

The behavior of $g(r)$ in crystalline solids is very different from that for gases at low density. In a gas atoms move freely throughout the container, atoms interact primarily through binary collisions, and only weak local

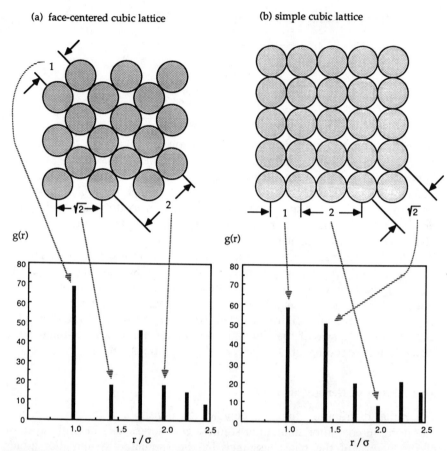

FIGURE 6.20 Use of the radial distribution function to discriminate between crystal lattice structures. (*a*) Structure of a close-packed ($\rho\sigma^3 = 1.414$) fcc lattice computed for 108 atoms and periodic boundary conditions. (*b*) Structure of a close-packed ($\rho\sigma^3 = 1.953$) simple cubic lattice computed using 125 atoms and periodic boundary conditions. The sketches above the plots show one plane from each three-dimensional lattice.

structures form around any one atom. Specifically, the low-density limit of $g(r)$ is given by [2, 3]

$$\lim_{\rho \to 0} g(r) = \exp\left[-\frac{u(r)}{kT}\right] \tag{6.103}$$

where $u(r)$ is the pair potential energy function. A plot of this low-density limit appears in Figure 6.21.

For liquids and amorphous solids the behavior of $g(r)$ is intermediate between crystal and gas: liquids exhibit *short-range* order similar to that in

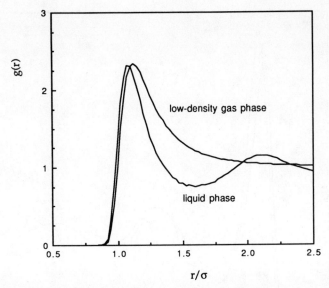

FIGURE 6.21   Comparison of radial distribution functions for gas and liquid phases of the Lennard-Jones fluid, both at $\langle T^* \rangle = 1.178$. The gas phase curve was obtained from the low-density limit (6.103), while the liquid phase curve was obtained from a molecular dynamics simulation at $\rho^* = 0.7$.

crystals, but *long-range* disorder like that in gases. A typical liquid $g(r)$, from a simulation of the Lennard-Jones fluid, is also shown in Figure 6.21. Goldman [30] has fit the Lennard-Jones $g(r)$ to an empirical equation in density, temperature, and radial separation r.

For simulations near fluid–solid phase boundaries, the behavior of $g(r)$ can be used to help identify the phase of the simulated system. For crystalline solids, $g(r)$ contains deeper valleys and higher narrower peaks (particularly after the first peak) than does $g(r)$ for fluids. In addition, for a partially crystallized substance, $g(r)$ may contain secondary peaks not found in $g(r)$ for a fluid. Such additional peaks are caused by remnants of the lattice structure and appear in Figure 6.22 in the distribution function for a partially melted lattice.

6.5   SUMMARY

In this chapter we reviewed how static properties can be computed from equilibrium phase-space trajectories generated by molecular dynamics. From simulations of modest duration, simple thermodynamic properties and the radial distribution function are easily and reliably obtained. But the response functions and entropic properties are more challenging.

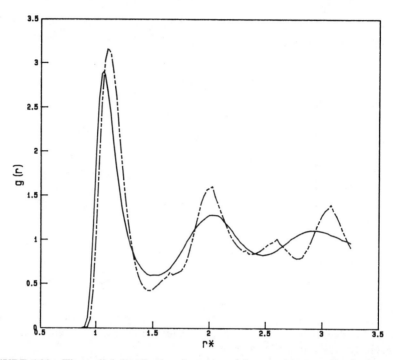

FIGURE 6.22 The radial distribution function $g(r)$ may often be used to distinguish a fluid state (continuous line) from a metastable solidlike state (broken line). These curves are from simulations that used the Lennard-Jones potential with 256 atoms at a density of $\rho\sigma^3 = 0.9$. The fluid state was at $kT/\varepsilon = 1.087$, while the metastable state was at $kT/\varepsilon = 0.80$. See Figure 5.7 for the location of these states on the Lennard-Jones phase diagram.

The response functions are difficult because they are proportional to fluctuations, which are less well behaved than simple thermodynamic quantities. When confronted with some response function χ, our strategy is to first consider the defining derivative

$$\chi = \left(\frac{\partial B}{\partial y}\right)_z \tag{6.104}$$

where B, y, and perhaps z are simple thermodynamic quantities. If B and y can be reliably obtained from simulations of modest length and if z is easily controlled during a run, then respectable values of χ can be obtained from a series of simulations at constant z. During those runs, B and y are computed, and afterward the derivative (6.104) is estimated. However, if these conditions on B, y, and z cannot be met, then we must evaluate the appropriate fluctuation formula for χ. This will require highly precise simulations using large numbers of molecules and extending over long durations.

A similar strategy pertains to entropic properties. We use thermodynamic integration if the integrand is some combination of simple thermodynamic properties and if the integration is along a path of constant z, where z is easily controlled during a run. Otherwise, we must resort to the test particle or coupling parameter schemes.

The general approach can be summarized in this way: our goal is to identify simple methods that give reliable results regardless of the computer time required. We should not go to extraordinary lengths to save computer time[†]; after all, it's the computer that's supposed to be the labor-saving device.

REFERENCES

[1] K. S. C. Freeman and I. R. McDonald, "Molecular Theory of Surface Tension," *Mol. Phys.*, **26**, 529, (1973).

[2] T. M. Reed and K. E. Gubbins, *Applied Statistical Mechanics*, McGraw-Hill, New York, 1973.

[3] D. A. McQuarrie, *Statistical Mechanics*, Harper & Row, New York, 1976.

[4] J. J. Nicolas, K. E. Gubbins, W. B. Streett, and D. J. Tildesley, "Equation of State for the Lennard-Jones Fluid," *Mol. Phys.*, **37**, 1429 (1979).

[5] F. H. Ree, "Analytic Representation of Thermodynamic Data for the Lennard-Jones Fluid," *J. Chem. Phys.*, **73**, 5401 (1980).

[6] Y. Adachi, I. Fijihara, M. Takamiya, and K. Nakanishi, "Generalized Equation of State for Lennard-Jones Fluids. I. Pure Fluids and Simple Mixtures," *Fluid Phase Equilibria*, **39**, 1 (1988).

[7] J. P. Boon and S. Yip, *Molecular Hydrodynamics*, McGraw-Hill, New York, 1980.

[8] G. Casanov and A. Levi in *Physics of Simple Liquids*, H. N. V. Temperley, J. S. Rowlinson, and G. S. Rushbrooke, Eds., North-Holland, Amsterdam, 1968.

[9] J. Bigeleisen, M. W. Lee, and F. Mandel, "Mean Square Force in Simple Liquids and Solids from Isotope Effect Studies," *Accts. Chem. Res.*, **8**, 179 (1975).

[10] S. M. Thompson, D. J. Tildesley, and W. B. Streett, "The $^{14}N_2 / ^{15}N_2$ and $^{14}N_2 / ^{14}N^{15}N$ Liquid–Vapour Isotope Separation Factor," *Mol. Phys.* **32**, 711 (1976).

[11] J. O. Hirschfelder, "Classical and Quantum Mechanical Hypervirial Theorems," *J. Chem. Phys.*, **33**, 1462 (1960).

[12] C. G. Gray, S. S. Wang, and K. E. Gubbins, "Monte Carlo Calculations of the Mean Squared Force in Molecular Liquids," *Chem. Phys. Lett.*, **26**, 610 (1974).

[13] J. L. Lebowitz, J. K. Percus, and L. Verlet, "Ensemble Dependence of Fluctuations with Application to Machine Computations," *Phys. Rev.*, **153**, 250 (1967).

[14] P. S. Y. Cheung, "On the Calculation of Specific Heats, Thermal Pressure Coefficients and Compressibilities in Molecular Dynamics Simulations," *Mol. Phys.*, **33**, 519 (1977).

[†]Do not take this as a license to write inefficient code; that is not the issue being addressed here.

[15] E. M. Pearson, T. Halicioglu, and W. A. Tiller, "Laplace-Transform Technique for Deriving Thermodynamic Equations from the Classical Microcanonical Ensemble," *Phys. Rev. A*, **32**, 3030 (1985).

[16] J. S. Rowlinson, *Liquids and Liquid Mixtures*, 2nd ed., Butterworths, London, 1969.

[17] K. Denbigh, *The Principles of Chemical Equilibrium*, 4th ed., Cambridge University Press, New York, 1981, pp. 96–97, 120.

[18] G. Jacucci and N. Quirke, "Free Energy Calculations for Crystals" in *Computer Simulation of Solids*, C. R. A. Catlow and W. C. Mackrodt, Eds., Springer-Verlag, Berlin, 1982.

[19] K. S. Shing and K. E. Gubbins, "A Review of Methods for Predicting Fluid Phase Equilibria: Theory and Computer Simulation," *ACS Adv. Chem. Ser.*, **204**, 73 (1983).

[20] D. Frenkel, "Free-Energy Computation and First-Order Phase Transitions" in *Molecular–Dynamics Simulation of Statistical – Mechanical Systems*, G. Ciccotti and W. G. Hoover, Eds., North-Holland, Amsterdam, 1986.

[21] M. Mezei and D. L. Beveridge, "Free Energy Simulations," *Ann. NY Acad. Sci.*, **482**, 1 (1986).

[22] B. Widom, "Some Topics in the Theory of Fluids," *J. Chem. Phys.*, **39**, 2808 (1963).

[23] J. G. Powles, W. A. B. Evans, and N. Quirke, "Non-destructive Molecular Dynamics Simulation of the Chemical Potential of a Fluid," *Mol. Phys.*, **46**, 1347 (1982).

[24] J. G. Powles, "The Liquid–Vapour Coexistence Line by Computer Simulation à la Widom," *Mol. Phys.*, **41**, 715 (1980).

[25] S. Romano and K. Singer, "Calculation of the Entropy of Liquid Chlorine and Bromine by Computer Simulation," *Mol. Phys.*, **37**, 1765 (1979).

[26] J. G. Kirkwood, "Statistical Mechanics of Fluid Mixtures," *J. Chem. Phys.*, **3**, 300 (1935); reprinted in *Theory of Solutions*, Z. W. Salsburg, Ed., John Gamble Kirkwood Collected Works, Gordon and Breach, New York, 1968.

[27] M. Mezei, "Direct Calculation of the Excess Free Energy of the Dense Lennard-Jones Fluid," *Mol. Simul.*, **2**, 201 (1989).

[28] L. Verlet, "Computer 'Experiments' on Classical Fluids. II. Equilibrium Correlation Functions," *Phys. Rev.*, **165**, 201 (1968).

[29] M. Dixon and P. Hutchinson, "A Method for the Extrapolation of Pair Distribution Functions," *Mol. Phys.*, **33**, 1663 (1977).

[30] S. Goldman, "An Explicit Equation for the Radial Distribution Function of a Dense Lennard-Jones Fluid," *J. Phys. Chem.*, **83**, 3033 (1979).

EXERCISES

6.1 What problems can you anticipate in attempting a molecular dynamics simulation at a liquid density using 32 atoms and the truncated Lennard-Jones potential?

6.2 A molecular dynamics simulation of the Lennard-Jones fluid at $\rho\sigma^3 = 0.7$ and $kT/\varepsilon = 0.96$ yields a pressure $\langle P\sigma^3/\varepsilon \rangle = -0.12 \pm 0.01$. This is the full pressure, not just the residual part. The code has been checked and verified at other state conditions. What is the significance of this negative pressure?

6.3 Define a dynamic quantity by $A(t) = W(\mathbf{r}^N)\exp[-\mathscr{U}(\mathbf{r}^N)/kT]$. Now use the fact that $\langle dA/dt \rangle = 0$ to derive the identity (6.23).

6.4 For the mean-square force (6.21) (a) derive the working expression (6.22) for the Lennard-Jones model and (b) derive the alternative form (6.25) with (6.26).

6.5 (a) Show that the pressure and internal energy are related by

$$\left(\frac{\partial U}{\partial V} \right)_{TN} = -P + T\left(\frac{\partial P}{\partial T} \right)_{VN}$$

(b) Now show that the long-range corrections to U and P for the Lennard-Jones fluid, (6.9) and (6.15), are thermodynamically inconsistent: they fail to obey the relation in (a).

6.6 In estimating the long-range corrections (6.9) and (6.15) to the Lennard-Jones internal energy and pressure, we approximated $g(r)$ as unity beyond the cutoff distance r_c. This is generally not a good assumption, especially at high densities. Nevertheless, the estimates for U_{LR} and P_{LR} are generally reliable. Explain.

6.7 Test the mean-field approximations (6.9) and (6.15) for the long-range corrections to the internal energy and pressure by using simulation results for $g(r)$ at $r_c < r < \frac{1}{2}L$ to compute improved estimates for the corrections. What are the magnitudes of the systematic errors introduced by using (6.9) and (6.15)? How do those errors change with density?

6.8 From the general expression for the adiabatic compressibility (6.40), derive the form (6.43) that applies for the Lennard-Jones substance.

6.9 Derive the following expressions for thermodynamic response functions: (a) Eq. (6.49) for the isothermal compressibility, (b) Eq. (6.50) for the isobaric heat capacity, (c) Eq. (6.51) for the volume expansivity, and (d) Eq. (6.52) for the Joule–Thomson coefficient.

6.10 A simulation of 256 Lennard-Jones atoms at $\rho\sigma^3 = 0.7$ produces the following:

$$\langle kT/\varepsilon \rangle = 1.1912 \pm 0.0006 \qquad \langle \mathscr{U}/N\varepsilon \rangle = -4.766 \pm 0.001$$

$$\langle P\sigma^3/\varepsilon \rangle = 0.634 \pm 0.004$$

The run also yields the following fluctuations:

$$\left\langle \left(\delta kT / \varepsilon \right)^2 \right\rangle = 0.287 \pm 0.008 \qquad \left\langle \left(\delta \mathcal{U} / \varepsilon \right)^2 / N \right\rangle = 0.643 \pm 0.02$$

$$\left\langle \left(\delta P \sigma^3 / \varepsilon \right)^2 \right\rangle = 0.042 \pm 0.001$$

$$\left\langle \left(\delta kT / \varepsilon \right) \left(\delta P \sigma^3 / \varepsilon \right) \right\rangle = -0.0058 \pm 0.0002$$

The uncertainties are statistical, quoted as twice the standard deviations of the averages. Compute the following response functions and their *statistical uncertainties* (see Appendix E): (a) residual isometric heat capacity, (b) adiabatic compressibility, (c) thermal pressure coefficient, and (d) isothermal compressibility.

6.11 A sequence of Lennard-Jones simulations produced the following ⊠ results for the configurational internal energy. All runs were performed at the same density and total energy.

N	$\langle \mathcal{U} / N \varepsilon \rangle$
108	-5.37 ± 0.015
256	-5.34 ± 0.01
500	-5.34 ± 0.007
864	-5.37 ± 0.005

(a) Use an unweighted least-squares fit to obtain an estimate for $\langle \mathcal{U} / N \varepsilon \rangle$ in the thermodynamic limit. Also estimate the uncertainty in your limiting value.

(b) Repeat (a) but use a weighted fit, as suggested in Appendix E. Also compute the uncertainty in this limiting value.

(c) Can you both quantitatively and qualitatively distinguish between the results obtained in (a) and (b)?

6.12 Ree has fit simulation *PVT* data for the Lennard-Jones fluid to the function [F. H. Ree, *J. Chem. Phys.*, **73**, 5401 (1980)]

$$\frac{P}{\rho kT} = 1 + Ax^{10} + \sum_{i=1}^{5} x^i \left[B_i - \frac{iC_i}{\sqrt{T^*}} + \frac{D_i}{T^*} \right]$$

where $x = \rho^* / T^{*1/4}$, $\rho^* = \rho \sigma^3$, $T^* = kT / \varepsilon$, and $A = 2.17619$. The equation applies for temperatures in the range $0.76 \leq T^* \leq 2.698$ and densities in the range $0.05 \leq \rho^* \leq 0.96$. Values for the other constants are as follows:

i	B	C	D
1	3.629	5.3692	-3.4921
2	7.2641	6.5797	18.6980
3	10.4924	6.1745	-35.5049
4	11.4590	-4.2685	31.8151
5	0	1.6841	-11.1953

☒ (a) Write a program that uses this Ree equation to compute pressures at specified densities and temperatures.

◧ (b) From this equation, derive expressions for the density and temperature dependence of each of the following and add the results to your program:

    (i) Configurational internal energy (see the expression in Exercise 6.5)

    (ii) Isometric heat capacity

    (iii) Thermal pressure coefficient

    (iv) Isothermal compressibility

    (v) Adiabatic compressibility

    (vi) Residual chemical potential

6.13 Use the Ree equation of state (Exercise 6.12) to compute the vapor–liquid saturation curve for the Lennard-Jones fluid. Use one of the following (equivalent) methods:

(a) Maxwell's equal-area construction or

(b) at specified subcritical temperatures, set the liquid- and vapor-phase chemical potentials equal, set the pressures equal, and solve the resulting two equations simultaneously for the liquid- and vapor-phase densities. Compare your result with that shown in Figure 5.7.

6.14 For any dynamic quantity A having mean-square fluctuation

$$\left\langle (\delta A)^2 \right\rangle = \left\langle (\langle A \rangle - A)^2 \right\rangle$$

verify that an alternative form is $\langle (\delta A)^2 \rangle = \langle A^2 \rangle - \langle A \rangle^2$.

6.15 For an isolated system prove that the mean-square fluctuations in temperature $\langle (\delta T)^2 \rangle$ must be proportional to the mean-square fluctuations in the configurational internal energy $\langle (\delta \mathscr{U})^2 \rangle$.

6.16 Consider any equilibrium property A that is evaluated in a simulation of a finite-size system using

$$\langle A \rangle = \langle A \rangle_{MD} + A_{LR}$$

where A_{LR} is a long-range correction that is constant independent of time. Determine whether or not A_{LR} contributes to the mean-square fluctuation $\langle (\delta A)^2 \rangle = \langle (\langle A \rangle - A)^2 \rangle$.

6.17 (a) For any dynamic quantity A verify that the time average of its reciprocal $\langle A^{-1} \rangle$ is related to the reciprocal of its time average by [E. M. Pearson, T. Halicioglu, and W. A. Tiller, *Phys. Rev. A.*, **32**, 3030 (1985)]

$$\langle A^{-1} \rangle = \frac{1}{\langle A \rangle} \left\langle \frac{1}{1 + \delta A / \langle A \rangle} \right\rangle$$

$$= \frac{1}{\langle A \rangle} \left[1 + \frac{\langle (\delta A)^2 \rangle}{\langle A \rangle^2} + \cdots \right]$$

where $\delta A = A - \langle A \rangle$ measures the fluctuation of A about its average.

(b) Now show that for an isolated system in the large-N limit, (6.36) for C_v^R in terms of the reciprocal kinetic energy is identical to (6.37) for C_v^R in terms of fluctuations in the configurational internal energy.

6.18 (a) How are instantaneous values of C_v distributed about their average?

(b) Derive a fluctuation expression that would enable you to compute, from a single run, the isometric temperature derivative of C_v.

6.19 Consider a time-dependent quantity $M(t)$ that should be distributed in a Gaussian about the time average $\langle M \rangle$. During a simulation a time-dependent error $\varepsilon(t)$ attaches itself to $M(t)$. The error randomly fluctuates about zero with its average $\langle \varepsilon \rangle = 0$.

(a) What error does $\varepsilon(t)$ cause in the mean-square fluctuation $\langle (\delta M)^2 \rangle$?

(b) If $\varepsilon(t)$ is distributed in a Gaussian about zero, will the error cause the distribution of $M(t)$ to deviate from a Gaussian?

(c) If $\varepsilon(t)$ is distributed uniformly about zero, will the error cause the distribution of $M(t)$ to deviate from a Gaussian?

6.20 Perform a simulation of the Lennard-Jones substance, then use the ⊠ results for the radial distribution function $g(r)$ to make the following consistency checks:

(a) The number of atoms N used in the run should be given by (6.97) in the form

$$4\pi \rho \int_0^\infty g(r) r^2 \, dr = N - 1 \approx N$$

(b) The configurational internal energy should be obtained from (6.5),

$$\frac{U_c}{N} = 2\pi\rho \int_0^\infty u(r)g(r)r^2\,dr$$

(c) The pressure should be obtained from the generalization of (6.14)

$$\frac{P}{\rho kT} = 1 - \frac{2\pi\rho}{3kT} \int_0^\infty r\frac{du(r)}{dr}g(r)r^2\,dr$$

6.21 How would you estimate uncertainties in simulation results for $g(r)$?

6.22 Write subroutine(s) to compute the distribution function $g(r)$ in one
⊠ dimension and insert them into the one-dimensional Lennard-Jones
program in Appendix I. Test your implementation by performing the
consistency checks from Exercise 6.20.

6.23 The isothermal compressibility is related to the radial distribution
function by (D. A. McQuarrie, *Statistical Mechanics*, Harper & Row,
New York, 1976, Chapter 13)

$$\rho kT\kappa_T = 1 + 4\pi\rho \int_0^\infty [g(r) - 1]r^2\,dr$$

What difficulties are encountered in trying to use simulation results for
$g(r)$ to compute κ_T from this equation?

6.24 Analogous to the radial distribution function $g(r)$, define a one-particle distribution function $k(r)$ that is proportional to the probability of
finding an atomic center at a distance $r \pm \frac{1}{2}\Delta r$ from a space-fixed
origin. The origin could, for example, be located at the center of the
primary cell. Thus, analogous to (6.94), we define

$$\rho k(r) = \left\langle \sum_i^N \delta[|r - r_i|] \right\rangle$$

(a) Derive the expression, analogous to (6.98), that gives the true
probability for finding any atomic center at position r relative to
the fixed origin.

(b) Derive the working expression, analogous to (6.102), that would
enable you to compute $k(r)$ during a simulation.

(c) Qualitatively sketch $k(r)$ versus r for a uniform homogeneous
fluid, taking $r = 0$ as the center of the container. Would computing
$k(r)$ during a simulation provide any diagnostic check on the
code?

6.25 The program MDSS in Appendix L provides the option of running
☒ either the shifted-force or the truncated Lennard-Jones potential. If
we implement the shift at the distance of the minimum in the potential, then we have a simulation of soft spheres that interact with purely
repulsive forces. Perform a sequence of parallel runs using the truncated potential in one series and the purely repulsive model in the
other. Use the results to study the effects of attractive forces on (a)
simple thermodynamic properties, (b) thermodynamic response functions, and (c) the distribution function $g(r)$.

6.26 In Exercises 1.12 and 1.13 procedures analogous to Monte Carlo and
to molecular dynamics were suggested for measuring the depth of a
body of water. In both schemes advantage was taken of the correlation
of depth with position on the water's surface. Extending these ideas,
the determination of entropy is analogous to measuring the area of the
surface. Explain why simply sampling the surface at discrete points
does *not* enable us to measure its area.

6.27 Derive the test particle expression for the residual chemical potential
of a pure substance when N, V, and T fix the thermodynamic state.

6.28 Write the necessary subroutines to implement the test particle method
☒ (6.72) for determining the residual chemical potential and insert them
▣ into the Lennard-Jones program MDSS in Appendix L.

    (a) Test your implementation by comparing results with values from
either the Ree or Nicolas equation of state.

    (b) Perform simulations to compare test particle results with results
from thermodynamic integration, as described in Section 6.3.1.

6.29 Consider a quantity M that is computed from a sequence of values of
$f(x)$, each of which is determined from a separate simulation. The
sequence of simulations spans the range $[a, b]$ for an independent
variable x. At the conclusion of the simulation sequence, M is computed by

$$M = \int_a^b f(x)\, dx$$

For example, M could be ΔS from (6.59), or S_{res} from (6.78), or ΔS
from (6.84). Each value for $f(x)$ has some uncertainty in it; the
problem is to estimate the accumulated uncertainty in M (see Appendix E). We assume any uncertainties in x are negligible.

    (a) If the uncertainties δf in $f(x)$ are constant, independent of x,
then show that the uncertainty in M is given by

$$\delta M = |b - a|\, \delta f$$

(b) If the uncertainties in f change with x, but nevertheless if the uncertainties are mutually independent (as they usually are when the f-values are from separate runs), then show that a reasonable estimate of δM is

$$\delta M = \Delta x \sqrt{\sum_{k}^{N} (\delta f_k)^2}$$

where N values of $f(x)$ have been generated at equal increments Δx on $[a, b]$.

6.30 A colleague proposes the following scheme for using molecular dynamics to obtain the chemical potential for a pure substance in an isolated system. He starts with the fundamental equation (6.56) in the form

$$\mu \, dN = dE - T \, dS + P \, dV$$

The proposal is to perform two simulations at the same volume V and same total energy E. The first run is to use a number of atoms N; the second $(N + 1)$. Thus, $\Delta N = 1$ and the fundamental equation reduces to

$$\mu = -\int T \, dS$$

From the two runs we cannot directly determine the entropy, but the temperature changes when we change the density, so the proposal is to compute

$$\int T \, dS = \int T \left(\frac{\partial S}{\partial T} \right)_V dT = \int_{T_1}^{T_2} T \left(\frac{C_v}{T} \right) dT \approx C_v (T_2 - T_1)$$

Hence,

$$\mu \overset{?}{=} - C_v (T_2 - T_1)$$

Are there any problems with this scheme?

6.31 Use a Maxwell relation to show that another form of (6.60) is

$$S(\rho_2) - S(\rho_1) = - \int_{\rho_1}^{\rho_2} \frac{\gamma_v}{\rho^2} \, d\rho$$

where γ_v is the thermal pressure coefficient (6.44) and here S is intensive.

6.32 Develop a detailed plan for using molecular dynamics to compute the vapor–liquid saturation curve of the Lennard-Jones substance (cf. Exercise 6.13). What are the difficulties in trying to directly simulate two phases in equilibrium? Your plan should include estimates of the number of production runs to be done, the length of each run, and the anticipated accuracy and precision of the final results. What problems do you expect to encounter and what checks can you apply to the results?

6.33 What changes in your Lennard-Jones code would have to be made to simulate N atoms in vacuo rather than have them confined by periodic boundary conditions? For an introduction to the uses that can be made of such cluster simulations, see R. S. Berry, *Sci. Am.*, **263**(2), 68 (1990).

6.34 What changes in your Lennard-Jones code would have to be made to simulate a fluid in contact with one flat wall? Assume the wall repels the atoms with an r^{-12} potential. In this situation, is there any advantage in using rectangular rather than cubic cells?

7

DYNAMIC PROPERTIES

From the static properties of the previous chapter, we now turn to dynamic properties: time correlation functions, thermal transport coefficients, and dynamic structure. Time correlation functions measure how the value of some dynamic quantity $A(t)$ may be related to the value of some other quantity $B(t)$. Such correlation functions can be computed directly from molecular dynamics data, and methods for doing so are described in Section 7.1.

Certain time correlation functions can be used to compute thermal transport coefficients such as viscosity, thermal conductivity, and diffusion coefficients. The computations can be done either by invoking Green–Kubo formulas, in which a correlation function is integrated over time, or by invoking Einstein relations, in which some other correlation function (a mean-square displacement) is differentiated with respect to time. In either case, transport coefficients are equilibrium properties of a substance and can be obtained from equilibrium molecular dynamics, as described in Section 7.2.

Lastly we consider dynamic structure—the dynamic analog of the radial distribution function—as measured by the space–time correlation function $G(r,t)$. This function separates into a self part and a distinct part; both are discussed in Section 7.3.

7.1 TIME CORRELATION FUNCTIONS

Imagine we are engaged in remote clandestine monitoring of enemy activities. Occasionally, we radio reports to headquarters. At the conclusion of one such transmission, we intercept an enemy transmission to *his* headquarters.

Could the enemy have intercepted our signal, coded it, and sent it on? Can we test whether his signal contains our message *without* breaking his code? If the enemy has not been too clever, we can.

7.1.1   Definitions and Limiting Behavior

Let $A(t)$ and $B(t)$ represent two time-dependent signals and define the time correlation function $C(t)$ by

$$C(t) = \lim_{\tau \to \infty} \frac{1}{\tau} \int_0^\tau A(t_0) B(t_0 + t) \, dt_0 \tag{7.1}$$

This integral represents an average accumulated over many time origins t_0, with each origin taken from a system at equilibrium. The quantity A is sampled at the time origin and B is sampled after a delay time t. Thus, the correlation function C depends on the length of the delay, and because it is an equilibrium property, it is independent of the time origin. Such correlation functions are said to be *stationary*. For the time average in (7.1) we can write the usual angular bracket notation (2.41),

$$C(t) = \langle A(t_0) B(t_0 + t) \rangle \tag{7.2}$$

The function $C(t)$ measures the correlation between the value of A at t_0 and that of B at $t_0 + t$. If A and B differ and are unrelated, then they are uncorrelated, and C reduces to the product of the averages (see Exercise 7.5),

$$C \to \langle A \rangle \langle B \rangle \tag{7.3}$$

On the other hand, if $A(t_0)$ causes or contributes to $B(t_0 + t)$, then A and B are correlated, and C *may* differ from the simple product $\langle A \rangle \langle B \rangle$. Unfortunately, the converse is not true: values of C different from $\langle A \rangle \langle B \rangle$ do not prove that A and B are related. They may be correlated by coincidence rather than by causal connection.

When A and B are different quantities, C is called a *cross-correlation* function. For the systems considered in this book the Hamiltonians are invariant (i.e., of even parity) under spatial inversion ($\mathbf{r}_i \to -\mathbf{r}_i$, $\mathbf{p}_i \to -\mathbf{p}_i$). In such cases, if $A(t)$ and $B(t)$ are of opposite parities, then their time correlation function is zero for all times. When A and B are the same quantity, C is called an *autocorrelation* function, and then C measures how the value of A at $t_0 + t$ is correlated with its value at t_0.

Returning now to our problem of the enemy transmission, let $A(t)$ represent our signal and $B(t)$ represent the enemy's; see Figure 7.1. In some cases we can test whether A and B are correlated by computing the cross-correlation function $\langle A(t_0) B(t_0 + t) \rangle$ and comparing it with the auto-

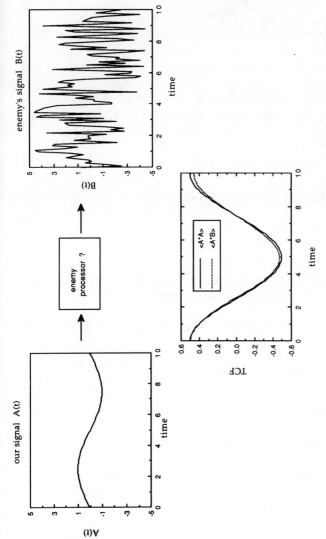

FIGURE 7.1 Time correlation functions (TCFs) can be used to test whether an output signal $B(t)$ is correlated with an input signal $A(t)$. Here the input was $A(t) = \sin(2\pi t/10)$. The output $B(t)$ was formed by adding to $A(t)$ a random noise that was uniformly distributed on $[-4, 4]$. The test for correlation, shown in the TCF plot, is whether the autocorrelation $\langle A(0)A(t)\rangle$ (solid line) equals the cross correlation $\langle A(0)B(t)\rangle$ (broken line). The slight disagreement in the two correlation functions was caused by a small nonuniformity in the random noise added to $A(t)$ and to lower statistical precision in the calculation for $\langle A(0)B(t)\rangle$ when $t > 4$.

correlation of our signal $\langle A(t_0)A(t_0 + t)\rangle$. For example, assume the enemy has simply superimposed on our signal a random noise $R(t)$ that is uniformly distributed about zero; so

$$B(t) = A(t) + R(t) \tag{7.4}$$

Then the cross-correlation function will equal the autocorrelation function,

$$\langle A(t_0)B(t_0 + t)\rangle = \langle A(t_0)A(t_0 + t)\rangle \tag{7.5}$$

because the noise $R(t)$ is unrelated to our signal $A(t)$ and the average $\langle R\rangle = 0$. In fact, this is the situation depicted in Figure 7.1, where we have used for our signal

$$A(t) = \sin\left(\tfrac{2}{10}\pi t\right) \tag{7.6}$$

Then the definition of C (7.1) leads to the autocorrelation

$$\langle A(t_0)A(t_0 + t)\rangle = \tfrac{1}{2}\cos\left(\tfrac{2}{10}\pi t\right) \tag{7.7}$$

Thus, in this situation, the autocorrelation function is itself periodic. In Figure 7.1, the cross-correlation $\langle A(t_0)B(t_0 + t)\rangle$ is compared with the rhs of (7.7), leading us to conclude that the enemy's transmission $B(t)$ does in fact contain our own $A(t)$.

The stationary property of correlation functions $C(t)$ means that C is invariant under translations of the time origin,

$$C(t) = \langle A(t_0)B(t_0 + t)\rangle = \langle A(t_0 + s)B(t_0 + s + t)\rangle \tag{7.8}$$

An important special case occurs if we let $s = -t$, for then (7.8) gives

$$\langle A(t_0)B(t_0 + t)\rangle = \langle A(t_0 - t)B(t_0)\rangle \tag{7.9}$$

When C is an autocorrelation function, then (7.9) reduces to

$$\langle A(t_0)A(t_0 + t)\rangle = \langle A(t_0 - t)A(t_0)\rangle \tag{7.10}$$

For substances involving continuous intermolecular forces, (7.10) can be satisfied only if the autocorrelation $C(t)$ is an *even* function of time. For discontinuous forces, such as appear in hard-body substances, time derivatives of C may not be well defined and then C may be either an even or an odd function of time.

We can distinguish between two time derivatives of a correlation function $C(t)$: that wrt time origins t_0 and that wrt delay times t. The derivative wrt time origins must vanish because of stationarity:

$$\frac{dC(t)}{dt_0} = \frac{d}{dt_0} \langle A(t_0) B(t_0 + t) \rangle = 0 \tag{7.11}$$

or in general,

$$\langle \dot{A}(t_0) B(t_0 + t) \rangle = -\langle A(t_0) \dot{B}(t_0 + t) \rangle \tag{7.12}$$

For autocorrelation functions, (7.12) takes the form

$$\langle \dot{A}(t_0) A(t_0 + t) \rangle = -\langle A(t_0) \dot{A}(t_0 + t) \rangle = -\langle A(t_0 - t) \dot{A}(t_0) \rangle \tag{7.13}$$

The rhs of (7.12) is proportional to the derivative wrt the delay time,

$$\frac{dC(t)}{dt} = \frac{d}{dt} \langle A(t_0) B(t_0 + t) \rangle = \langle A(t_0) \dot{B}(t_0 + t) \rangle \tag{7.14}$$

In the limit of no delay time, $C(0)$ defines a *static* correlation function,

$$C(0) = \lim_{\tau \to \infty} \frac{1}{\tau} \int_0^\tau A(t_0) B(t_0) \, dt_0 \tag{7.15}$$

$$= \langle A(t_0) B(t_0) \rangle \equiv \langle AB \rangle \tag{7.16}$$

In the other extreme, the long-time limit, the behavior of $C(t)$ depends on whether A and B are periodic. Here, we are interested in nonperiodic functions, and in those cases A and B usually become uncorrelated after long delay times,

$$\lim_{t \to \text{large}} C(t) = \langle A \rangle \langle B \rangle \tag{7.17}$$

Thus, for the quantities of interest to us, the time correlation functions decay over time from $\langle AB \rangle$ to $\langle A \rangle \langle B \rangle$. Sometimes it is convenient to normalize $C(t)$ by its static correlation

$$\hat{C}(t) = \frac{\langle A(t_0) B(t_0 + t) \rangle}{\langle AB \rangle} \tag{7.18}$$

and then $\hat{C}(t)$ decays from unity to its uncorrelated limit $\langle A \rangle \langle B \rangle / \langle AB \rangle$. For autocorrelation functions, $\hat{C}(t)$ can never be greater than unity because, before a delay occurs, $A(t_0)$ is already completely correlated with itself.

It is straightforward to obtain the short-time behavior of autocorrelation functions. For times near t_0, expand $A(t_0)A(t_0 + t)$ in a Taylor series about

$t = 0$

$$A(t_0)A(t_0 + t) = A^2(t_0) + t\left[A(t_0)\dot{A}(t_0 + t)\right]_{t=0}$$

$$+ \frac{1}{2!}t^2\left[A(t_0)\ddot{A}(t_0 + t)\right]_{t=0} + \cdots \quad (7.19)$$

Now average over time origins to obtain the correlation function,

$$\langle A(t_0)A(t_0 + t)\rangle = \langle A^2(t_0)\rangle + t\langle A(t_0)\dot{A}(t_0)\rangle + \frac{1}{2!}t^2\langle A(t_0)\ddot{A}(t_0)\rangle + \cdots$$

$$(7.20)$$

Because of the time derivatives (7.11) and (7.14), the first-order term in (7.20) can be written as

$$\langle A(t_0)\dot{A}(t_0)\rangle = \frac{1}{2}\frac{d}{dt_0}\langle A^2(t_0)\rangle \quad (7.21)$$

and the rhs of (7.21) is zero because of stationarity; hence, the Taylor expansion (7.20) reduces to

$$\langle A(t_0)A(t_0 + t)\rangle = \langle A^2(t_0)\rangle + \tfrac{1}{2}t^2\langle A(t_0)\ddot{A}(t_0)\rangle + \cdots \quad (7.22)$$

Thus, for small delays, the autocorrelation function is quadratic in time: it must depart from its static limit with zero slope. The general features of autocorrelation functions are illustrated in Figure 7.2.

7.1.2 Single-Particle Correlations

The autocorrelation functions divide into two classes: the one-particle functions, in which the dynamic quantity $A(t)$ is a property of individual molecules, and the collective functions, in which $A(t)$ depends on the accumulated contributions from all molecules in a system. A typical member of the first class is the velocity autocorrelation function $\Psi(t)$, defined by

$$\Psi(t) = \langle \mathbf{v}_i(t_0) \cdot \mathbf{v}_i(t_0 + t)\rangle \quad (7.23)$$

where $\mathbf{v}_i$ is the velocity of an atom i. The static limit of Ψ is just the mean-square velocity

$$\Psi(0) = \langle \mathbf{v}_i^2\rangle \quad (7.24)$$

In the special case of the one-dimensional harmonic oscillator (ODHO), $\Psi(t)$

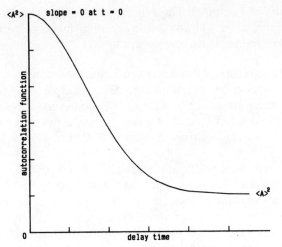

FIGURE 7.2 An autocorrelation function $\langle A(0)A(t)\rangle$ is initially equal to $\langle A^2\rangle$, but as the delay time grows, it decays, initially with zero slope, to $\langle A\rangle^2$.

can be evaluated analytically. Thus, using the expressions for velocity $v(t)$ and position $x(t)$ given in Exercise 2.2, we find

$$\Psi(t) = \langle v^2\rangle\cos\omega t \qquad (7.25)$$

Because the motion of the ODHO is periodic, $\Psi(t)$ fails to approach a constant long-time limit.

However, in simulations of N atoms, atomic motions are chaotic, not periodic, and then the long-time limit is the square of the average velocity, $\Psi(t) \to \langle v\rangle^2$. For nonflowing systems, $\langle v\rangle = 0$, and so

$$\lim_{t \to \text{large}} \Psi(t) = 0 \qquad (7.26)$$

Because the velocity autocorrelation function is a single-particle function and all atoms are indistinguishable in a pure substance, the statistical precision of a calculation for $\Psi(t)$ can be improved by averaging over all N atoms in a system,

$$\Psi(t) = \frac{1}{N}\left\langle \sum_{i}^{N} \mathbf{v}_i(t_0)\cdot\mathbf{v}_i(t_0+t) \right\rangle \qquad (7.27)$$

Now the static limit is proportional to the temperature,

$$\Psi(0) = \frac{3kT}{m} \qquad (7.28)$$

where m is the atomic mass. Thus, for nonperiodic motion in an equilibrium system of N atoms, the velocity autocorrelation function starts from a value proportional to temperature and decays to zero.

Calculational Algorithms. Time correlation functions can be computed either during a simulation or afterward. On the principle that two moderately complicated codes are easier to maintain than one very complicated code, we prefer to do the computation after a run. During a run we save the necessary dynamic quantity $A(t)$ by writing to tape or disk at regular intervals, say every 10 time-steps.

Consider then the velocity autocorrelation function $\Psi(t)$ and assume we have the velocities of N atoms stored on a magnetic tape at L discrete times. This is a huge amount of data, even for a modest calculation: if at 1000 discrete times we have stored three components of each velocity for 108 atoms, then we have 3.24×10^5 double-precision numbers from which to extract one velocity autocorrelation function. From this data we approximate the time average (7.27) for $\Psi(t)$ by

$$\Psi(t) = \frac{1}{MN} \sum_{k}^{M} \sum_{i}^{N} \mathbf{v}_i(t_k) \cdot \mathbf{v}_i(t_k + t) \tag{7.29}$$

Here M is the total number of available time origins; its value changes with the delay time t,

$$M = L - \frac{t}{\Delta t} \tag{7.30}$$

and Δt is the time increment at which velocities have been stored. The calculation of $\Psi(t)$ involves several nested loops, and the efficiency of the calculation depends on how we handle the large amount of data involved. In order of increasing efficiency and sophistication, we discuss three algorithms.

Algorithm I. The least efficient scheme implements (7.29) directly; that is, since in (7.29) the delay time is obviously the independent variable, we are tempted to loop over the delay time, calculating the rhs of (7.29) on each pass:

```
1) set value for the delay time t
2) loop over time origins t_k, {k=1,M}
3) at each origin, loop over atoms to read and store
   velocities v_i(t_k), {i=1,N}
4) from origin t_k, read down tape through delay t
5) at delay t, loop over the atoms to read velocities
   v_i(t_k+t), {i=1,N} and accumulate the integrand
   [v_i(t_k)·v_i(t_k+t)], {i=1,N}
6) rewind the tape
7) goto step 1
```

For a given delay time, each pass through the loop over time origins t_k shifts the origins further down the tape. This translation of origins is illustrated in Figure 7.3.

Since, according to (7.30), the number of terms M in the sum (7.29) decreases with t, statistical uncertainties in $\Psi(t)$ increase as the delay time grows large. Moreover, for small systems, uncertainties at long delay times are compounded by periodicity introduced by periodic boundary conditions (Section 2.9 and Table 2.2).

Algorithm II. Reflecting on Figure 7.3 leads us to realize that the calculation can be made more efficient by rearranging the order in which the loops are nested. In particular, $\Psi(t)$ is invariant under an interchange of the loops over delay time and time origin, and in so doing, we obtain the following algorithm:

```
1) read down tape to next time origin tₖ, {k=1,M}
2) loop over atoms to read and store velocities vᵢ(tₖ),
   {i=1,N}
3) loop over delay times t
4) at each delay t, loop over atoms to read velocities
   vᵢ(tₖ+t), {i=1,N} and accumulate the integrand
   [vᵢ(tₖ)·vᵢ(tₖ+t)], {i=1,N}
5) rewind the tape
6) goto step 1
```

Algorithm III. The second algorithm executes faster than the first, but it is still slow because it requires the tape to be rewound on completing the calculations for each time origin. If sufficient data can be held in fast-access memory, then these rewinds can be eliminated; that is, the calculation can be performed using only one pass through the tape. Typically, the amount of fast memory required is modest. For example, a respectable run duration is 50,000 time-steps, and with velocities stored at intervals of $\Delta t = 10$ time-steps ≈ 0.1 psec, the number of time origins available on tape would be $L = 5000$. But the largest delay time to which the autocorrelation function is to be calculated might typically be only $t_{max} = 5$ psec, so the corresponding number of time-steps J is

$$J = \frac{t_{max}}{\Delta t} + 1 \tag{7.31}$$

For our example, $J = 51$, which is certainly much less than the total number of steps ($L = 5000$) on the tape. Most machines will readily allow storing

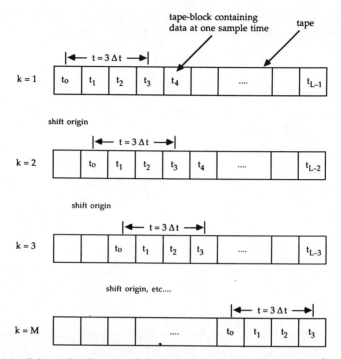

FIGURE 7.3 Schematic of a tape of dynamic data stored at L times, t_i, $\{i = 0, L-1\}$, showing shifts in sampling as the calculation of a time correlation function loops over time origins, t_k, $\{k = 1, M\}$. Here the calculation is for the one delay time $t = 3\,\Delta t$ and the total number of available origins M is related to L and t by (7.30).

$3 \times N \times J$ double-precision numbers in fast memory. The algorithm is then:

```
1) read velocities for N atoms over the first J time-
   steps into fast memory, [[v_i(t_k)], i=1,N], k=1,J
2) take first step in fast memory as time origin k=1
3) loop over delay times t
4) for each delay time t, accumulate [v_i(t_k)·v_i(t_k+t)],
   {i=1,N}
5) at end of loop over t, shift locations of all veloc-
   ities in fast memory, [[v_i(t_k)←v_i(t_k+1)], k=1,J-1],
   i=1,N
6) read from tape into fast memory N velocities for the
   new Jth time-step, v_i(t_J), i=1,N
7) goto step 2
```

As is often the case when analyzing large amounts of data, the efficiency of this algorithm results from execution speed combined with use of fast-access memory.

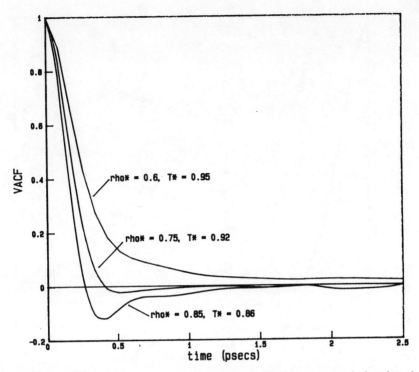

FIGURE 7.4 Effect of density on the normalized velocity autocorrelation function for 108 Lennard-Jones atoms. These results are averages formed over 500 time origins, with sample times separated by intervals of 10 integration time-steps. The dimensional time was obtained using the molecular weight and potential parameters for argon: $\varepsilon / k = 120$ K, $\sigma = 3.405$ Å. The results at $\rho^* = 0.75$ and 0.85 were smoothed by quadratic interpolation.

Sample results for the normalized velocity autocorrelation function (vacf) $\hat{\Psi}(t)$ are shown in Figure 7.4. At low density $\Psi(t)$ remains positive for all times and decays monotonically to zero, but at liquid densities $\Psi(t)$ becomes negative and passes through a minimum before approaching zero. At low densities collisions tend to merely scatter atoms without reversing their trajectories; hence, Ψ remains positive. However, at high densities, atoms are closely packed, so that rebounding collisions are more numerous than scattering collisions, and the many rebounds cause Ψ to change sign. At the middle density in the figure, $\Psi(t)$ has decayed to zero by 1 psec; however, at both lower and higher densities, longer delay times are required for a velocity to become uncorrelated with an earlier value.

Uncertainties. There are several ways for estimating the precision of results for $\Psi(t)$. One is to compute separately its three Cartesian contributions because, for a homogeneous isotropic substance, the three components

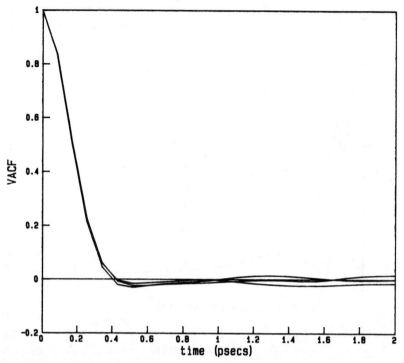

FIGURE 7.5 Cartesian components x, y, and z of the normalized (unsmoothed) velocity autocorrelation function for 108 Lennard-Jones atoms at $\rho^* = 0.75$, $T^* = 0.92$. Calculational details as in Figure 7.4. For times $t > 1$ psec, the variation among the components is ± 0.02. Often, as here, the delay times are simply not sampled at intervals sufficiently small to show the correlation function departing from $t = 0$ with zero slope.

should be equal,

$$\langle \dot{x}_i(t_0) \dot{x}_i(t_0 + t) \rangle = \langle \dot{y}_i(t_0) \dot{y}_i(t_0 + t) \rangle = \langle \dot{z}_i(t_0) \dot{z}_i(t_0 + t) \rangle \quad (7.32)$$

In Figure 7.5 the three components are shown for the moderately dense state from Figure 7.4. For delays $t < 0.3$ psec the three components coincide, but for delays $t > 1$ psec they deviate by ± 0.02.

A second way for estimating statistical uncertainties is prescribed by Zwanzig and Ailawadi [1]. Assuming that the dynamic quantity [here, the velocity $\mathbf{v}(t)$] is a random variable with a Gaussian distribution, the uncertainty is estimated by

$$\varepsilon(t) \approx \pm \sqrt{\frac{2\tau}{t_{max}}} \left[1 - \hat{\Psi}(t) \right] \quad (7.33)$$

where t_{max} is the duration over which the normalized correlation function $\hat{\Psi}(t)$ is computed. Thus, to halve the uncertainty, the duration of the simulation must be increased by a factor of 4. In (7.33), τ is a relaxation time defined by

$$\tau = 2 \int_0^\infty \left[\hat{\Psi}(t) \right]^2 dt \tag{7.34}$$

If $\Psi(t)$ decayed exponentially, which means $\Psi(t) \sim \exp(-t/\tau)$, then τ would be the time required for $\hat{\Psi}$ to decrease from unity to $e^{-1} = 0.368$. For the velocity autocorrelation function in Figure 7.5, numerical integration of (7.34) gives a relaxation time $\tau = 0.34$ psec, which is 40 integration time-steps. Since here Ψ was computed over 5000 time-steps, we have $\tau/t_{max} = \frac{1}{125}$; so at long times when $\Psi \to 0$, the uncertainty given by (7.33) is $\varepsilon(\infty) = \pm 0.13$.

This estimate of the statistical uncertainty is too high because for single-particle correlation functions such as $\Psi(t)$, (7.33) neglects the statistical improvement gained by averaging over all N atoms. By the law of large numbers (see Appendix M), we expect the statistical precision to vary as $N^{1/2}$. Thus, since here $N = 108$, the statistical uncertainty at long times is roughly $\varepsilon(\infty) \approx \pm 0.01$, which is the same order of magnitude as obtained from the variations in Figure 7.5.

In addition to sampling statistics, time correlation functions are adversely affected by periodic boundary conditions. This is because correlation functions measure relations between events occurring at different times, and those events are influenced by the artificial periodicity imposed by the boundaries. The problem is discussed in Section 2.9. Unrealistic correlations can occur when the delay time exceeds the periodic correlation time τ_{pbc}, which, according to (2.115), can be estimated from the velocity of sound. At a specified system density, we can increase periodic correlation times by increasing the system size; thus as shown in Table 2.2, τ_{pbc} increases as the cube root of the number of atoms.

Long-Time Tails. Of the many findings made using molecular dynamics, one of the first was discovery of long-time tails in certain autocorrelation functions. According to the Boltzmann and Enskog equations of kinetic theory, the tail (long-time portion) of the velocity autocorrelation function $\Psi(t)$ should decay exponentially; however, molecular dynamics results [2, 3] on hard-disc ($d = 2$) and hard-sphere ($d = 3$) fluids show that $\Psi(t)$ decays as a power law, $t^{-d/2}$.

This long-lived correlation of the velocity of one atom i is attributed [3] to collective effects involving the many atoms that surround atom i. As atom i moves through a fluid, it creates a vortex motion in neighboring atoms; see Figure 7.6. At low densities where the mean free path is long, this vortex pushes atom i along its path, giving rise in $\Psi(t)$ to an extended region of positive values. Such a region is starting to develop in the low-density result

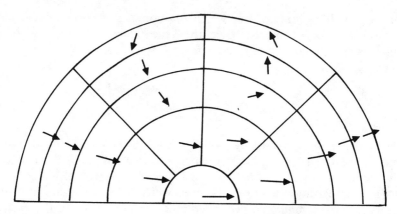

FIGURE 7.6 Upper half of a vortex velocity field created about one disk in a two-dimensional hard-disk simulation. The smallest semicircle represents half of the central disk. Arrows are average velocities for other disks in the circular sections shown; each arrow originates from the center of its section. Lengths of arrows are proportional to the magnitudes of the velocities. The lower half-plane (not shown) is, by symmetry, analogous to this upper half. The simulation used 224 hard disks and was performed at an area of twice that at close packing. At low density the backflow of the vortex pushes the central disk along its trajectory. At high density the central disk will soon reverse its velocity due to a collision and then the vortex backflow will retard the reversed velocity. (This figure is from Alder and Wainwright [3] and is used with permission.)

for $\Psi(t)$ appearing in Figure 7.4. At high densities the mean free path is short, atom i travels a distance, the vortex develops, but then i undergoes a collision that reverses its velocity. After the collision, the motion of atom i is retarded by the vortex backflow, creating an extended region of negative values in the autocorrelation function $\Psi(t)$.

An analytic expression for the tail can be obtained from a hydrodynamic analysis of the long-time behavior of $\Psi(t)$ [4–6],

$$\lim_{t \to \text{large}} \Psi(t) = \frac{2kT}{3\rho m}\left[4\pi\left(D + \frac{\eta}{\rho m}\right)t\right]^{-3/2} \tag{7.35}$$

where m is the atomic mass, D is the self-diffusion coefficient, and η is the shear viscosity. Explicit study of long-time tails by simulation is a delicate problem because the effects are small, and therefore long runs are needed to obtain adequate statistical precision. Moreover, the problem is aggravated by the necessity of using large systems to circumvent the recurrence correlations introduced by using periodic boundaries (see the previous section on uncertainties).

Sample results are shown in Figure 7.7 for the long-time tail of the velocity autocorrelation function. These results are taken from the work of Levesque

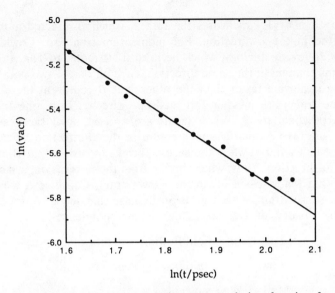

ln(t/psec)

FIGURE 7.7 Long-time tail in the velocity autocorrelation function for 4000 soft repulsive atoms at $\rho^* = 0.45$, $T^* = 2.17$. The time span here is 5 psec $\leq t \leq 7.8$ psec. Solid line has slope $= -\frac{3}{2}$. (This figure is from Levesque and Ashurst [7] and is used with permission.)

and Ashurst [7], who performed two runs, each using 4000 atoms, over a combined duration of 42,600 time-steps. In the figure the simulation data are to be compared with the solid line, which has slope $-\frac{3}{2}$.

7.1.3 Collective Correlations

The velocity autocorrelation function is straightforward to evaluate because the relevant dynamic quantity, the velocity, is purely kinetic and does not depend explicitly on intermolecular forces. Moreover, because it is a single-particle function, its statistical precision can be improved by averaging over all atoms. However, because it is a single-particle function, the velocity autocorrelation function is time consuming to compute. We now turn to more easily but less accurately computed correlation functions—the collective functions.

As a source for a representative collective function, consider the microscopic stress tensor **J**, given by

$$\mathbf{J} = \begin{bmatrix} J_{xx} & J_{xy} & J_{xz} \\ J_{yx} & J_{yy} & J_{yz} \\ J_{zx} & J_{zy} & J_{zz} \end{bmatrix} \tag{7.36}$$

in which element $J_{\alpha\beta}$ measures the rate at which β-directed momentum is transported in the α-direction. For momentum transport, Cowling [8] has given the following analogy, which he in turn attributes to Tait. Imagine two long trains running in the same direction, each on one of two parallel tracks; one train is moving faster than the other. Let β represent the direction in which the trains are moving and let α be directed from one train to the other, orthogonal to β. When trains are abreast, mail sacks are thrown between the two. A sack has momentum in the β-direction because of the motion of its train, but when the sack is thrown, it carries that β-momentum in the $\pm \alpha$-direction. Thus, when thrown from the faster train, a sack gives an impulse that tends to speed up the slower train. Conversely, when thrown from the slower train, a sack gives an impulse that tends to slow the faster train. The analogy involves the following correspondences:

$$\text{Fast train} \Leftrightarrow \text{fluid flowing in center of a pipe}$$

$$\text{Slow train} \Leftrightarrow \text{fluid flowing near pipe walls} \tag{7.37}$$

$$\text{Mail sacks} \Leftrightarrow \text{molecules}$$

Thus, when molecules in one region of fluid move to another region, they carry along momentum. On entering regions of slow-moving fluid, fast molecules tend to increase the average momentum in that region, and conversely. This momentum transport occurs in stationary substances as well as flowing substances, but of course, in nonflowing materials the time-average momentum transport may be zero.

In the train analogy a mail sack thrown to a train contributes in two ways to the train's motion: its own momentum contributes (molecular motion) and the impulse on collision also contributes (molecular forces). Thus, each element of the stress tensor is composed of a kinetic part and a potential part,[†]

$$J_{\alpha\beta} = m \sum_i^N v_{i\alpha} v_{i\beta} + \frac{1}{2} \sum_{i \neq j}^N r_{ij\beta} F_{ij\alpha} \tag{7.38}$$

where m is the atomic mass, $v_{i\alpha}$ is the α-component of the velocity of atom i, $r_{ij\beta}$ is the β-component of the vector $\mathbf{r}_{ij}$ separating atoms i and j, and $F_{ij\alpha}$ is the α-component of the force exerted on atom i by atom j. For a homogeneous isotropic substance, the stress tensor is symmetric; moreover, the time average of the trace of the stress tensor gives, according to the virial theorem, the pressure (see Appendix B),

$$\frac{1}{3} \left\langle \sum_{\alpha = x, y, z} J_{\alpha\alpha} \right\rangle = PV \tag{7.39}$$

[†]Some authors call the object defined by (7.36) and (7.38) the pressure tensor and take its negative to be the stress tensor.

while the time average of each off-diagonal element vanishes,

$$\langle J_{\alpha\beta} \rangle = 0 \qquad \text{for all } \alpha \neq \beta \tag{7.40}$$

The stress autocorrelation function is formed from the off-diagonal elements,

$$\eta(t) = \frac{\rho}{3kT} \frac{1}{N} \sum \langle J_{\alpha\beta}(t_0) J_{\alpha\beta}(t_0 + t) \rangle \tag{7.41}$$

where the sum accumulates three terms given by cyclic permutations of the indices $\alpha\beta(= xy, yz, zx)$. Because $J_{\alpha\beta}$ is a sum of two terms, the correlation function $\eta(t)$ has three contributions: (i) a kinetic term, which measures the correlation of momentum transport caused by atomic motions; (ii) a potential term, which measures the correlation of momentum transport caused by interatomic forces; and (iii) a cross term, which measures the coupling of atomic motions and forces. The kinetic term dominates $\eta(t)$ at gas densities, while the potential term dominates at liquid densities.

If, during a simulation, the atomic positions and velocities are saved periodically to disk, the subsequent analysis for $\eta(t)$ will be time consuming because part of the stress tensor involves intermolecular forces, which must be evaluated by a double sum over all atoms. The computation of $\eta(t)$ would then require time proportional to that used for the simulation itself. However, the calculation can be made more efficient if, during the simulation, we compute and save the three off-diagonal elements of the stress tensor, $J_{\alpha\beta}$. The calculation of the stress tensor adds nothing to the time needed for the simulation because the time-consuming task, the calculation of the force F_{ij}, is already being done. With elements of the stress tensor stored on disk, analysis for $\eta(t)$ from (7.41) is now done simply by

$$\eta(t) = \frac{\rho}{3kT} \frac{1}{NM} \sum \sum_{n}^{M} J_{\alpha\beta}(t_n) J_{\alpha\beta}(t_n + t) \tag{7.42}$$

where M is the number of time origins sampled and the outer sum is over the three cyclic permutations of $\alpha\beta$. Thus, many collective correlations are *easier* to compute than single-particle correlations because neither atomic positions nor velocities need to be stored or subsequently analyzed. However, collective correlations are less precisely computed than single-particle correlations because the precision cannot be improved by summing over N atoms.

Just as for the velocity autocorrelation function, the static limit of the stress autocorrelation function (7.41) is related to thermodynamic quantities and the relation provides a consistency check. For the stress correlation the static limit $\eta(0)$ is the shear modulus: Sitting in a canoe on Lake Wobegon, you strike the water's surface with the blade of a paddle. The water's *initial* response to the sudden shock is a resistance to deformation much like that of

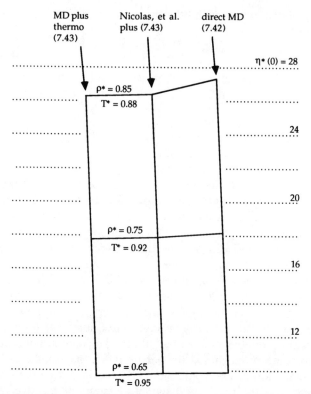

FIGURE 7.8   Comparison of results for the shear modulus $\eta^*(0) = \eta(0)\sigma^3/\varepsilon$ of the Lennard-Jones fluid. The simulation results were obtained from runs using 108 atoms with averages using (7.42) computed from 500 time origins and adjacent origins separated by 10 integration time-steps.

a solid, resistance characterized by two elastic constants[†]: the bulk modulus (the inverse isothermal compressibility) and the shear modulus (the coefficient of rigidity). However, if you push the paddle quietly into the water, the resistance to deformation is that of a viscous fluid, resistance now characterized by two other quantities: the bulk viscosity and the shear viscosity. It is because of water's elastic resistance to a suddenly applied force that, if thrown hard and at low angle, a stone will skip across the lake's surface.

For the Lennard-Jones substance the shear modulus $\eta(0)$ is related to thermodynamics by [9]

$$\eta^*(0) = 3P^* - \tfrac{24}{5}\rho^* U_c^* - 2\rho^* T^* \tag{7.43}$$

[†]W. Thomson, Lecture XX, p.210: "If you make a vibration in glycerine quick enough it will act like a perfectly elastic solid."

where U_c is the configurational internal energy. Figure 7.8 shows a two-way plot (see Appendix D) in which results for $\eta(0)$ obtained directly from (7.42) are compared with values given by the thermodynamic relation (7.43). For the thermodynamic quantities in (7.43) we used direct simulation values; we also used values obtained from the Nicolas et al. empirical equation [10]. If we accept that the simulation results using (7.43) are the most reliable of the three methods, then at the states shown in the figure the Nicolas equation consistently overestimates $\eta(0)$ by 1% while the direct formula (7.42) overestimates $\eta(0)$ by 2–4%.

Figure 7.9 shows stress autocorrelation functions computed from a simulation of 500 Lennard-Jones atoms at a modest density. For delay times $t > 1.5$ psec, the figure shows that the uncertainty in $\eta(t)$ is about ± 0.04. On increasing the number of time origins by a factor of 10 (from 2000 to 20,000), the figure confirms the Zwanzig–Ailawadi estimate (7.33) in that the precision increases by about a factor of 3. Comparing Figure 7.9 for the stress with Figure 7.5 for the velocity indicates that more atoms and longer runs are

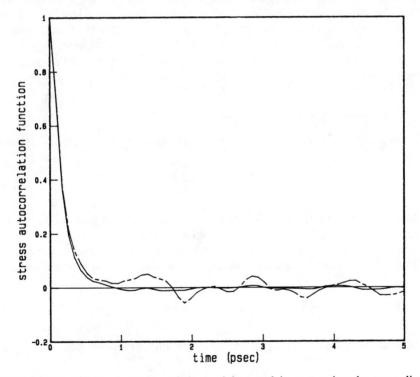

FIGURE 7.9 Effect of number of time origins used in computing the normalized stress autocorrelation function for 500 Lennard-Jones atoms at $\rho^* = 0.72$, $T^* = 1.00$. Broken line is from 2000 time origins; solid line is from 20,000 origins. Adjacent origins were separated by 10 integration time-steps.

generally needed to achieve high statistical precision for a collective correlation function.

7.2 TRANSPORT COEFFICIENTS

When mass, energy, or momentum is transferred through a system, the transport is described, to first order, by a phenomenological relation of the form

$$\text{Flux} = -\,\text{coefficient} \times \text{gradient} \tag{7.44}$$

The flux measures the transfer per unit area in unit time, the gradient provides the driving force for the flux, and the coefficient characterizes the resistance to flow. Examples of (7.44) include Newton's law of viscosity, Fick's law of diffusion, Fourier's law of heat conduction, and Ohm's law of electrical conduction. We normally think of these laws applied to nonequilibrium situations: we apply a temperature gradient to a pot of water, the pot offers a resistance in terms of a thermal conductivity, and the result is a heat flux that boils the water. But in addition to nonequilibrium situations, the linear transport equations (7.44) also apply to microscopic fluctuations that occur in a system at equilibrium [11]. Thus, the transport coefficients, which are properties of matter, can be extracted from equilibrium molecular dynamics simulations.

7.2.1 Generalized Einstein Relations

To illustrate, consider one-dimensional diffusion as described by Fick's law,

$$N\dot{x} = -\,D\,\frac{\partial N}{\partial x} \tag{7.45}$$

where $N = N(x,t)$ is the number of atoms per unit volume (i.e., length) located at position x at time t, $\dot{x}$ is the local velocity at (x,t), and D is the diffusion coefficient. Then $(N\dot{x})$ is the flux. From a material balance on a differential element of fluid, we obtain the equation for continuity of mass,

$$\frac{\partial N}{\partial t} + \frac{\partial(N\dot{x})}{\partial x} = 0 \tag{7.46}$$

and combining these two equations gives the diffusion equation

$$\frac{\partial N}{\partial t} = D\,\frac{\partial^2 N}{\partial x^2} \tag{7.47}$$

For a set of initial conditions, the diffusion equation can be solved for the temporal and spatial evolution of $N(x,t)$. For example, if N_0 atoms were concentrated at the origin $x = 0$ at time $t = 0$, then the solution to (7.47) is [12]

$$N(x,t) = \frac{N_0}{2\sqrt{\pi Dt}} \exp\left[\frac{-x^2}{4Dt}\right] \qquad (7.48)$$

Thus, at any time $t > 0$ the atoms are spatially distributed in a Gaussian about the origin, and as time evolves, atoms diffuse away from the origin, causing the Gaussian to collapse. Imagine one drop of ink added to a pan of water; the water is left undisturbed—no stirring—so, at the point of insertion, the local concentration slowly decreases as the ink diffuses throughout the container.

At any time $t > 0$ the second moment of the distribution gives the mean-square displacement of atoms

$$\left\langle [x(t) - x(0)]^2 \right\rangle = \frac{1}{N_0} \int x^2 N(x,t)\, dx \qquad (7.49)$$

Putting (7.48) into (7.49) and performing the integration, we find that the mean-square displacement is simply related to the diffusion coefficient,

$$\left\langle [x(t) - x(0)]^2 \right\rangle = 2Dt \qquad (7.50)$$

This result applies when the time t is large compared to the average time between collisions of atoms. The three-dimensional analog of (7.50) is

$$\lim_{t \to \infty} \frac{\left\langle [\mathbf{r}(t) - \mathbf{r}(0)]^2 \right\rangle}{6t} = D \qquad (7.51)$$

In realizing (7.50) and (7.51) from simulation, the brackets would be interpreted as averages over time origins, as in (7.1). Since at a specified state condition the diffusion coefficient is a constant, (7.51) implies that the mean-square displacement grows linearly at large delay times. A relation analogous to (7.51) applies to each of the thermal transport coefficients [13]; see Table 7.1.

Equation (7.50) was first obtained by Einstein [14] and is one of a family of fluctuation–dissipation equations that relate transport properties to time or ensemble averages performed on systems at equilibrium. The validity of such relations rests on a proposal of Onsager's [15] called the *fluctuation–regression hypothesis*. Consider a system initially isolated at density $\rho = N/V$ and total energy E. From these initial conditions, the microscopic equations of motion, over time, carry the macroscopic state to the equilibrium state. This

TABLE 7.1   Generalized Einstein and Green–Kubo Formulas for Self-Diffusion Coefficient, Shear, and Longitudinal Viscosity

$$K = \lim_{t \to \infty} \langle [A(t) - A(0)]^2 \rangle / 2t = \int_0^\infty d\tau \langle \dot{A}(\tau) \dot{A}(0) \rangle$$

K	$A(t)$	$\dot{A}(t)$
Self-diffusion coefficient D	$x_i(t)$	$\dot{x}_i(t)$
Shear viscosity $\eta \, VkT$	$m \sum_i^N \dot{x}_i(t) y_i(t)$	$m \sum_i^N \dot{x}_i(t) \dot{y}_i(t) + \sum_{i<j}^N y_{ij}(t) F_{ijx}(t)$
Longitudinal viscosity[a] $\left(\frac{4}{3}\eta + \zeta\right) VkT$	$m \sum_i^N \dot{x}_i(t) x_i(t)$	$m \sum_i^N \dot{x}_i(t) \dot{x}_i(t) + \sum_{i<j}^N x_{ij}(t) F_{ijx}(t) - PV$

[a] ζ is the bulk viscosity.

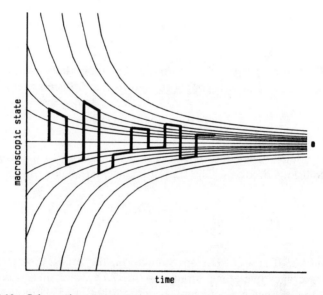

FIGURE 7.10   Schematic representation of relaxation from spontaneous fluctuations that occur in an isolated system. The equilibrium state is represented by the central horizontal line marked with a closed circle at the extreme right. The darkened path represents states visited during fluctuations about equilibrium. These fluctuations merely move the system among a family of trajectories that all eventually converge to the equilibrium state. (Adapted from van Kampen [16] and used with permission.)

convergence to equilibrium occurs regardless of the positions and velocities assigned initially to the molecules, as illustrated schematically in Figure 7.10.

Now, equilibrium is dynamic in the sense that the state of the system fluctuates about the equilibrium state, but those fluctuations only move the system from one convergent path to another. On average, relaxation from a fluctuation tends to carry the system back toward the equilibrium state; this is the point of Figure 7.10. Compared to macroscopically imposed deviations from equilibrium, spontaneous microscopic fluctuations are weak and therefore they can only be linear in the driving forces. As Zwanzig has remarked [11], "Any exception to this would be astonishing indeed." Thus, in addition to describing a system's response to many macroscopically imposed departures from equilibrium, the linear transport laws also describe, on the average, the response to spontaneous microscopic fluctuations.

Consider now the particular case of diffusion: even though a system is at equilibrium, molecules locally aggregate and redisperse, creating local density fluctuations. According to the fluctuation–regression hypothesis, these spontaneous fluctuations tend to be proportional to local gradients, they dissipate (on average) according to Fick's law, and therefore (7.51) provides a means for obtaining the diffusion coefficient from equilibrium molecular dynamics.

7.2.2 Green–Kubo Relations

The Einstein formulas are remarkable because they relate, in simple ways, macroscopic transport coefficients to molecular quantities; moreover, each relation can be reformulated in terms of a time correlation function [11, 17–19]. In the reformulation that follows we start with an arbitrary dynamic quantity $A(t)$ to emphasize that the procedure is independent of the kind of transport being considered. For diffusion, the atomic position $x(t)$ would play the role of $A(t)$.

Consider then $A(t)$, which has a total time derivative

$$\dot{A}(t) = \frac{dA}{dt} \tag{7.52}$$

At time t the displacement of A from its value at $t = 0$ is

$$A(t) - A(0) = \int_0^t dt'\, \dot{A}(t') \tag{7.53}$$

Squaring both sides and averaging over time origins gives the mean-square displacement as

$$\text{msd} = \left\langle [A(t) - A(0)]^2 \right\rangle = \int_0^t dt'' \int_0^t dt' \left\langle \dot{A}(t')\dot{A}(t'') \right\rangle \tag{7.54}$$

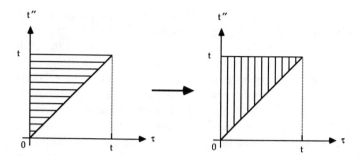

FIGURE 7.11   Effect of interchanging the order of integration in (7.57) is to change the evaluation of the area from an accumulation of horizontal strips to an accumulation of vertical strips. The area and the value of the integral are unaffected.

Since the integrand is symmetric in t' and t'' and since the integration is over a square in $t'-t''$ space, we get the same value for the integral if we integrate over only one diagonal half of the square and double that result,

$$\text{msd} = 2\int_0^t dt'' \int_0^{t''} dt' \left\langle \dot{A}(t')\dot{A}(t'') \right\rangle \qquad (7.55)$$

Stationary correlation functions are unaffected by shifting the time origin to t', so

$$\left\langle \dot{A}(t')\dot{A}(t'') \right\rangle = \left\langle \dot{A}(t''-t')\dot{A}(0) \right\rangle \qquad (7.56)$$

Now introduce the change of variables $\tau = t'' - t'$, so $dt' = -d\tau$ and the inner integral takes lower limit $\tau = t''$ and upper limit $\tau = 0$. Reversing those limits changes the sign of the integral, leaving

$$\text{msd} = 2\int_0^t dt'' \int_0^{t''} d\tau \left\langle \dot{A}(\tau)\dot{A}(0) \right\rangle \qquad (7.57)$$

This double integral accumulates horizontal strips over the upper diagonal of a square, as in Figure 7.11; however, the integral is invariant if we interchange the two integrations and accumulate vertical strips. Then (7.57) becomes

$$\text{msd} = 2\int_0^t d\tau \left\langle \dot{A}(\tau)\dot{A}(0) \right\rangle \int_\tau^t dt'' = 2t\int_0^t d\tau \left\langle \dot{A}(\tau)\dot{A}(0) \right\rangle \left(1-\frac{\tau}{t}\right) \qquad (7.58)$$

or

$$\frac{\left\langle [A(t)-A(0)]^2 \right\rangle}{2t} = \int_0^t d\tau \left\langle \dot{A}(\tau)\dot{A}(0) \right\rangle \left(1-\frac{\tau}{t}\right) \qquad (7.59)$$

Taking the long-time limit, we find

$$\lim_{t \to \infty} \frac{\left\langle [A(t) - A(0)]^2 \right\rangle}{2t} = \int_0^\infty d\tau \left\langle \dot{A}(\tau) \dot{A}(0) \right\rangle \qquad (7.60)$$

This establishes the general relation between the mean-square displacement of a dynamic quantity and an integral over a time correlation function. The time correlation function integrals were derived by Green [20] for thermal transport coefficients and by Kubo [21] for electrical phenomena, so the relations are commonly called Green–Kubo formulas, though neither Green nor Kubo were the first to obtain such relations [11]. For the self-diffusion coefficient the Green–Kubo formula contains the velocity autocorrelation function, for shear viscosity the integration is over the stress autocorrelation function, and for bulk viscosity, the autocorrelation involves diagonal elements of the stress tensor. These relations are summarized in Table 7.1; analogous expressions for other transport coefficients, including thermal conductivity, are given by Hansen and McDonald [22]. The Einstein and Green–Kubo formulas are general in that they do not depend explicitly on intermolecular forces: they apply to materials composed of hard spheres, soft spheres, and nonspherical molecules.

With a mean-square displacement related by (7.59) to a time correlation function, we may evaluate short-time behavior. Substituting into (7.59) the Taylor series for $\left\langle \dot{A}(\tau) \dot{A}(0) \right\rangle$, equivalent to (7.22), we find for small delay times,

$$\left\langle [A(t) - A(0)]^2 \right\rangle = 2 \int_0^t d\tau (t - \tau) \left[\left\langle \dot{A}^2(0) \right\rangle + \cdots \right] \qquad (7.61)$$

$$= t^2 \left\langle \dot{A}^2(0) \right\rangle + \cdots \qquad (7.62)$$

Thus, a mean-square displacement should increase quadratically from $t = 0$, initially with zero slope. The quadratic (7.62) will not necessarily apply to hard-body substances because derivatives of $A(t)$ for discontinuous potentials may be ill-defined.

An example of (7.62) is presented in Figure 7.12; specifically, the figure shows mean-square displacements in atomic positions for the Lennard-Jones fluid at two densities. At the lower density $\rho^* = 0.1$ (solid line), the quadratic behavior near $t = 0$ is pronounced, and the diffusion coefficient (slope) is large, characteristic of gases. At the higher density $\rho^* = 0.75$, the linear

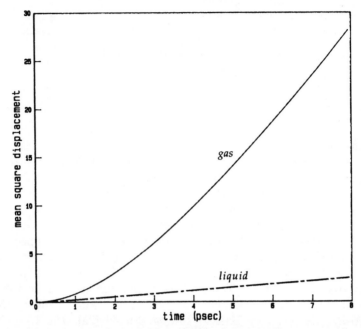

FIGURE 7.12 Examples of positional mean-square displacements for Lennard-Jones fluids. Solid line is for a gaseous state, $N = 32$, $\rho^* = 0.1$, $T^* = 1.475$, and shows pronounced quadratic behavior at short time, linear behavior at long time. Dashed line is for a liquid state, $N = 108$, $\rho^* = 0.75$, $T^* = 1.37$. The mean-square displacement is plotted here in units of σ^2; time has been made dimensional using potential parameters characteristic of argon.

region develops quickly, obscuring the quadratic behavior at small times, and the diffusion coefficient is small, characteristic of liquids.

7.2.3 Diffusion Coefficients

Because it is proportional to single-particle correlations, the self-diffusion coefficient in pure fluids is the transport coefficient most reliably extracted from molecular dynamics: systems of modest size and runs of modest length are sufficient. Moreover, in addition to the Einstein and Green–Kubo formulas, the self-diffusion coefficient can be evaluated in several other ways. For example, the Einstein equation (7.51) can be rewritten as (see Exercise 7.12)

$$D = \frac{1}{6N} \lim_{t \to \infty} \frac{d}{dt} \left\langle \sum_i^N \left[\mathbf{r}_i(t) - \mathbf{r}_i(0) \right]^2 \right\rangle \tag{7.63}$$

which shows that D is proportional to the slope of the mean-square displacement at long times. The form (7.63) is preferred, particularly at low densities, over the original Einstein form (7.51). Otherwise, the Einstein and Green–Kubo formulas yield consistent results for D, and there appears to be little advantage to one over the other.

If we choose to evaluate D from the mean-square displacement, then the atomic positions must be unbounded; otherwise, periodic boundaries will prevent development of the proper linear behavior. In the jargon of simulation, the mean-square displacement must be evaluated using positions from which the periodic boundary conditions have been "unfolded." A simple way to achieve this unfolding is to carry, in the simulation, an additional set of position vectors to which periodic boundary conditions are never applied. Aside from the unfolding, computation of a mean-square displacement from simulation data is algorithmically the same as that for computing an autocorrelation function: we sample a quantity at two different times, combine the samples algebraically, and average over many sample times. Therefore, mean-square displacements can be computed using the same procedures used to compute autocorrelation functions, procedures such as Algorithm III given in Section 7.1.2.

Alder et al. [23] have reported simulation results for the diffusion coefficient in hard-sphere fluids and have compared those results to predictions given by the Enskog theory. For the Lennard-Jones fluid, Levesque and Verlet [24] have reported extensive results, which are correlated by

$$D \frac{\sqrt{m/\varepsilon}}{\sigma} = \frac{0.0445T^*}{\rho^{*2}} - 0.194\rho^* + 0.1538 \qquad (7.64)$$

(The parameter values here differ by a factor of $\sqrt{48}$ from those in ref. 24 because of a difference in the choice of the unit of time.) The fit (7.64) applies for densities $\rho^* > 0.65$; it reproduces the data within a few percent, except for $T^* < 1$, where the deviations may reach 5%. While the Alder et al. hard-sphere results were corrected for both $1/N$ and long-time tail effects, (7.64) includes neither correction and so (7.64) should be used only as a guide to the state-dependent behavior of D.

More recently, Erpenbeck [25] has performed careful simulations of the Lennard-Jones fluid and evaluated the diffusion coefficient by both Green–Kubo and nonequilibrium molecular dynamics. He concludes that for evaluating the self-diffusion coefficient, the equilibrium Green–Kubo method is preferred; some of his results [26] are shown in Figure 7.13. The figure indicates that at least at this state condition the N-dependent correction for 108 atoms contributes about 10% of the diffusion coefficient. Erpenbeck uses (7.35) to account for the long-time tail in the velocity autocorrelation function; this increases the diffusion coefficient by an additional 6.7%, and so his final estimate is $D^* = 0.03044$ at $\rho^* = 0.8442$, $T^* = 0.722$.

For mixtures, evaluation of diffusion coefficients from simulation is less straightforward than for pure fluids. For a binary mixture of components 1

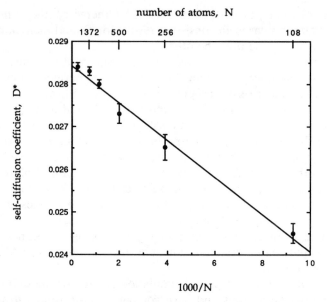

FIGURE 7.13 Effect of system size on the self-diffusion coefficient D^* for the Lennard-Jones liquid at $\rho^* = 0.8442$, $T^* = 0.722$, as reported by Erpenbeck [26]. The line is a least-squares fit. $D^* = D(m/\varepsilon)^{1/2}/\sigma$.

and 2 with mole fractions $x_1 = N_1/N$ and $x_2 = N_2/N$, the Green–Kubo expression for the mutual diffusion coefficient is [27]

$$D_{12} = \frac{1}{3} \left(\frac{\partial^2 (G/NkT)}{\partial x_1^2} \right)_{PT} \int_0^\infty d\tau \langle \mathbf{A}(\tau) \cdot \mathbf{A}(0) \rangle \qquad (7.65)$$

Here G is the mixture Gibbs free energy and $\mathbf{A}(t)$ is defined by

$$\mathbf{A}(t) = x_2 \sum_{i=1}^{N_1} \mathbf{v}_i(t) - x_1 \sum_{i=N_1+1}^{N} \mathbf{v}_i(t) \qquad (7.66)$$

where $\mathbf{v}_i$ is the velocity of atom i. Thus, in general, we need the composition dependence of the mixture free energy $G(x_1)$. To evaluate $G(x_1)$ from simulation, we must apply either the test particle method or the coupling-parameter method; see Section 6.3. Those methods would have to be applied to the mixture over a range of compositions. However, in the special case of a binary ideal solution $G(x_1)$ is simply

$$\frac{G}{NkT} = x_1 \frac{G_{pure\,1}}{NkT} + x_2 \frac{G_{pure\,2}}{NkT} + x_1 \ln x_1 + x_2 \ln x_2 \qquad (7.67)$$

where the mixture and both pure components are all at the same T and P. Then the derivative in (7.65) becomes

$$\left(\frac{\partial^2(G/NkT)}{\partial x_1^2}\right)_{PT} = \frac{1}{x_1 x_2} \tag{7.68}$$

Aside from the complication of the free energy, note that the mutual diffusion coefficient D_{12} is a collective quantity: the dynamic quantity $A(t)$ in (7.66) is already accumulated over all atoms. Thus, simulation results for D_{12} are less precise than those for self-diffusion coefficients obtained from runs of the same duration. However, for simple mixtures the cross-correlation in (7.65) is often negligible, and then the Green–Kubo integrand in (7.65) decomposes into a sum of single-particle correlations [27]. Because of these inherent difficulties, few simulation results are available for mixture diffusion coefficients [28, 29].

7.2.4 Shear Viscosity

To illustrate a transport property related to collective correlations, we consider the shear viscosity η. As shown in Table 7.1, the viscosity can be evaluated from both Green–Kubo and generalized Einstein formulas. For soft-body potentials, most simulators have used the Green–Kubo equation, which is an integral over the stress autocorrelation function,

$$\eta_{xy} = \frac{1}{VkT} \int_0^\infty \langle J_{xy}(t) J_{xy}(0) \rangle \, dt \tag{7.69}$$

Unlike self-diffusion, which is a single-particle phenomenon, a shear stress necessarily involves the entire system; consequently, we cannot improve the statistical precision of results for viscosity by averaging over the N particles in the system. However, for stationary, homogeneous, uniform fluids, the statistical precision can be improved somewhat by averaging over all six terms that result from the stress tensor,

$$\eta = \tfrac{1}{6}(\eta_{xy} + \eta_{xz} + \eta_{yx} + \eta_{yz} + \eta_{zx} + \eta_{zy}) \tag{7.70}$$

The standard deviation of these components from their average provides a guide to the precision of the calculation; but this deviation is only a guide that tends to underestimate the statistical error.

We noted in Section 7.1.3 that the stress autocorrelation function is composed of three parts; likewise, the viscosity divides into a kinetic part, a potential part, and a kinetic–potential part. According to hydrodynamic theory [4, 6], the kinetic part contains a long-time tail,

$$\lim_{t \to \text{large}} \langle J_{xy}(t)J_{xy}(0) \rangle = \frac{V(kT)^2}{120\pi^{3/2}} t^{-3/2} \left[\frac{7}{(2\eta/\rho m)^{3/2}} + \frac{1}{\Gamma^{3/2}} \right] \quad (7.71)$$

where

$$\Gamma = \left(\frac{C_p}{C_v} - 1 \right) \frac{\lambda}{\rho m C_p} + \frac{4\eta}{3\rho m} + \frac{\zeta}{\rho m} \quad (7.72)$$

Here ζ is the bulk viscosity and λ is the thermal conductivity. There is some evidence that long-time tails may also exist in the potential and kinetic–potential parts of the stress correlation [30, 31].

The situation regarding the shear viscosity in Lennard-Jones fluids has been confused because of an erroneous result published in 1973 [32]. The source of the error remains unresolved, though it has been conjectured that the simulation was partially metastable [33] or, perhaps, merely that the statistical error was severely underestimated [26]. In any case, the erroneous value was initially interpreted as demonstrating that the Green–Kubo method was not a reliable procedure for obtaining viscosities and that attitude, in turn, stimulated development of nonequilibrium molecular dynamics methods for obtaining transport properties. More recently, new equilibrium simulations have been performed using the Green–Kubo formula for viscosity, and those results are in statistically good agreement with one another and with nonequilibrium results. Some of those results are presented in Figure 7.14.

The figure shows that, as expected, the statistical precision is not as good as for the diffusion coefficient. However, the N-dependence is small, perhaps small enough to ignore for $N \geq 500$ [34]. More troubling is that results from Lennard-Jones and repulsive soft-sphere simulations [33] suggest that viscosity is sensitive to the long-range part of the intermolecular potential. Therefore, to obtain the full viscosity for the complete Lennard-Jones fluid, it may be that sizable corrections to η must be applied to results obtained from truncated Lennard-Jones runs [26].

For hard spheres the Green–Kubo expression for viscosity is not directly usable because elements of the stress tensor (7.38) involve the force, and forces among hard bodies are discontinuous. Therefore, we appeal to the Einstein relation given in Table 7.1,

$$\eta_{xy} = \frac{1}{VkT} \lim_{t \to \infty} \frac{1}{2t} \left\langle [A_{xy}(t) - A_{xy}(0)]^2 \right\rangle \quad (7.73)$$

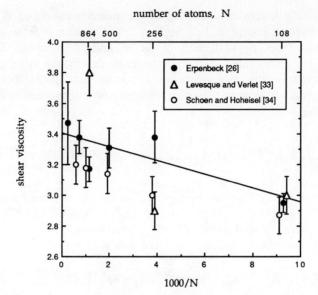

FIGURE 7.14 Effect of system size on the shear viscosity $\eta^* = \eta\sigma^3/\varepsilon\tau^*$ of Lennard-Jones fluids at $\rho^* = 0.8442$, $T^* = 0.722$. The line is a simple least-squares fit to Erpenbeck's [26] data. For clarity, some points have been shifted slightly to the left or right. τ^* is the time unit $\sigma(m/\varepsilon)^{1/2}$.

where

$$A_{xy}(t) = m\sum_{i}^{N}\dot{x}_i(t)y_i(t) \tag{7.74}$$

Recall that because of the equivalence between Green–Kubo and Einstein formulas, the time derivative of A_{xy} is an element of the stress tensor (7.38),

$$\dot{A}_{xy}(t) = \mathsf{J}_{xy}(t) \tag{7.75}$$

The Einstein approach to viscosity is appealing because A_{xy} is readily calculated during a run and at each time-step the computation produces a single number that is to be stored for subsequent determination of a mean-square displacement.

Unfortunately, in periodic systems the Einstein relation (7.73) for viscosity cannot be used directly [35]. In periodic systems Newton's second law is

$$m\frac{d\dot{\mathbf{r}}_i}{dt} = \sum_{\alpha}\sum_{j}\mathbf{F}(\mathbf{r}_{ij} - \alpha L) \tag{7.76}$$

where L is the length of one edge of the primary cube and α is the cell translation vector introduced in Section 2.9. The sum over α accumulates contributions from $\mathbf{r}_{ij}$ and each of its 26 images (in three dimensions). However, for a truncated potential the minimum image criterion (Section 5.1.3) selects from that sum the term having the smallest distance $|\mathbf{r}_{ij} + \alpha L|$; all other terms will be zero. Then an element of the stress tensor is computed by [cf. (7.38)]

$$J_{xy}(t) = m \sum_i \dot{x}_i \dot{y}_i + \frac{1}{2} \sum_\alpha \sum \sum_{j \neq i} (y_{ij} - \alpha_y L) F_x(\mathbf{r}_{ij} - \alpha L) \qquad (7.77)$$

To test whether the Einstein relation (7.73) applies, we must show that the quantity A_{xy} given by (7.74) is an integral of J_{xy} given by (7.77); that is, we must test whether (7.74) and (7.77) are related by (7.75).

To make the test, take the time derivative of (7.74),

$$\dot{A}_{xy}(t) = m \sum_i \dot{x}_i \dot{y}_i + m \sum_i y_i \ddot{x}_i \qquad (7.78)$$

For the x-component of the acceleration, substitute the x-component of the second law; this gives

$$\dot{A}_{xy}(t) = m \sum_i \dot{x}_i \dot{y}_i + \sum_\alpha \sum \sum_{j \neq i} y_i F_x(\mathbf{r}_{ij} - \alpha L) \qquad (7.79)$$

Now symmetrize the second term on the rhs with respect to atoms i and j by writing

$$\dot{A}_{xy}(t) = m \sum_i \dot{x}_i \dot{y}_i + \frac{1}{2} \sum_\alpha \sum \sum_{j \neq i} [y_i F_x(\mathbf{r}_{ij} - \alpha L) + y_j F_x(\mathbf{r}_{ji} + \alpha L)] \qquad (7.80)$$

and applying Newton's third law,

$$F_x(\mathbf{r}_{ij} - \alpha L) = -F_x(\mathbf{r}_{ji} + \alpha L) \qquad (7.81)$$

We obtain

$$\dot{A}_{xy}(t) = m \sum_i \dot{x}_i \dot{y}_i + \frac{1}{2} \sum_\alpha \sum \sum_{j \neq i} y_{ij} F_x(\mathbf{r}_{ij} - \alpha L) \qquad (7.82)$$

where $y_{ij} = y_i - y_j$. Thus we find that the time derivative of A_{xy} in (7.82) generally differs from J_{xy} in (7.77): the two expressions do not provide the equivalence between the Green–Kubo and Einstein equations for viscosity, as demanded by (7.75). When the minimum image criterion is used, we can readily see that (7.82) is wrong because then the image distance used to compute the force F_x in (7.82) will only sometimes (namely, only when $\alpha = \{0,0,0\}$) correspond to the distance y_{ij} that also appears in (7.82).

Note that the inconsistency in (7.82) between y_{ij} and $F_x(\mathbf{r}_{ij} - \alpha L)$ cannot be rectified by simply using positions from which the periodic boundary conditions have been "unfolded." For example, if we compute y_{ij} from positions $\mathbf{r}_i$ and $\mathbf{r}_j$ to which periodic boundary conditions are never applied, as suggested for diffusion in Section 7.2.3, then at long times the distance y_{ij} will definitely *not* correspond to the distance used to compute the force $F_x(\mathbf{r}_{ij} - \alpha L)$ and the equivalence between the Green–Kubo and Einstein forms, represented by (7.75), will *still* not be satisfied. This failure of a direct implementation of an Einstein relation in periodic systems also occurs for the bulk viscosity.

Thus, to evaluate viscosity from the Einstein relation, we proceed indirectly. For continuous intermolecular potentials, we may simply compute elements of the stress tensor via (7.77) and then numerically integrate to obtain the displacement [36],

$$\Delta A_{xy}(t) \equiv A_{xy}(t_0 + t) - A_{xy}(t_0) = \int_{t_0}^{t_0 + t} d\tau\, J_{xy}(\tau) \qquad (7.83)$$

Then

$$\left\langle \left[A_{xy}(t_0 + t) - A_{xy}(t_0) \right]^2 \right\rangle = \left\langle \left[\Delta A_{xy}(t) \right]^2 \right\rangle \qquad (7.84)$$

For hard spheres, we seek numerical forms of (7.83) that avoid explicit evaluation of the discontinuous force that appears in J_{xy}. For example, write (7.77) in the form [23]

$$J_{xy}(t) = \dot{A}_{xy}(t) = m \sum_i^N \dot{x}_i(t)\,\dot{y}_i(t) + m \sum\sum_{i<j} y_{ij}(t) \frac{d\dot{x}_i(t)}{dt} \qquad (7.85)$$

In writing (7.85) we do not explicitly show the periodic images, which are no longer an issue. Now approximate (7.83) by

$$\Delta A_{xy}(t) \approx \sum_{\Delta\tau}^{t} \dot{A}_{xy}(t)\, \Delta\tau \qquad (7.86)$$

where $\Delta\tau$ is of the order of a collision time. Putting (7.85) into (7.86) gives the approximation

$$\Delta A_{xy}(t) \approx m \sum_{\Delta\tau}^{t} \Delta\tau \sum_i^N \dot{x}_i(t)\,\dot{y}_i(t) + m \sum_{\Delta\tau}^{t} \Delta\tau \sum\sum_{i<j} y_{ij}(t) \frac{d\dot{x}_i(t)}{dt} \qquad (7.87)$$

Now the acceleration $d\dot{x}_i/dt = 0$ except when dt spans a collision involving sphere i; therefore, we choose $\Delta\tau = dt$, the time between successive collisions. Then (7.87) reduces to

$$\Delta A_{xy}(t) \approx m \sum_{\Delta\tau}^{t} \Delta\tau \sum_{i}^{N} \dot{x}_i(t)\dot{y}_i(t) + m \sum_{\Delta\tau}^{t} \Delta\dot{x}_i(t)y_{ij}(t) \qquad (7.88)$$

In the second term on the rhs of (7.88), i and j are the two colliding spheres, the change in velocity occurs over the collision, and the double sum has been dropped because all other terms are zero. The mean-square displacement is now obtained by using (7.88) on the rhs of (7.84).

Besides (7.88), other approximations can be generated for hard-sphere viscosity [30]. Similar strategies must be employed for other transport properties of hard bodies if the appropriate Green–Kubo formula explicitly involves the intermolecular force.

7.3 DYNAMIC STRUCTURE

Imagine arriving at the bottom of stairs that empty into the great hall of a railway station; it is Friday afternoon, rush hour. From your position at the staircase you must negotiate your way through the crowd, first to the ticket counter, and then to the platform. Your train departs in 5 minutes. The ticket counter lies a distance r_a from the staircase and you estimate a time t_a minutes to walk from the stairs to the ticket counter. The train lies a distance r_b from the stairs and you estimate that another t_b minutes will be needed to buy the ticket and walk to the platform. Because of the crowd, your progress through the station will be nonlinear. You wonder (a) what is the probability $P_d(r_a, t_a)$ that another person will be at the ticket counter when you arrive there and (b) what is the probability $P_s(r_b, t_a + t_b)$ that you will arrive at the platform in $t_a + t_b < 5$ minutes.

The quantities P_s and P_d reflect information about the structure of the crowd; in particular, these probabilities depend on how a local density $\rho(\mathbf{r}, t)$, at position $\mathbf{r}$ and time t, is related to a local density $\rho(0, 0)$ at another position $\mathbf{r} = 0$ at some earlier time $t = 0$. For matter at the molecular level, these structural probabilities are related to a density–density time correlation function $G(\mathbf{r}, t)$, which is defined by

$$G(\mathbf{r}, t) \equiv \frac{1}{\rho}\langle\rho(\mathbf{r}, t)\rho(0, 0)\rangle = \frac{1}{N}\left\langle \sum_i \sum_j \delta[\mathbf{r} + \mathbf{r}_i(0) - \mathbf{r}_j(t)]\right\rangle \qquad (7.89)$$

Here δ is the Dirac delta symbol and the angular brackets represent the usual time average over a system at equilibrium. The quantity $G(\mathbf{r}, t)$ is called the space–time correlation function; it is the classical analog of the van Hove function used to characterize dynamic structure measured in inelastic neutron scattering experiments [37]. Note from (7.89) that G has units of density and therefore the δ-symbol in (7.89) must have units of V^{-1}; consequently, $G(\mathbf{r}, t)$ is not itself a probability, it is proportional to a probability,

$G(\mathbf{r}, t) \propto$ probability that an atom is at position $\mathbf{r}$ at time t given that an atom was at the origin $\mathbf{r} = 0$ at initial time $t = 0$

The space–time function is a dynamic quantity because it depends on the delay t; nevertheless, it is an equilibrium property obtained by averaging over an equilibrium trajectory in phase space.

It is evident from (7.89) that the space–time correlation function separates into two parts. Terms having $i = j$ yield the *self* correlation function $G_s(\mathbf{r}, t)$, for which the atom at $(\mathbf{r}, t)$ is the same atom that occupied the origin at $(\mathbf{r} = 0, t = 0)$. Terms having $i \neq j$ yield the *distinct* correlation function $G_d(\mathbf{r}, t)$ for which the atom at $(\mathbf{r}, t)$ differs from the one that occupied the origin. Thus, (7.89) can be written as

$$G(\mathbf{r}, t) = G_s(\mathbf{r}, t) + G_d(\mathbf{r}, t) \tag{7.90}$$

The self function is discussed in the next section and the distinct function is discussed in Section 7.3.2.

7.3.1 Self Space–Time Correlation Function

The self correlation function is formally defined from (7.89) as

$$G_s(\mathbf{r}, t) = \frac{1}{N} \left\langle \sum_i \sum_{j=i} \delta[\mathbf{r} + \mathbf{r}_i(0) - \mathbf{r}_j(t)] \right\rangle \tag{7.91}$$

For homogeneous uniform substances, G_s depends only on the scalar distance r, so we may write (7.91) as

$$G_s(r, t) = \frac{1}{N} \left\langle \sum_i \delta[r - |\mathbf{r}_i(0) - \mathbf{r}_i(t)|] \right\rangle \tag{7.92}$$

Physically, G_s measures a probability,

$$G_s(r, t)\, d\mathbf{r} = \begin{cases} \text{probability that at time } t \text{ an atom is a distance } r \\ \text{from an origin } (r = 0) \text{ given that the same atom} \\ \text{was at the origin at the initial time } t = 0 \end{cases} \tag{7.93}$$

and so it is normalized at each instant by

$$\int G_s(r,t)\, d\mathbf{r} = 1 \tag{7.94}$$

At the initial time, G_s must be a delta symbol,

$$G_s(r,0) = \delta \tag{7.95}$$

while at long times and large distances, the position of the atom is unrelated to its initial position, the atom is equally likely to be located anywhere in the container of volume V, and G_s is a constant. The value of that constant is found from the normalization (7.94) to be

$$\lim_{r \to \infty} G_s(r,t) = \lim_{t \to \infty} G_s(r,t) = \frac{1}{V} \approx 0 \tag{7.96}$$

Therefore, as the delay time lengthens, we expect G_s to start from a delta symbol and to collapse to nearly zero. During that collapse, we might guess that G_s is approximately Gaussian. Further, since G_s describes the motion of an individual atom, we expect G_s to be related to the single-particle mean-square displacement. Indeed, a reliable approximation is (Exercise 7.24)

$$G_s(r,t) \approx \frac{1}{\left[\frac{4}{6}\pi \langle \Delta r^2(t)\rangle\right]^{3/2}} \exp\left[-\frac{r^2}{\frac{4}{6}\langle \Delta r^2(t)\rangle}\right] \tag{7.97}$$

where $\langle \Delta r^2(t)\rangle$ is the mean-square displacement at delay time t. In fluids, the Gaussian (7.97) is a reliable approximation for short and long times: at short times, before collisions occur, an atom diffuses as a free particle, the trajectories are straight lines, $\mathbf{r} = \mathbf{v}t$, and $\langle \Delta r^2(t)\rangle$ is given by the first term in a Taylor series (7.62),

$$\langle \Delta r^2(t)\rangle = \frac{3kT}{m}t^2 \tag{7.98}$$

At long times, after many collisions, atomic motion is much like a random walk, the Einstein relation (7.51) is obeyed, and we have

$$\langle \Delta r^2(t)\rangle = 6Dt \tag{7.99}$$

where D is the self-diffusion coefficient. At intermediate times, the Gaussian approximation (7.97) is least reliable; the duration of the intermediate period depends on state condition [38, 39]. A sample G_s is shown for the Lennard-Jones fluid in Figure 7.15. In the figure, points are simulation results and the lines were computed from the Gaussian (7.97) using values for the mean-

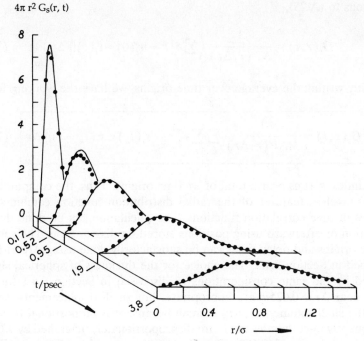

4π r² G_s(r, t)

t/psec

r/σ ⟶

FIGURE 7.15   Self space–time correlation function for 256 Lennard-Jones atoms at $\rho^* = 0.75$, $T^* = 1.012$. Points are simulation results obtained from averages over 2000 time origins, each origin separated by 10 time-steps. Lines are from the Gaussian approximation (7.97) using values of the mean-square displacement obtained from the simulation.

square displacement taken from the simulation. For the state condition of the figure, the Gaussian approximation is not closely obeyed for delay times from about 0.5 to 2 psec.

To obtain an expression by which G_s can be evaluated from simulation data, consider the probability (7.93) written for a specified time and a finite increment in r,

$$G_s(r,t)V(r,\Delta r) = \frac{\langle N(r,t)\rangle_{\Delta r}}{N} \tag{7.100}$$

Here, as in Section 6.4.1, the notation $\langle \cdots (r)\rangle_{\Delta r}$ means a time average over a spherical shell of radius r, thickness Δr, and volume $V(r, \Delta r)$. The numerator in (7.100) is the average number of atoms found at time t lying in a spherical shell of radius r and thickness Δr centered on the atom's position at time $t = 0$. The averaging is over time origins, as in a time correlation function. The number $N(r,t)$ is obtained by sampling all N atoms; thus, analogous to (A.25),

$$G_s(r,t) = \frac{1}{NV(r,\Delta r)}\left\langle \sum_i \delta[r - |\mathbf{r}_i(0) - \mathbf{r}_i(t)|]\,\Delta \mathbf{r}\right\rangle_{\Delta r} \quad (7.101)$$

Explicitly writing the average over time origins, we have the working form

$$\boxed{G_s(r,t) = \frac{1}{MNV(r,\Delta r)}\sum_k \sum_i \delta[r - |\mathbf{r}_i(t_k) - \mathbf{r}_i(t_k + t)|]\,\Delta \mathbf{r}} \quad (7.102)$$

where index k runs over a total of M time origins. Thus, the computation of $G_s(r,t)$ involves features of the radial distribution function combined with features of time correlation functions. The calculation can be done during a simulation or afterward using positions stored on disk or tape. The sampling of time origins and delay times can be satisfactorily done using Algorithm III described in Section 7.1.2. The choice for the thickness of spherical shells is dictated by the same considerations as discussed in Section 6.4.1 for $g(r)$; values near $\Delta r = 0.025\sigma$ are appropriate. For small displacements, say $r < 0.2\sigma$, the shell volume $V(r,\Delta r)$ is small because it is proportional to r^2, and consequently uncertainties in G_s are disproportionately magnified by $V(r,\Delta r)$ in the denominator of (7.102). Therefore, rather than report $G_s(r,t)$ directly, we report results in the form $[r^2 G_s(r,t)]$, as in Figure 7.15.

7.3.2 Distinct Space–Time Correlation Function

For homogeneous uniform substances, the distinct correlation function is formally defined from (7.89) as

$$G_d(r,t) = \frac{1}{N}\left\langle \sum_i \sum_{j \neq i} \delta[r - |\mathbf{r}_i(0) - \mathbf{r}_j(t)|]\right\rangle \quad (7.103)$$

and it measures the probability

$$\frac{1}{N}G_d(r,t)\,d\mathbf{r} = \begin{cases}\text{probability that at time } t \text{ an atom is a distance} \\ r \text{ from an origin } (r = 0) \text{ given that some other} \\ \text{atom was at the origin at the initial time } t = 0\end{cases} \quad (7.104)$$

The normalization is

$$\frac{1}{N}\int G_d(r,t)\,d\mathbf{r} = 1 - \frac{1}{N} \quad (7.105)$$

In words, (7.105) means that accumulating G_d over the available volume merely counts the $N-1$ atoms other than atom i in the container. At the

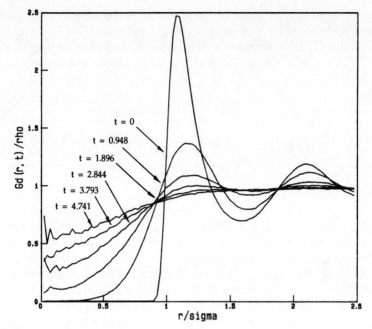

FIGURE 7.16 Distinct space–time correlation function $G_d(r,t)/\rho^*$ from a simulation of 256 Lennard-Jones atoms at $\rho^* = 0.75$, $T^* = 1.012$. Averages were computed over 2000 time origins, each origin separated by 10 time-steps. The delay times t are shown in picoseconds and were made dimensional using potential parameters characteristic of argon.

initial time, the static function $G_d(r,0)$ is proportional to the radial distribution function [cf. (7.104) with (6.98)]

$$G_d(r,0) = \rho g(r), \qquad (7.106)$$

while at long times and large separations the position of one atom is unrelated to the earlier position of another atom, so

$$\lim_{r \to \infty} G_d(r,t) = \lim_{t \to \infty} G_d(r,t) = \rho \qquad (7.107)$$

Therefore, as the delay time lengthens, we expect G_d to collapse from $\rho g(r)$ to the constant ρ at all pair separations r. This behavior is illustrated in Figure 7.16 for the Lennard-Jones fluid. The figure shows that in 1 psec the first peak falls by 50% (from 2.5 to ~ 1.25); that is, at this particular state condition the cage of nearest neighbors is short-lived. This behavior is consistent with the negligible negative region in the velocity autocorrelation function that occurs for this state condition; see Figure 7.4.

To obtain G_d from simulation data, a working expression may be derived in exactly the same way as the corresponding expressions for $g(r)$ and G_s. The result is

$$G_d(r,t) = \frac{1}{M(N/2)V(r,\Delta r)} \sum_k \sum_i \sum_{j>i} \delta\big[r - |\mathbf{r}_i(t_k) - \mathbf{r}_j(t_k + t)|\big]\, \Delta\mathbf{r}$$

(7.108)

This expression is within a factor of ρ of that for evaluating the radial distribution function, except that the positions of atoms i and j are sampled at different times; when the delay time $t = 0$, (7.108) reduces to (6.102) for $[\rho g(r)]$. Note that because of the additional sum over atoms in (7.108) compared to (7.102) for G_s, the calculation of G_d requires about a factor of N more computer time than does the calculation of G_s. Thus, for 256 atoms, if (7.102) requires 1 CPU minute to yield G_s, then (7.108) will require 4 CPU hours to yield G_d. Just as for G_s, the shell volume in the denominator of (7.108) magnifies small uncertainties at small r-values. Such magnification is evident in Figure 7.16.

7.4 SUMMARY

Dynamic properties are principal results provided by molecular dynamics. The time correlation functions and thermal transport coefficients discussed in this chapter are typical: we considered one single-particle function—the velocity autocorrelation function—which leads to the self-diffusion coefficient, and one collective function—the stress autocorrelation function—which leads to the shear viscosity. Of course, many other correlation functions and transport coefficients can be determined by simulation, but the ones discussed here embody the prominent problems that will be encountered in any computation.

The space–time correlation function $G(r,t)$ characterizes the dissipation of local density fluctuations; as such, it measures correlations in both space and time, correlations that molecular dynamics is able to address easily. Indeed, in many situations molecular dynamics can access those correlations more cleanly—less ambiguously—than either theory or experiment. Nevertheless, most simulators concentrate on relatively simple problems involving correlations in either space or time: in many applications the full power of molecular dynamics remains unexploited.

REFERENCES

[1] R. Zwanzig and N. K. Ailawadi, "Statistical Error Due to Finite Time Averaging in Computer Experiments," *Phys. Rev.*, **182**, 280 (1969).

[2] B. J. Alder and T. E. Wainwright, "Velocity Autocorrelations for Hard Spheres," *Phys. Rev. Lett.*, **18**, 988 (1967).

[3] B. J. Alder and T. E. Wainwright, "Decay of the Velocity Autocorrelation Function," *Phys. Rev. A*, **1**, 18 (1970).

[4] M. H. Ernst, E. H. Hauge, and J. M. J. van Leeuwen, "Asymptotic Time Behavior of Correlation Functions. I. Kinetic Terms," *Phys. Rev. A*, **4**, 2055 (1971).

[5] J. R. Dorfman and E. G. D. Cohen, "Velocity-Correlation Functions in Two and Three Dimensions: Low Density," *Phys. Rev. A*, **6**, 776 (1972).

[6] Y. Pomeau and P. Résibois, "Time Dependent Correlation Functions and Mode–Mode Coupling Theories," *Phys. Rpt.*, **19**, 63 (1975).

[7] D. Levesque and W. T. Ashurst, "Long-Time Behavior of the Velocity Autocorrelation Function for a Fluid of Soft Repulsive Particles," *Phys. Rev. Lett.*, **33**, 277 (1974).

[8] T. G. Cowling, *Molecules in Motion*, Hutchinson's University Library, London, 1950, p. 53.

[9] R. Zwanzig and R. D. Mountain, "High-Frequency Elastic Moduli of Simple Fluids," *J. Chem. Phys.*, **43**, 4464 (1965).

[10] J. J. Nicolas, K. E. Gubbins, W. B. Streett, and D. J. Tildesley, "Equation of State for the Lennard-Jones Fluid," *Mol. Phys.*, **37**, 1429 (1979).

[11] R. Zwanzig, "Time-Correlation Functions and Transport Coefficients in Statistical Mechanics," *Ann. Rev. Phys. Chem.* **16**, 67 (1965).

[12] D. A. McQuarrie, *Statistical Mechanics*, Harper & Row, New York, 1976, p. 387.

[13] E. Helfand, "Transport Coefficients from Dissipation in a Canonical Ensemble," *Phys. Rev.*, **119**, 1 (1960).

[14] A. Einstein, "On the Movement of Small Particles Suspended in a Stationary Liquid Demanded by the Molecular-Kinetic Theory of Heat," *Ann. Phys.* (*Leipzig*), **17**, 549 (1905); English translation in A. Einstein, *Investigations on the Theory of the Brownian Movement*, Dover, New York, 1956.

[15] L. Onsager, "Reciprocal Relations in Irreversible Processes. I.," *Phys. Rev.* **37**, 405 (1931); "Reciprocal Relations in Irreversible Processes. II.," *Phys. Rev.*, **38**, 2265 (1931).

[16] N. G. van Kampen, *Stochastic Processes in Physics and Chemistry*, North-Holland, Amsterdam, 1984, p. 270.

[17] D. A. McQuarrie, *Statistical Mechanics*, Harper & Row, New York, 1976, p. 515.

[18] J. P. Boon and S. Yip, *Molecular Hydrodynamics*, McGraw-Hill, New York, 1980, pp. 47–48.

[19] J. P. Hansen and I. R. McDonald, *Theory of Simple Liquids*, 2nd ed., Academic, London, 1986, p. 201.

[20] M. S. Green, "Markoff Random Processes and the Statistical Mechanics of Time-dependent Phenomena. II. Irreversible Processes in Fluids," *J. Chem. Phys.*, **22**, 398 (1954).

[21] R. Kubo, "Statistical Mechanical Theory of Irreversible Processes. I. General Theory and Simple Applications to Magnetic and Conduction Problems," *J. Phys. Soc. Jpn.*, **12**, 570 (1957).

[22] J. P. Hansen and I. R. McDonald, *Theory of Simple Liquids*, 2nd ed., Academic, London, 1986, p. 283.

[23] B. J. Alder, D. M. Gass, and T. E. Wainwright, "Studies in Molecular Dynamics. VIII. The Transport Coefficients for a Hard-Sphere Fluid," *J. Chem. Phys.*, **53**, 3813 (1970).

[24] D. Levesque and L. Verlet, "Computer 'Experiments' on Classical Fluids. III. Time-dependent Self-Correlation Functions," *Phys. Rev. A*, **2**, 2514 (1970).

[25] J. J. Erpenbeck, "Comparison of Green–Kubo and Nonequilibrium Calculations of the Self-Diffusion Constant of a Lennard-Jones Fluid," *Phys. Rev. A*, **35**, 218 (1987).

[26] J. J. Erpenbeck, "Shear Viscosity of the Lennard-Jones Fluid near the Triple Point: Green–Kubo Results," *Phys. Rev. A*, **38**, 6255 (1988).

[27] J. P. Hansen and I. R. McDonald, *Theory of Simple Liquids*, 2nd ed., Academic, London, 1986, p. 289.

[28] G. Jacucci and I. R. McDonald, "Structure and Diffusion in Mixtures of Rare-Gas Liquids," *Physica*, **80A**, 607 (1975).

[29] M. Schoen and C. Hoheisel, "The Mutual Diffusion Coefficient D_{12} in Binary Liquid Model Mixtures. Molecular Dynamics Calculations Based on Lennard-Jones (12-6) Potentials. I. The Method of Determination," *Mol. Phys.*, **52**, 33 (1984); "II. Lorentz–Berthelot Mixtures," *Mol. Phys.*, **52**, 1029 (1984).

[30] J. J. Erpenbeck and W. W. Wood, "Molecular Dynamics Calculations of Shear Viscosity Time-Correlation Functions for Hard Spheres," *J. Stat. Phys.*, **24**, 455 (1981).

[31] T. R. Kirkpatrick, "Large Long-Time Tails and Shear Waves in Dense Classical Liquids," *Phys. Rev. Lett.*, **53**, 1735 (1984).

[32] D. Levesque, L. Verlet, and J. Kürkijarvi, "Computer 'Experiments' on Classical Fluids. IV. Transport Properties and Time-Correlation Functions of the Lennard-Jones Liquid near Its Triple Point," *Phys. Rev. A*, **7**, 1690 (1973).

[33] D. Levesque and L. Verlet, "Molecular Dynamics Calculations of Transport Coefficients," *Mol. Phys.*, **61**, 143 (1987).

[34] M. Schoen and C. Hoheisel, "The Shear Viscosity of a Lennard-Jones Fluid Calculated by Equilibrium Molecular Dynamics," *Mol. Phys.*, **56**, 653 (1985).

[35] J. J. Erpenbeck, private communication, 1991.

[36] C. Hoheisel and R. Vogelsang, "Thermal Transport Coefficients for One- and Two-Component Liquids from Time Correlation Functions Computed by Molecular Dynamics," *Comp. Phys. Rep.*, **8**, 1 (1988).

[37] L. van Hove, "Correlations in Space and Time and Born Approximation Scattering in Systems of Interacting Particles," *Phys. Rev.*, **95**, 249 (1954).

[38] A. Rahman, "Correlations in the Motion of Atoms in Liquid Argon," *Phys. Rev.*, **136**, 405 (1964).

[39] W. A. Steele and W. B. Streett, "Computer Simulations of Dense Molecular Fluids. I. Time-dependent Statistical Properties of Single Diatomic Molecules," *Mol. Phys.*, **39**, 279 (1980).

EXERCISES

7.1 Demonstrate whether or not the autocorrelation of $A(t) = \sin 2\pi t$ is stationary.

7.2 Show that $\langle \sin \omega t_0 \sin \omega(t_0 + t) \rangle = \langle \cos \omega t_0 \cos \omega(t_0 + t) \rangle$. Thus, two different dynamic quantities may have the same autocorrelation function.

7.3 Show that the autocorrelation of the sum $A(t) + B(t)$ is not necessarily equal to the sum of the autocorrelations of A and B.

7.4 A function $f(x)$ is square integrable on $[a, b]$ if the following integral exists:

$$\int_a^b |f(x)|^2 \, dx$$

The absolute value sign allows for the possibility that f might be complex. If both $f(x)$ and $g(x)$ are square integrable on $[a, b]$, then notice that we must have

$$\int_a^b \int_a^b |f(x)g(y) - f(y)g(x)|^2 \, dx \, dy \geq 0 \qquad \text{(E7.4)}$$

(a) Starting from (E7.4) and considering f and g to simply be real, derive the *Schwarz inequality*,

$$\left| \int_a^b f(x)g(x) \, dx \right| \leq \left(\int_a^b f(x)f(x) \, dx \right)^{1/2} \left(\int_a^b g(y)g(y) \, dy \right)^{1/2}$$

(b) Use the Schwarz inequality to prove that the magnitude of any autocorrelation function cannot exceed its static limit; that is, prove that

$$|\langle A(t_0) A(t_0 + t) \rangle| \leq \langle A^2(t_0) \rangle$$

7.5   For a time correlation function $C(t) = \langle A(t_0)B(t_0 + t) \rangle$, first write A and B each as fluctuations about their means,

$$A(t) = \langle A \rangle + \delta A(t) \qquad B(t) = \langle B \rangle + \delta B(t)$$

Then show that if A and B are uncorrelated, $C(t) = \langle A \rangle \langle B \rangle$.

7.6   A dynamic quantity $A(t)$ evaluated during a simulation has superimposed on it a random error $\varepsilon(t)$ whose time average is zero. How would the presence of this error affect the value computed for the autocorrelation of $A(t)$?

7.7   Derive (7.25), that is, derive the expression for the velocity autocorrelation function of the one-dimensional harmonic oscillator.

7.8   Use the analytic solution for the ODHO given in Exercise 2.2 to illustrate (7.12). That is, show that the one-dimensional harmonic oscillator has

$$\langle x(t_0)v(t_0 + t) \rangle = -\langle v(t_0)x(t_0 + t) \rangle$$

7.9   For one-dimensional diffusion with N_0 atoms concentrated at the origin $x = 0$ at time $t = 0$, the solution to the diffusion equation (7.47) is the Gaussian distribution (7.48). At any time $t > 0$, the second moment of that distribution is

$$\langle x^2(t) \rangle = \frac{\int_{-x(t)}^{x(t)} x^2(t) N(x,t) \, dx}{\int_{-x(t)}^{x(t)} N(x,t) \, dx} \qquad \text{(E7.9a)}$$

(a) Since the average position at any time is the origin, that is, $\langle x(t) \rangle = \langle x(0) \rangle = 0$, show that the lhs of (E7.9a) can be written as

$$\langle x^2(t) \rangle = \langle [x(t) - x(0)]^2 \rangle \qquad \text{(E7.9b)}$$

(b) Since at any time $t > 0$ no atoms appear at $x > x(t)$ or at $x < -x(t)$, we can safely extend the integration limits in (E7.9a) to infinity. Now put the distribution (7.48) plus (E7.9b) into (E7.9a), and so derive the Einstein relation (7.50).

7.10   Helfand has noted that not only the second moment but also higher moments of the Gaussian (7.48) can be used to obtain the one-dimensional self-diffusion coefficient [E. Helfand, *Phys. Rev.*, **119**, 1 (1960)].

Show that the fourth moment can be used in the form

$$D^2 = \frac{1}{12 \, t^2} \langle [x(t) - x(0)]^4 \rangle$$

If both the second and fourth moments produce consistent values for D, how must the two moments be related?

7.11 Start with the Einstein relation (7.51) and show that, for an ideal gas, the self-diffusion coefficient is bounded only by the walls of the container; that is, diffusion is not well defined for ideal gases. Repeat the argument starting from the Green–Kubo formula for D.

7.12 (a) Use the definition of a derivative to prove Leibnitz's rule in the form

$$\frac{d}{dx} \int_0^{b(x)} f(x, z) \, dz = f(x, b) \frac{db(x)}{dx} + \int_0^{b(x)} \frac{\partial f(x, z)}{\partial x} \, dz$$

(b) Now use Leibnitz's rule along with (7.59) to prove (7.63); that is, show that

$$\lim_{t \to \infty} \frac{d}{dt} \langle [A(t) - A(0)]^2 \rangle = \lim_{t \to \infty} \frac{1}{t} \langle [A(t) - A(0)]^2 \rangle$$

7.13 Combine (7.43) with (6.20) to show that, for a Lennard-Jones substance, the shear modulus can be written as

$$\eta^*(0) = \rho^* T^* + \tfrac{24}{5} \rho^* \left[6 \langle r^{*-12} \rangle - \langle r^{*-6} \rangle \right]$$

7.14 Consider N hard spheres of diameter σ and mass m forming a dilute gas in a volume V, and so having number density $\rho = N/V$. The gas is at temperature T, so the spheres move with mean speed $v_0 = (kT/m)^{1/2}$. On average, the spheres are uniformly distributed throughout V. Imagine constructing two parallel planes, S and S', separated by distance h. We want to estimate the number of collisions a sphere suffers in moving through h.

Initially, assume in plane S that target-spheres are stationary; then, there are ρh targets per unit area of plane S. Since a collision occurs whenever a moving sphere's center is at a distance σ from a target's center, each target presents an effective area of $\pi \sigma^2$ for collision. On average, then, the fractional area of S occupied by targets is $\pi \sigma^2 \rho h$; in other words, the probability of a moving sphere colliding with a target in distance h is, on average, $\pi \sigma^2 \rho h$ per unit area of S.

Now the time needed for a sphere to cover distance h is h/v_0, so the collision frequency is

$$\left. \begin{array}{l} \text{number of collisions per unit time for one} \\ \text{sphere moving among stationary targets} \end{array} \right\} = \frac{\pi \sigma^2 \rho h}{h/v_0} = \pi \sigma^2 \rho v_0$$

But the target spheres are themselves moving with speed v_0, and their motion shortens the time between collisions, so the speed of approach for two spheres is not v_0, but the root-mean-square speed of the two,

$$\sqrt{v_0 + v_0} = v_0 \sqrt{2}$$

Thus, the collision frequency is

$$\left. \begin{array}{l} \text{number of collisions per unit time for one} \\ \text{sphere moving among moving targets} \end{array} \right\} = \pi \sigma^2 \rho v_0 \sqrt{2}$$

The reciprocal collision frequency is the mean time between collisions for one sphere,

$$t_c = \frac{1}{\sqrt{2}\,\pi \sigma^2 \rho v_0}$$

and the mean distance one sphere travels between collisions, l, is the mean free path (A rigorous derivation of this expression for the hard-sphere mean free path is given in D. A. McQuarrie, *Statistical Mechanics*, Harper & Row, New York, 1976, pp. 370–372, 375):

$$l = t_c v_0 = \frac{1}{\sqrt{2}\,\pi \sigma^2 \rho}$$

Consider a gas at 300 K, 0.05 g/cm^3 with molecular weight $M = 50$. Assume the molecules to be nearly spherical with 4 Å diameters.
(a) Estimate the hard-sphere mean free path in Å.
(b) Estimate the hard-sphere mean speed, v_0, in cm/sec.
(c) Estimate the average collision frequency per sphere.

7.15 Estimate the self-diffusion coefficient for a dilute hard-sphere gas. Start with the Green–Kubo expression for diffusion in the form

$$D = v_0^2 \int_0^\infty \hat{\Psi}(t)\, dt$$

where v_0 is the mean speed defined in Exercise 7.14. Then assume the normalized velocity autocorrelation function decays exponentially,

$$\hat{\Psi}(t) = e^{-t/\tau}$$

where τ is a relaxation time. Estimate this relaxation time by the time required for a sphere, moving with mean speed v_0, to cover the mean distance between collisions, l, where l is the mean free path given in Exercise 7.14. You should find that the diffusion coefficient D is proportional to $\sqrt{T}$, inversely proportional to ρ, inversely proportional to $\sqrt{m}$, and inversely proportional to σ^2.

7.16 (a) Use the Taylor series (7.22) along with (7.14) to express the short-time behavior of the velocity autocorrelation function in the form

$$\langle \mathbf{v}(t_0) \cdot \mathbf{v}(t_0 + t) \rangle = \langle v^2(t_0) \rangle - \frac{t^2}{2\,m^2} \langle F^2 \rangle + \cdots$$

where $\langle F^2 \rangle$ is the mean square force on an atom (see Section 6.1.3).

(b) From your result in (a), show that the short-time behavior of the normalized velocity autocorrelation function can be written as

$$\hat{\Psi}(t) = \exp \frac{-t^2 \langle F^2 \rangle}{6mkT}$$

(c) Use the expression from (b) in the Green–Kubo integral to derive the following approximate relation between the self-diffusion coefficient and the mean square force

$$D \approx kT \sqrt{\frac{3\pi kT}{2m\langle F^2 \rangle}}$$

7.17 In a system of N atoms, the dynamic quantity for obtaining the shear viscosity from the generalized Einstein relation is

$$A(t) = m \sum_i^N \dot{x}_i(t) y_i(t)$$

Show that for a nonflowing system the value of $A(t)$ is independent of the location of the origin used to measure the positions $y_i(t)$.

7.18   (a) Prove that $\langle A(t) \rangle = -\langle B(t) \rangle$, where

$$A(t) = \sum_i^N \dot{x}_i(t) y_i(t) \qquad \text{and} \qquad B(t) = \sum_i^N x_i(t) \dot{y}_i(t)$$

(b) Now prove that $\langle [A(t) - A(0)]^2 \rangle = \langle [B(t) - B(0)]^2 \rangle$.

7.19   For a nonflowing system of N atoms at equilibrium, show that the time average is zero for any off-diagonal element of the stress tensor, J_{xy},

$$J_{xy} = \frac{dA(t)}{dt}$$

where $A(t)$ is defined in Exercise 7.17.

7.20   For any dynamic quantity $A(t)$, show that its mean-square displacement obeys

$$\langle [A(t) - A(0)]^2 \rangle = -2\langle A(0)[A(t) - A(0)] \rangle$$

7.21   In the generalized Einstein expression for viscosity, the dynamic quantity $A(t)$ is defined in Exercise 7.17. Show that A can also be written in the symmetrized form

$$A(t) = m \sum_{i<j}^N \sum^N \dot{x}_{ij}(t) y_{ij}(t)$$

where $\dot{x}_{ij}$ is a relative velocity, $\dot{x}_{ij} = \dot{x}_i - \dot{x}_j$, and y_{ij} is a relative position, $y_{ij} = y_i - y_j$.

7.22   Estimate the shear viscosity for a dilute hard-sphere gas. For hard spheres, the Green–Kubo expression for the shear viscosity (7.69) reduces to

$$\eta = \frac{m^2}{kTV} \int_0^\infty dt \left\langle \left(\sum_i^N v_{ix}(t) v_{iy}(t) \right) \left(\sum_j^N v_{jx}(0) v_{jy}(0) \right) \right\rangle$$

As in Exercise 7.15, assume $\eta(t)$ decays exponentially,

$$\eta = \eta(0) \int_0^\infty dt\, e^{-t/\tau}$$

where τ is the relaxation time for the $x - y$ element of the stress tensor and

$$\eta(0) = \frac{m^2}{kTV}\left\langle\left(\sum_i^N v_{ix}(0)v_{iy}(0)\right)\left(\sum_j^N v_{jx}(0)v_{jy}(0)\right)\right\rangle$$

(a) Using the fact that at any instant the velocities of atoms i are uncorrelated with those of atoms $j \neq i$ and, further, that the x-component of an atomic velocity is independent of its y-component, show that this expression reduces to $\eta(0) = \rho kT$.

(b) Now make the same assumption for τ as used in Exercise 7.15, and thereby show that the hard-sphere gas viscosity is independent of the number density ρ, proportional to $\sqrt{m}$, proportional to $\sqrt{T}$, and inversely proportional to σ^2.

7.23 Use the normalization conditions for $G_s(r,t)$ (7.94) and for $G_d(r,t)$ (7.105) to obtain the value of

$$\int G(r,t)\, d\mathbf{r}$$

at any instant t. Give a physical explanation of your result.

7.24 The single-particle correlation function $G_s(r, t)$ can be obtained from a more general single-particle space–momentum–time distribution function by integrating over the velocities,

$$G_s(r,t) = \int f(\mathbf{r},\mathbf{v},t)\, d\mathbf{v} \qquad (E7.24a)$$

At short times, before collisions, an atom translates freely, so $\mathbf{r} = \mathbf{v}t$, and the distribution function f separates into

$$f(\mathbf{r},\mathbf{v},t) = f_m(\mathbf{v})\, \delta[\mathbf{v} - \mathbf{r}/t] \qquad (E7.24b)$$

where f_m is Maxwell's velocity distribution (2.54). Use (E7.24b) in (E7.24a) to show that

$$G_s(r,t) = C \exp\left[\frac{-mr^2}{2kTt^2}\right]$$

Then use the normalization condition (7.94) to evaluate the constant C, and so derive the short-time Gaussian approximation (7.97) for G_s.

7.25 (a) Modify the code GENTCF from Appendix J to compute the velocity autocorrelation function (7.29).

$\boxtimes$ (b) Further modify GENTCF to compute the mean-square displacement.

7.26 Use the hard-sphere code MDHS from Appendix K to perform a
⊠ simulation using 108 spheres at packing fraction $\eta = 0.3$. Every 50
collisions, store the sphere positions and velocities to disk.

   (a) Use your modified code GENTCF from Exercise 7.25 to compute
   the velocity autocorrelation function and the mean-square displacement from your simulation data.

   (b) With your results from (a) for the velocity autocorrelation function
   (vacf), use (7.59) to compute the mean-square displacement and
   compare with the mean-square displacement (msd) obtained in (a).

7.27 **(a)** Apply the Green–Kubo formula to the hard sphere vacf obtained
⊠ in Exercise 7.26 to obtain the self-diffusion coefficient $D^* = D/\sigma(kT/m)^{1/2}$.

   (b) Apply the Einstein formula (7.63) to the msd obtained in Exercise
   7.26 to compute D^*.

   (c) Compare your results for D^* from (a) and (b) with the Enskog
   expression for hard spheres (see D. A. McQuarrie, *Statistical
   Mechanics*, Harper & Row, New York, 1976, Chapter 19):

$$D^* = \frac{\sqrt{\pi}/4}{P/\rho kT - 1}$$

   (d) Repeat the test of the Enskog theory by doing several simulations
   at various packing fractions $\eta < 0.47$.

7.28 Modify code GENTCF as required in Exercise 7.25 to compute the vacf
⊠ and the msd. Apply your modified GENTCF code to data generated
using the one-dimensional Lennard-Jones simulation code MDODSS
from Appendix I. In particular, (a) study how the vacf and msd each
respond to changes in density at fixed temperature and (b) how each
respond to changes in temperature at fixed density. (c) How does the
diffusion coefficient respond to the changes performed in (a) and (b)?

7.29 Using the code GENTCF as a guide, write a program that analyzes
⊠ simulation data to obtain the self space–time correlation function
$G_s(r,t)$. Apply your program to simulation data for the one-dimensional Lennard-Jones substance obtained using code MDODSS from
Appendix I. Compare your results with the one-dimensional form of
the Gaussian (7.97).

7.30 Using the code GENTCF as a guide, write a program that analyzes
⊠ simulation data to obtain the distinct space–time correlation function
$G_d(r,t)$. Apply your program to simulation data for the one-dimensional Lennard-Jones substance obtained using code MDODSS from
Appendix I.

APPENDIX A

THE DELTA SYMBOL

Molecular simulations are particularly useful in providing spatial distribution functions—how atoms or molecules arrange themselves throughout the available volume. The computation of such functions involves a counting process, and this appendix discusses the notation used to represent the counting.

As a concrete example, consider a closed, one-dimensional system of length L bounded on $[a, b]$ and containing N atoms, as in Figure A.1. The bulk number density (atoms per unit length) is $\rho = N/L$. We might want to characterize how the atoms are distributed on L, and therefore we define a local density $\rho(x)$ by

$$\rho(x) = \frac{N(x)}{V(x)} = \lim_{\Delta x \to 0} \frac{N(x, \Delta x)}{V(x, \Delta x)} \tag{A.1}$$

where $N(x, \Delta x)$ is the number of atomic centers lying in a length element centered at x and having thickness Δx. Thus, $V(x, \Delta x) = \Delta x$ is the "volume" (i.e., length) of the element. If we accumulate this local density over the entire system, we should obtain the total number of atoms present,

$$\int_a^b \rho(x) \, dx = N \tag{A.2}$$

At one instant during a simulation we could compute $\rho(x)$ by dividing L into elements, each of finite thickness Δx, and counting the number of

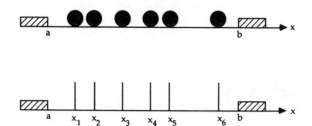

FIGURE A.1 Two representations of a one-dimensional system confined to a line segment $L = b - a$. The schematic version on the upper axis is reinterpreted on the lower axis as a sequence of δ-symbols (A.4).

atomic centers in each element. How shall we symbolically represent this counting operation? The atomic centers occupy discrete positions x_i $\{i = 1, 2, \ldots, N\}$ on $[a, b]$ and so we represent those locations by spikes (vertical lines of infinite height or spectral lines, if you like), as in Figure A.1. We now represent the local density $\rho(x)$ by the notation

$$\rho(x) = \sum_i^N \delta(x - x_i) \tag{A.3}$$

where the δ-*symbol* takes values

$$\delta(x - x_i) = \begin{cases} \infty & \text{if center of atom } i \text{ is located at } x \\ 0 & \text{if center of atom } i \text{ is not at } x \end{cases} \tag{A.4}$$

Note that for dimensional consistency in (A.3), $\delta(x)$ must have "units" of L^{-1}, though since δ can be only zero or infinite, it hardly matters numerically what its units may be. Putting (A.3) into (A.2), we see that

$$\int_a^b \sum_i^N \delta(x - x_i)\, dx = N \tag{A.5}$$

Interchanging the order of integration and summation in (A.5) leads to the normalization

$$\int_a^b \delta(x - x_i)\, dx = 1 \tag{A.6}$$

which merely indicates that atom i is located somewhere in the system.

Although (A.3) is a convenient representation for $\rho(x)$, it cannot be used to compute the local density. Implicit in (A.3) is a limiting process; that is, at a specified point x there can be at most only one atomic center, so the local density at a point is either 1 or 0. Thus, $\rho(x)$ can be computed only for a region of finite size Δx. To obtain a useful form for computing $\rho(x)$, we first substitute (A.3) into (A.2),

$$\int_a^b \rho(x)\, dx = \int_a^b \sum_i^N \delta(x - x_i)\, dx \qquad (A.7)$$

Then we write the integrals as sums over finite sample elements,

$$\sum_{\Delta x} \rho(x)\, \Delta x = \sum_{\Delta x} \sum_i^N \delta(x - x_i)\, \Delta x \qquad (A.8)$$

and note that the inner sum on the rhs is $N(x, \Delta x)$, the number of atomic centers at x within thickness Δx,

$$\boxed{N(x, \Delta x) = \sum_i^N \delta(x - x_i)\, \Delta x} \qquad (A.9)$$

So (A.8) becomes

$$\sum_{\Delta x} \rho(x)\, \Delta x = \sum_{\Delta x} N(x, \Delta x) \qquad (A.10)$$

This equation should be satisfied term by term, so we obtain the computationally useful form

$$\rho(x) = \frac{N(x, \Delta x)}{\Delta x} \qquad (A.11)$$

which is, of course, (A.1). *The important conceptual relation here is (A.9), which represents the counting operation.* The lesson is that to count the number of objects N, we must apply (A.9) to accumulate areas, each of which, by (A.6), is unity. We have obtained (A.11) for one instant during the simulation; if an equilibrium average is desired, then (A.11) would be averaged over time.

In addition to deriving expressions involving counting, the δ-symbol notation is helpful in deriving forms for quantities that are weighted averages of the density distribution. That is, the notation helps in evaluating integrals of the form

$$I = \int_a^b f(x)\rho(x)\, dx \qquad (A.12)$$

As an example, consider evaluating the location of the center of mass in our one-dimensional system,

$$\bar{x}_{com} = \frac{\int_a^b x\rho(x)\,dx}{\int_a^b \rho(x)\,dx} \tag{A.13}$$

Using (A.2) and (A.3) in (A.13) leaves

$$\bar{x}_{com} = \frac{1}{N}\int_a^b x \sum_i^N \delta(x - x_i)\,dx \tag{A.14}$$

Interchanging the order of summation and integration gives

$$\bar{x}_{com} = \frac{1}{N}\sum_i^N \int_a^b x\,\delta(x - x_i)\,dx \tag{A.15}$$

Now consider one term I_i in the sum on the rhs of (A.15); that is, write (A.15) as

$$\bar{x}_{com} = \frac{1}{N}\sum_i^N I_i \tag{A.16}$$

For a given atom i, the δ-symbol in I_i is zero unless $x = x_i$; so we can reduce the integration limits in I_i to within a small amount ε of x_i,

$$I_i = \int_{x_i-\varepsilon}^{x_i+\varepsilon} x\,\delta(x - x_i)\,dx \tag{A.17}$$

Over the interval $[x_i - \varepsilon, x_i + \varepsilon]$, x is nearly constant at the value x_i, and it is more nearly constant as we make ε smaller; so

$$I_i = x_i \int_{x_i-\varepsilon}^{x_i+\varepsilon} \delta(x - x_i)\,dx \tag{A.18}$$

According to the normalization condition (A.6), the integral on the rhs is unity, (A.18) reduces to

$$I_i = x_i \tag{A.19}$$

and putting this back into (A.16), we find the anticipated expression for locating the center of mass,

$$\bar{x}_{com} = \frac{1}{N} \sum_{i}^{N} x_i \tag{A.20}$$

The important concept here is what happened in passing from (A.17) to (A.19); that is, in general,

$$\boxed{\int_a^b f(x)\, \delta(x-c)\, dx = f(c)} \tag{A.21}$$

provided $f(x)$ is continuous and c lies on $[a,b]$. The result (A.21) is a representation for selecting from a function $f(x)$ the particular value of f at $x = c$. Equation (A.21) is the motivation for introducing the δ-symbol, and it serves as a formal definition. For example, the normalization condition (A.6) is just a special case of (A.21). The δ-symbol is commonly called the Dirac delta function, but this is a misnomer—it is not a function[†] and the usual defining characteristics (A.4) and (A.6) are contradictory. The δ-symbol is merely a notation by which we represent the selection process in (A.21).

In three dimensions the local density $\rho(\mathbf{r})$ is defined, analogous to (A.1), as

$$\rho(\mathbf{r}) = \frac{N(\mathbf{r})}{V(\mathbf{r})} = \lim_{\Delta r \to 0} \frac{N(\mathbf{r}, \Delta r)}{V(\mathbf{r}, \Delta r)} \tag{A.22}$$

Now $N(\mathbf{r}, \Delta r)$ is the number of atomic centers in a spherical shell of radius r and thickness Δr centered on an origin and $V(\mathbf{r}, \Delta r)$ is the volume of the shell. The δ-symbol form for $\rho(\mathbf{r})$ is

$$\rho(\mathbf{r}) = \sum_{i}^{N} \delta(\mathbf{r} - \mathbf{r}_i) \tag{A.23}$$

and the three-dimensional analog to the normalization condition (A.6) is

$$\int \delta(\mathbf{r} - \mathbf{r}_i)\, d\mathbf{r} = 1 \tag{A.24}$$

Note that for dimensional consistency, $\delta(\mathbf{r})$ must have units of V^{-1}. Proceeding through a sequence of steps analogous to (A.7)–(A.11) leads to a computationally useful form for evaluating the local density at any instant,

$$\rho(\mathbf{r}) = \frac{N(\mathbf{r}, \Delta r)}{V(\mathbf{r}, \Delta r)} = \frac{\sum_{i}^{N} \delta(\mathbf{r} - \mathbf{r}_i)\, \Delta r}{V(\mathbf{r}, \Delta r)} \tag{A.25}$$

[†]R. V. Churchill, *Operational Mathematics*, 3rd ed., McGraw-Hill, New York, 1972, pp. 33–34.

APPENDIX B

PRESSURE

To relate the pressure to molecular quantities, we usually rely on the virial theorem from classical dynamics [e.g., see 1], but the derivation of the virial theorem appears to fail for a system with periodic boundaries [2]. The difficulty can be summarized in the following way. For a system of N atoms, the derivation of the virial theorem starts with the quantity

$$I = \tfrac{1}{2} \sum_i^N m\mathbf{r}_i \cdot \mathbf{r}_i \tag{B.1}$$

which is a measure of the shape of the object formed by the N atoms. Taking two time derivatives gives

$$\frac{d^2 I}{dt^2} = \sum_i^N m\dot{\mathbf{r}}_i^2 + \sum_i^N m\ddot{\mathbf{r}}_i \cdot \mathbf{r}_i \tag{B.2}$$

Then using Newton's second law on the rhs and forming time averages give

$$\left\langle \frac{d^2 I}{dt^2} \right\rangle = m \left\langle \sum_i^N \dot{\mathbf{r}}_i^2 \right\rangle + \left\langle \sum_i^N \mathbf{F}_i \cdot \mathbf{r}_i \right\rangle \tag{B.3}$$

Now in the usual development either external fields or container walls preserve the shape of the N-particle object (the shape measured by I), if not at each instant, then at least on a time average. Consequently, the lhs equals zero and the force F_i on the rhs divides into internal and external terms. But

332

in a system with periodic boundaries there are neither walls nor external forces to preserve I, and the usual derivation of the virial theorem appears problematic.

This appendix gives an alternative derivation in which the pressure is interpreted as a momentum flux. The derivation is given in some detail because the general strategy applies not only to the pressure but also to other macroscopic quantities [3].

B.1 GENERAL FORM

Consider N atoms that, at one instant, occupy a cubic region of space having volume V and side L (in molecular dynamics this is the usual shape of the primary cell). The collection of atoms is at equilibrium with zero total linear momentum. Imagine a planar surface of area $A = L^2$ inserted into the system and oriented perpendicular to the x-axis, as in Figure B.1. The pressure can be defined as the force per unit area acting normal to the surface,

$$P_x = \frac{F_x}{A} \tag{B.4}$$

Using Newton's second law, this can be written as

$$P_x = \frac{1}{A} \frac{d(mv_x)}{dt} \tag{B.5}$$

Thus, pressure is a momentum flux; it is the amount of momentum that crosses a unit area of the surface in unit time. In general, this flux is composed of two parts: (a) P_m, the momentum carried by the atoms themselves as they cross the area during dt, and (b) P_f, the momentum transferred as a result of forces acting between atoms that lie on different sides of the surface [4, 5],

$$P_x = P_{mx} + P_{fx} \tag{B.6}$$

B.2 MOMENTUM FLUX CAUSED BY ATOMIC MOTION

The part of the pressure determined by atomic motion is the part usually derived in kinetic theory. We give an abbreviated derivation here; detailed derivations can be found in any good text on kinetic theory [e.g., 6]. On the

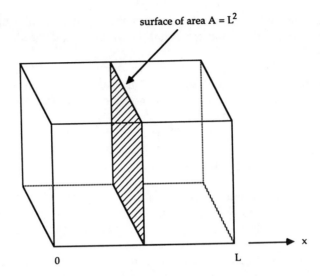

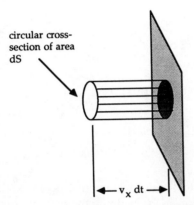

FIGURE B.1 *Top:* Imaginary plane inserted perpendicular to the *x*-axis across which the momentum flux will be evaluated. *Bottom:* Those atoms having velocities in dv_x about v_x and that cross an element dS on the plane A during interval dt must have been in the cylinder $\{v_x \, dt \, dS\}$ at some point during the time interval.

planar surface of area A, construct a circular element of area dS. Let

$$N(v_x)\, dv_x = \begin{array}{l}\text{number of atoms having velocities}\\ \text{within } dv_x \text{ of } v_x \text{ and that cross}\\ \text{element } dS \text{ in } +x \text{ direction during } dt\end{array} \qquad (B.7)$$

$$= \begin{array}{l}\text{volume of cylinder containing}\\ \text{those atoms that have the}\\ \text{specified velocities and that}\\ \text{approach } dS \text{ during } dt\end{array} \times \begin{array}{l}\text{density of}\\ \text{atoms having}\\ \text{the specified}\\ \text{velocities}\end{array} \qquad (B.8)$$

The volume of the cylinder is $\{v_x\,dt\,dS\}$ (see Figure B.1), while the density of atoms having the specified velocities is proportional to the velocity distribution $f(v_x, t)$. Recall from Section 2.7.1 that the velocity distribution is

$$f(v_x, t)\,dv_x = \begin{array}{l}\text{fraction of } N \text{ atoms that have, at}\\ \text{time } t, \text{ velocities within } dv_x \text{ of } v_x\end{array} \qquad (2.54)$$

So (B.8) becomes

$$N(v_x)\,dv_x = [v_x\,dt\,dS] \times \left[\frac{N}{V}f(v_x, t)\,dv_x\right] \qquad (B.9)$$

Then the total component of momentum normal to dS carried by the atoms that cross dS (in both directions) per unit area in unit time is

$$P_{mx} = \int m v_x v_x \frac{N}{V} f(v_x, t)\,dv_x \qquad (B.10)$$

This integral is, according to (2.56), proportional to the average of v_x^2,

$$P_{mx} = \frac{Nm}{V}\overline{v_x^2} \qquad (B.11)$$

Averaging over time produces

$$\langle P_{mx}\rangle = \frac{Nm}{V}\langle \overline{v_x^2}\rangle = \frac{2N}{V}\langle E_{kx}\rangle \qquad (B.12)$$

where E_k is the kinetic energy per atom. After repeating this procedure in the y- and z-directions, we can form the total convective contribution to the pressure and thereby obtain the ideal-gas law,

$$\langle P_m\rangle = \frac{2N}{3V}\left[\langle E_{kx}\rangle + \langle E_{ky}\rangle + \langle E_{kz}\rangle\right] = \frac{2N}{3V}\langle E_k\rangle \qquad (B.13)$$

B.3   MOMENTUM FLUX CAUSED BY INTERMOLECULAR FORCES

While kinetic theory gives (B.13) as a standard result, kinetic theory is rarely used to obtain an expression for P_f, the contribution resulting from intermolecular forces. We therefore give the following derivation in some detail. An analogous derivation is given by Hill [7], except that he immediately obtains the virial in terms of the radial distribution function.

At any instant, the second term in (B.7), P_{fx}, is the total force (per unit area) acting normal to the surface A, where the forces are caused by atoms

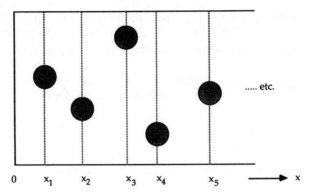

FIGURE B.2 Labeling of atoms to help evaluation of (B.15), which involves the average over all positions of the imagined surface (dotted lines) inserted perpendicular to the x-axis.

on one side of the surface interacting with atoms on the other side. Assuming the intermolecular forces are pairwise additive, P_{fx} can be written as

$$P_{fx} = \frac{1}{A} \sum_i' \sum_j'' \mathbf{F}_{ij} \cdot \hat{\mathbf{x}} \tag{B.14}$$

where $\hat{\mathbf{x}}$ is the unit vector in the $+x$-direction. The primes on the summation signs are to remind us that neither sum is over all N atoms; specifically, index i runs over all atoms on one side of the surface, while index j runs over all atoms on the other side. Averaging over all possible locations of the imaginary surface that is positioned along the x-axis, we obtain

$$\bar{P}_{fx} = \frac{1}{AL} \int_0^L \sum_i' \sum_j'' F_{xij} \, dx \tag{B.15}$$

Now label the atoms sequentially from 1 to N as their x-position increases from zero to L; that is, as in Figure B.2, label the atoms so that

$$x_1 \le x_2 \le x_3 \le \cdots \le x_N$$

and divide the edge of length L along the x-axis into $N+1$ intervals of unequal widths Δx_k, where

$$\Delta x_k = x_{k+1} - x_k \equiv x_{k,k+1} \tag{B.16}$$

We use a sum over these intervals Δx_k to approximate the integral in (B.15),

$$\bar{P}_{fx} = \frac{1}{V} \sum_{k=0}^{N} \sum_i{}' \sum_j{}'' F_{xij} \Delta x_k \qquad (B.17)$$

Thus, the distances between adjacent atoms serve as the integration intervals Δx (recall the analysis so far is all at one instant). Note that the first and last intervals Δx_0 and Δx_N do not contribute to the momentum flux because, in both cases, all N atoms lie on one side of the surface A; so

$$\bar{P}_{fx} = \frac{1}{V} \sum_{k=1}^{N-1} \sum_i{}' \sum_j{}'' F_{xij} x_{k,k+1} \qquad (B.18)$$

Further, when the surface lies between atoms k and $k+1$, index i runs from 1 to k, while index j runs from $k+1$ to N; thus

$$\bar{P}_{fx} = \frac{1}{V} \sum_{k=1}^{N-1} \sum_{i=1}^{k} \sum_{j=k+1}^{N} F_{xij} x_{k,k+1} \qquad (B.19)$$

Note also that if two atoms happen to have the same x-position, then that interval $\Delta x = 0$; that is, those two atoms cannot fall on different sides of the surface A, and so their interaction does not contribute to a force in the $\pm x$-direction.

Let us examine the first few terms in (B.19):

$$\bar{P}_{fx} = \frac{1}{V} \left[x_{12}(F_{x12} + F_{x13} + F_{x14} + \dots) \right.$$

$$+ x_{23}(F_{x13} + F_{x14} + \dots + F_{x23} + F_{x24} + \dots)$$

$$\left. + x_{34}(F_{x14} + \dots + F_{x24} + \dots + F_{x34} + \dots) + \dots \right] \qquad (B.20)$$

These can be rearranged to give

$$\bar{P}_{fx} = \frac{1}{V} \left[F_{x12} x_{12} + F_{x13}(x_{12} + x_{23}) + F_{x14}(x_{12} + x_{23} + x_{34}) \right.$$

$$\left. + \dots + F_{x23} x_{23} + F_{x24}(x_{23} + x_{34}) + \dots + F_{x34} x_{34} + \dots \right] \qquad (B.21)$$

Or in general,

$$\bar{P}_{fx} = \frac{1}{V} \sum_{i=1}^{N-1} \sum_{j=i+1}^{N} F_{xij} \sum_{k=i}^{j} x_{k,k+1} \qquad (B.22)$$

But

$$\sum_{k=i}^{j} x_{k,k+1} = x_{ij} \tag{B.23}$$

For example, $x_{12} + x_{23} + x_{34} = x_{14}$. Thus,

$$\bar{P}_{fx} = \frac{1}{V} \sum_{i=1}^{N-1} \sum_{j=i+1}^{N} F_{xij} x_{ij} \tag{B.24}$$

This is still at one instant; now average over time,

$$\langle P_{fx} \rangle = \frac{1}{V} \left\langle \sum_{i=1}^{N-1} \sum_{j=i+1}^{N} F_{xij} x_{ij} \right\rangle \tag{B.25}$$

Repeating this procedure for surfaces oriented perpendicular to the y- and z-axes produces analogous expressions for P_{fy} and P_{fz}. Then

$$\langle P_f \rangle = \frac{1}{3V} \left[\left\langle \sum_{i<j} \sum F_{xij} x_{ij} \right\rangle + \left\langle \sum_{i<j} \sum F_{yij} y_{ij} \right\rangle + \left\langle \sum_{i<j} \sum F_{zij} z_{ij} \right\rangle \right] \tag{B.26}$$

or

$$\langle P_f \rangle = \frac{1}{3V} \left\langle \sum_{i<j} \sum \mathbf{F}_{ij} \cdot \mathbf{r}_{ij} \right\rangle \tag{B.27}$$

which is the virial. Combining (B.13) and (B.27) gives the complete expression for the pressure,

$$P = \frac{2N}{3V} \langle E_k \rangle + \frac{1}{3V} \left\langle \sum_{i<j} \sum \mathbf{F}_{ij} \cdot \mathbf{r}_{ij} \right\rangle \tag{B.28}$$

where E_k is the kinetic energy per atom. For systems with periodic boundaries this becomes

$$P = \frac{2N}{3V} \langle E_k \rangle + \frac{1}{3V} \left\langle \sum_{\alpha} \sum_{i<j} \sum (\mathbf{r}_{ij} - \boldsymbol{\alpha} L) \cdot \mathbf{F}_{ij}(\mathbf{r}_{ij} - \boldsymbol{\alpha} L) \right\rangle \tag{B.29}$$

where $\boldsymbol{\alpha}$ is the cell translation vector defined in Section 2.9.

REFERENCES

[1] H. Goldstein, *Classical Mechanics*, 2nd ed., Addison-Wesley, Reading, MA, 1980, pp. 82–84.

[2] J. J. Erpenbeck and W. W. Wood, "Molecular Dynamics Techniques for Hard-Core Systems" in *Statistical Mechanics*, Part B, B. J. Berne, Ed., Plenum, New York, 1977.

[3] W. G. Hoover, *Computational Statistical Mechanics*, Elsevier, Amsterdam, 1991.

[4] E. H. Kennard, *Kinetic Theory of Gases*, McGraw-Hill, New York, 1938, pp. 211–213.

[5] R. D. Present, *Kinetic Theory of Gases*, McGraw-Hill, New York, 1958, pp. 101–102.

[6] S. Chapman and T. G. Cowling, *The Mathematical Theory of Nonuniform Gases*, Cambridge University Press, Cambridge, 1961.

[7] T. L. Hill, *An Introduction to Statistical Thermodynamics*, Addison-Wesley, Reading, MA, 1960, pp. 304–305.

APPENDIX C

THE MICROCANONICAL ENSEMBLE

Standard texts on statistical mechanics usually avoid detailed discussion of the microcanonical ensemble and its connections to thermodynamics. Isolated systems are not experimentally interesting, and consequently statistical mechanics is typically developed in terms of more accessible situations,[†] such as the canonical ensemble (fixed N, V, and T). The resulting thermodynamic properties are then converted, via Legendre transforms [1, 2], to the microcanonical ensemble.

However, the microcanonical ensemble is central to the analysis of molecular dynamics trajectories computed for isolated systems. Therefore, it is instructive to develop thermodynamics directly in the microcanonical situation. We consider an isolated system in which the total number of molecules N, system volume V, and total energy E are constants; for a pure substance these variables are sufficient to fix the thermodynamic state. Moreover, we limit the development to spherical molecules, which are the only kinds considered in this book.

Some purists carefully distinguish the traditional microcanonical ensemble from the "molecular dynamics" ensemble. In the latter, not only are N, V, and E fixed, but so too is the total linear momentum $\mathscr{P}$. This issue is discussed in Section 2.10.

In the microcanonical ensemble, the phase-space volume Ω is given by

$$\Omega = \frac{1}{h^{3N}N!} \int d\mathbf{r}^N \, d\mathbf{p}^N \theta[E - \mathscr{H}] \tag{C.1}$$

[†] "...the microcanonical ensemble bears the most direct relation to mechanics but is the most difficult ensemble to relate to thermodynamics." [3]

where h is Planck's constant, θ is the unit step function, and $\mathscr{H}$ is the Hamiltonian,

$$\mathscr{H} = \mathscr{H}(\mathbf{r}^N, \mathbf{p}^N) = E_k(\mathbf{p}^N) + \mathscr{U}(\mathbf{r}^N) \tag{C.2}$$

with E_k the kinetic energy and $\mathscr{U}$ the potential energy. Recall, $\mathbf{r}^N$ represents the set of N position vectors and $\mathbf{p}^N$ is the set of N momentum vectors. Similarly, the density of states ω is given by

$$\omega = \left(\frac{\partial \Omega}{\partial E}\right)_{NV} = \frac{1}{h^{3N}N!} \int d\mathbf{r}^N \, d\mathbf{p}^N \delta[E - \mathscr{H}] \tag{C.3}$$

where δ is the delta symbol (see Appendix A). Then for any dynamic quantity $A(\mathbf{r}^N, \mathbf{p}^N)$, the microcanonical ensemble average is given by

$$\langle A \rangle = \frac{1}{h^{3N}N!\omega} \int d\mathbf{r}^N \, d\mathbf{p}^N A(\mathbf{r}^N, \mathbf{p}^N) \delta[E - \mathscr{H}] \tag{C.4}$$

In the microcanonical ensemble the fundamental connection to thermodynamics is via the entropy S,

$$S = k \ln \Omega \tag{C.5}$$

C.1   INTEGRATION OVER MOMENTA

In some ensembles the integrations over the momenta in Ω and ω can be done, giving the kinetic (i.e., ideal-gas) contributions to the phase-space volume and density of states, respectively. However, in the microcanonical ensemble the integrations cannot be immediately done because the positions $\mathbf{r}^N$ and momenta $\mathbf{p}^N$ are coupled through the step function in (C.1) and through the delta symbol in (C.3). The $\mathbf{r}^N$ and $\mathbf{p}^N$ can be decoupled by taking Laplace transforms [4]. The procedure involves three steps, and we illustrate it here by applying it to Ω:

(i) Take the Laplace transform of Ω with respect to the total energy E.
(ii) Integrate over the momenta.
(iii) Take the inverse transform of the integrated result.

Recall that the Laplace transform of the unit step function is

$$L[\theta(t-a)] = \int_0^\infty \theta[t-a]e^{-st}\, dt = \frac{1}{s}e^{-sa} \tag{C.6}$$

while that for a delta symbol is [5]

$$L[\delta(t-a)] = \int_0^\infty \delta[t-a]e^{-st}\,dt = e^{-sa} \tag{C.7}$$

Thus, the Laplace transform of Ω with respect to E is

$$L[\Omega] = \frac{1}{h^{3N}N!}\int d\mathbf{r}^N\,d\mathbf{p}^N \frac{1}{s}e^{-s\mathcal{H}} \tag{C.8}$$

Using (C.2) for the Hamiltonian, (C.8) can be written as

$$L[\Omega] = \frac{1}{h^{3N}N!}\int d\mathbf{r}^N\,d\mathbf{p}^N \frac{1}{s}\exp[-sE_k(\mathbf{p}^N)]\exp[-s\mathcal{U}(\mathbf{r}^N)] \tag{C.9}$$

Now the kinetic energy E_k does not depend explicitly on the molecular positions, so we can perform the integrations over the momenta in (C.9). The integral is

$$\int d\mathbf{p}^N \exp\left[-\frac{s}{2m}\sum_i \mathbf{p}_i^2\right] = \left(\frac{2\pi m}{s}\right)^{3N/2} \tag{C.10}$$

so (C.9) becomes

$$L[\Omega] = \left(\frac{2\pi m}{sh^2}\right)^{3N/2}\frac{1}{N!}\int d\mathbf{r}^N \frac{1}{s}\exp[-s\mathcal{U}(\mathbf{r}^N)] \tag{C.11}$$

We now want the inverse transform of (C.11). Note that the s-dependence in (C.11) is of the form $\exp[-sk]/s^n$, which has inverse transform [6]

$$L^{-1}\left(\frac{\exp[-sk]}{s^n}\right) = \theta[t-k]\frac{(t-k)^{n-1}}{\Gamma[n]} \tag{C.12}$$

where Γ is the gamma function

$$\Gamma[n+1] = n\Gamma[n] = n! \tag{C.13}$$

Thus, the inverse transform of (C.11) is

$$\Omega = L^{-1}\{L[\Omega]\}$$

$$= \left(\frac{2\pi m}{h^2}\right)^{3N/2}\frac{1}{N!\Gamma[3N/2+1]}\int d\mathbf{r}^N (E-\mathcal{U})^{3N/2}\theta[E-\mathcal{U}] \tag{C.14}$$

Equation (C.14) is the desired result: it expresses the microcanonical phase-space volume as a $3N$-fold integral over the molecular positions. By an analogous procedure, we find for the density of states,

$$\omega = L^{-1}\{L[\omega]\}$$

$$= \left(\frac{2\pi m}{h^2}\right)^{3N/2} \frac{1}{N!\Gamma[3N/2]} \int d\mathbf{r}^N (E - \mathcal{U})^{3N/2-1} \theta[E - \mathcal{U}] \quad (C.15)$$

Then for any quantity $A(\mathbf{r}^N)$ that depends only on configurations, the ensemble average (C.4) becomes

$$\langle A \rangle = \left(\frac{2\pi m}{h^2}\right)^{3N/2} \frac{1}{N!\Gamma[3N/2]} \frac{1}{\omega} \int d\mathbf{r}^N (E - \mathcal{U})^{3N/2-1} \theta[E - \mathcal{U}] A(\mathbf{r}^N)$$

$$(C.16)$$

or

$$\langle A \rangle = \frac{\int d\mathbf{r}^N (E - \mathcal{U})^{3N/2-1} \theta[E - \mathcal{U}] A(\mathbf{r}^N)}{\int d\mathbf{r}^N (E - \mathcal{U})^{3N/2-1} \theta[E - \mathcal{U}]} \quad (C.17)$$

In the remainder of this appendix we illustrate how (C.14)–(C.16) are used to obtain microcanonical expressions for thermodynamic properties. We consider one simple thermodynamic property (temperature), one response function (isometric heat capacity), and one entropic property (chemical potential). Relations for other properties are developed by Pearson et al. [4].

C.2 TEMPERATURE

The fundamental equation of thermodynamics for a pure substance is (6.56), and therefore the temperature is

$$\frac{1}{T} = \left(\frac{\partial S}{\partial E}\right)_{NV} \quad (C.18)$$

Substituting (C.5) for S leaves

$$\frac{1}{T} = k\left(\frac{\partial \ln \Omega}{\partial E}\right)_{NV} = \frac{k}{\Omega}\left(\frac{\partial \Omega}{\partial E}\right)_{NV} \quad (C.19)$$

Applying the derivative in (C.19) to (C.1) for Ω and recalling that the derivative of a step function θ is the delta symbol δ, we find

$$\left(\frac{\partial \Omega}{\partial E}\right)_{NV} = \omega = \frac{1}{h^{3N}N!} \int d\mathbf{r}^N d\mathbf{p}^N \delta[E - \mathscr{H}] \tag{C.20}$$

Combining (C.19) and (C.20) gives

$$kT = \frac{\Omega}{\omega} \tag{C.21}$$

Now substitute (C.14) for Ω and (C.15) for ω and use (C.13) to express the gamma function as

$$\Gamma\left[\tfrac{3}{2}N + 1\right] = \tfrac{3}{2}N\Gamma\left[\tfrac{3}{2}N\right] \tag{C.22}$$

Then (C.21) becomes

$$kT = \frac{2}{3N} \frac{\int d\mathbf{r}^N (E - \mathscr{U})^{3N/2 - 1} \theta[E - \mathscr{U}](E - \mathscr{U})}{\int d\mathbf{r}^N (E - \mathscr{U})^{3N/2 - 1} \theta[E - \mathscr{U}]} \tag{C.23}$$

Comparing (C.23) with (C.17) shows that (C.23) is in the form of an ensemble average in which we identify $E - \mathscr{U}$ as the kinetic energy E_k,

$$\boxed{kT = \frac{2}{3N}\langle E_k \rangle} \tag{C.24}$$

Thus we find, in agreement with kinetic theory, that the absolute temperature is proportional to the average kinetic energy.

We noted at the end of Section 2.1 that translational kinetic energy is the work needed to move an object from rest to some final velocity. Therefore, one interpretation of temperature is that it is a measure of this work averaged over many objects (molecules) and over time. An alternative, geometric interpretation of temperature is suggested by (C.24). First consider a two-dimensional space having orthogonal coordinates x and y. The distance d of any point from the origin satisfies $d^2 = x^2 + y^2$. Now consider the momentum part of phase space, which has $3N$ orthogonal coordinates $\{p_{ix}, p_{iy}, p_{iz}\}$ with $i = 1, 2, \ldots, N$. In this space, analogous to the two-dimensional $x-y$ space, define a measure of "distance" from the origin as $\sum_i^N \mathbf{p}_i^2$. Therefore, temperature measures the average distance that the phase point lies from the origin in momentum space.

C.3   ISOMETRIC HEAT CAPACITY

The isometric heat capacity is defined by

$$C_v = \left(\frac{\partial E}{\partial T} \right)_{NV} \tag{C.25}$$

We now seek an expression for C_v in terms of the phase-space volume Ω. The derivative of (C.18) with respect to the total energy E is

$$\frac{-1}{T^2} \left(\frac{\partial T}{\partial E} \right)_{NV} = \left(\frac{\partial^2 S}{\partial E^2} \right)_{NV} \tag{C.26}$$

Combining (C.25) and (C.26) gives

$$\frac{-1}{T^2 C_v} = \left(\frac{\partial^2 S}{\partial E^2} \right)_{NV} \tag{C.27}$$

Differentiating (C.5) twice with respect to E gives

$$\left(\frac{\partial^2 S}{\partial E^2} \right)_{NV} = \frac{-k}{\Omega^2} \left(\frac{\partial \Omega}{\partial E} \right)_{NV}^2 + \frac{k}{\Omega} \left(\frac{\partial^2 \Omega}{\partial E^2} \right)_{NV} \tag{C.28}$$

Using (C.20) and (C.27) in (C.28) leaves

$$\frac{-1}{T^2 C_v} = -k \left(\frac{\omega}{\Omega} \right)^2 + \frac{k}{\Omega} \left(\frac{\partial^2 \Omega}{\partial E^2} \right)_{NV} \tag{C.29}$$

Now (C.21) relates the temperature to the ratio ω / Ω, so we use (C.21) to eliminate that ratio from (C.29), rearrange, and multiply through by the number of molecules N; then (C.29) becomes

$$\frac{Nk}{C_v} = N - \frac{NkT}{\omega} \left(\frac{\partial^2 \Omega}{\partial E^2} \right)_{NV} \tag{C.30}$$

We must now evaluate the derivative remaining in (C.30). Equation (C.20) identifies ω as the first derivative of Ω with respect to E; hence, differentiating (C.20) gives the second derivative as

$$\frac{1}{\omega} \left(\frac{\partial^2 \Omega}{\partial E^2} \right)_{NV} = \frac{1}{\omega} \left(\frac{\partial \omega}{\partial E} \right)_{NV} \tag{C.31}$$

To evaluate the derivative on the rhs of (C.31), we use (C.15) for ω; this leads to

$$\frac{1}{\omega}\left(\frac{\partial^2 \Omega}{\partial E^2}\right)_{NV}$$

$$= \frac{1}{\omega}\left(\frac{2\pi m}{h^2}\right)^{3N/2} \frac{1}{N!\Gamma\left[\dfrac{3N}{2}\right]} \left\{\left(\left(\frac{3N}{2}-1\right)\int d\mathbf{r}^N (E-\mathscr{U})^{3N/2-2}\theta[E-\mathscr{U}]\right.\right.$$

$$\left.\left. + \int d\mathbf{r}^N (E-\mathscr{U})^{3N/2-1}\delta[E-\mathscr{U}]\right\} \quad (C.32)$$

The second term on the rhs is zero because the integration over the delta symbol selects only the value of $\mathscr{U}=E$, and the integrand then vanishes because $E-\mathscr{U}=E-E=0$. In the first term, identifying $E-\mathscr{U}$ as the kinetic energy E_k, we have

$$\frac{1}{\omega}\left(\frac{\partial^2 \Omega}{\partial E^2}\right)_{NV} = \left(\frac{3N}{2}-1\right)\frac{1}{\omega}\left(\frac{2\pi m}{h^2}\right)^{3N/2} \frac{1}{N!\Gamma[3N/2]}$$

$$\times \int d\mathbf{r}^N (E-\mathscr{U})^{3N/2-1}\theta[E-\mathscr{U}]E_k^{-1} \quad (C.33)$$

Comparing with (C.16), we see that the rhs of (C.33) is in the form of an ensemble average of the reciprocal kinetic energy,

$$\frac{1}{\omega}\left(\frac{\partial^2 \Omega}{\partial E^2}\right)_{NV} = \left(\frac{3N}{2}-1\right)\langle E_k^{-1}\rangle \quad (C.34)$$

Putting (C.34) into (C.30) and inverting both sides gives the final result,

$$\boxed{\frac{C_v}{Nk} = \left[N - NkT\left(\frac{3N}{2}-1\right)\langle E_k^{-1}\rangle\right]^{-1}} \quad (C.35)$$

C.4 CHEMICAL POTENTIAL: TEST PARTICLE EXPRESSION

In this section we develop the microcanonical expression for the chemical potential in the test particle formalism. The derivation follows that given by Frenkel [7]. The fundamental equation of thermodynamics (6.56) leads to an interpretation of the chemical potential as

$$\mu = -T\left(\frac{\partial S}{\partial N}\right)_{EV} \tag{C.36}$$

Substituting (C.5) for S yields

$$\mu = -kT\left(\frac{\partial \ln \Omega}{\partial N}\right)_{EV} \tag{C.37}$$

Now in the test particle approach, the derivative in (C.37) is approximated by an increase in the number of molecules from N to $N+1$, where the added molecule is the test particle; thus

$$\mu \approx -kT\left(\frac{\ln \Omega_{N+1} - \ln \Omega_N}{N+1-N}\right)_{EV} = -kT\ln\frac{\Omega_{N+1}}{\Omega_N} \tag{C.38}$$

The development that follows is simpler in terms of ω rather than Ω, so we use (C.21) to eliminate Ω from (C.38) in favor of ω,

$$\mu \approx -kT\ln\frac{\omega_{N+1}T_{N+1}}{\omega_N T_N} = -kT\ln\frac{\omega_{N+1}}{\omega_N} \tag{C.39}$$

The temperatures T_{N+1} and T_N in (C.39) are the same because the addition of the test particle does not affect the momenta of the N real molecules. The objective now is to express the ratio ω_{N+1}/ω_N as a microcanonical ensemble average.

Using (C.15), this ratio becomes

$$\frac{\omega_{N+1}}{\omega_N} = \left(\frac{2\pi m}{h^2}\right)^{3/2}\frac{1}{N(3N/2)^{3/2}}\frac{\int d\mathbf{r}^{N+1}(E - \mathcal{U}_{N+1})^{3(N+1)/2-1}\theta[E-\mathcal{U}]}{\int d\mathbf{r}^N(E - \mathcal{U}_N)^{3N/2-1}\theta[E-\mathcal{U}]} \tag{C.40}$$

In obtaining (C.40) we have used $N+1 \approx N$ for large N and

$$\frac{\Gamma[3N/2]}{\Gamma[3N/2+3/2]} \approx \frac{1}{(3N/2)^{3/2}} \tag{C.41}$$

which is also valid for large N [8].

Now $(E - \mathscr{U}_{N+1}) = (E - \mathscr{U}_N - U_t)$, where U_t is the test particle potential energy of interaction with the real molecules. Further, $(E - \mathscr{U}_N) = E_k$ is the kinetic energy of the N real molecules. So the integrand in the numerator of (C.40) can be written

$$(E - \mathscr{U}_N - U_t)^{3(N+1)/2-1} = E_k^{3(N+1)/2-1}\left(1 - \frac{U_t}{E_k}\right)^{3(N+1)/2-1} \tag{C.42}$$

$$= E_k^{3/2}(E - \mathscr{U}_N)^{3N/2-1}\left(1 - \frac{U_t}{E_k}\right)^{3(N+1)/2-1} \tag{C.43}$$

Then (C.40) becomes

$$\frac{\omega_{N+1}}{\omega_N} = \left(\frac{2\pi m}{h^2}\right)^{3/2}\frac{1}{N(3N/2)^{3/2}}$$

$$\times \frac{\int d\mathbf{r}^N \, d\mathbf{r}_t E_k^{3/2}(E - \mathscr{U}_N)^{3N/2-1}(1 - U_t/E_k)^{3(N+1)/2-1}\theta[E - \mathscr{U}]}{\int d\mathbf{r}^N(E - \mathscr{U}_N)^{3N/2-1}\theta[E - \mathscr{U}]}$$

$$\tag{C.44}$$

where $\mathbf{r}_t$ is the position of the test particle. Comparing with (C.17) shows that (C.44) has the form of an ensemble average,

$$\frac{\omega_{N+1}}{\omega_N} = \left(\frac{2\pi m}{h^2}\right)^{3/2}\frac{1}{N}\int d\mathbf{r}_t\left\langle\left(\frac{2E_k}{3N}\right)^{3/2}\left(1 - \frac{U_t}{E_k}\right)^{3(N+1)/2-1}\right\rangle \tag{C.45}$$

If the measurements of the positions $\mathbf{r}^N$ are made relative to an origin located on the test particle, then the remaining integral in (C.45) can be done. That integral merely gives the system volume V, and using $\rho = N/V$, (C.45) becomes

$$\frac{\omega_{N+1}}{\omega_N} = \left(\frac{2\pi m}{h^2}\right)^{3/2}\frac{1}{\rho}\left\langle\left(\frac{2E_k}{3N}\right)^{3/2}\left(1 - \frac{U_t}{E_k}\right)^{3(N+1)/2-1}\right\rangle \tag{C.46}$$

Now, for $|U_t| < E_k$,

$$\left(1 - \frac{U_t}{E_k}\right)^{3(N+1)/2-1} \approx \exp\left(-\frac{[3(N+1)/2-1]U_t}{E_k}\right) \tag{C.47}$$

and the instantaneous temperature T_{in} is related to the kinetic energy by the instantaneous analog of (C.24); that is,

$$kT_{in} = \frac{2E_k}{3N} \tag{C.48}$$

Then (C.47) can be written as

$$\left(1 - \frac{U_t}{E_k}\right)^{3(N+1)/2 - 1} \approx \exp\left(-\frac{U_t}{kT_{in}} - \frac{U_t}{3NkT_{in}}\right) \approx \exp\left(-\frac{U_t}{kT_{in}}\right) \tag{C.49}$$

where we neglect the term that is $O(1/N)$. Using (C.49) in (C.46) gives

$$\frac{\omega_{N+1}}{\omega_N} = \left(\frac{2\pi m}{h^2}\right)^{3/2} \frac{1}{\rho} \left\langle (kT_{in})^{3/2} \exp\left(-\frac{U_t}{kT_{in}}\right) \right\rangle \tag{C.50}$$

Introducing the thermal de Broglie wavelength Λ, where

$$\Lambda = \left(\frac{h^2}{2\pi mkT}\right)^{1/2} \tag{C.51}$$

(C.50) becomes

$$\frac{\omega_{N+1}}{\omega_N} = \frac{1}{\rho\Lambda^3 \langle kT_{in} \rangle^{3/2}} \left\langle (kT_{in})^{3/2} \exp\left[-\frac{U_t}{kT_{in}}\right] \right\rangle \tag{C.52}$$

Using (C.52) in (C.39), the test particle expression for the chemical potential is

$$\boxed{\mu = -kT \ln\left[\frac{1}{\rho\Lambda^3 \langle kT_{in} \rangle^{3/2}} \left\langle (kT_{in})^{3/2} \exp\left(-\frac{U_t}{kT_{in}}\right) \right\rangle\right]} \tag{C.53}$$

Here T is the average temperature, as given by (C.24), while T_{in} is the instantaneous value.

If we want the residual chemical potential μ_{res}, then we must subtract from (C.53) the ideal-gas value at the same density ρ and temperature T,

$$\mu_{res} = \mu - \mu_{ig} = \mu - \left(-kT \ln\frac{1}{\rho\Lambda^3}\right) \tag{C.54}$$

Then combining (C.53) and (C.54) gives

$$\mu_{res} = -kT \ln\left[\frac{1}{\langle kT_{in}\rangle^{3/2}}\left\langle (kT_{in})^{3/2}\exp\left(-\frac{U_t}{kT_{in}}\right)\right\rangle\right] \quad (C.55)$$

C.5   ONE-DIMENSIONAL SYSTEM OF HARD SPHERES

We now illustrate the use of results from previous sections by determining the properties of an isolated system of hard spheres constrained to move in only one dimension. The usual derivations for this situation are given in a constant-pressure system [9] rather than in one at constant energy. Consider N hard spheres (more properly, hard rods in one dimension) occupying a line of length L; the system is isolated, so N, L, and the total energy E are constants. For a one-dimensional system the accessible phase-space volume, given by (C.14), reduces to

$$\Omega = \left(\frac{2\pi m}{h^2}\right)^{N/2}\frac{1}{N!\Gamma[N/2+1]}\int dr^N (E-\mathscr{U})^{N/2}\theta[E-\mathscr{U}] \quad (C.56)$$

Hard-sphere forces can act only between nearest neighbors, so the potential energy function is of the form

$$\mathscr{U} = \sum_i^N u(r_i - r_{i-1}) \quad (C.57)$$

When $i=1, r_{i-1}=r_0$ locates the origin, fixed at one end of the line of length L; moreover, we number the spheres serially from one end of the line to the other end. For this case, the integrals in (C.56) can be written as

$$\Omega = \left(\frac{2\pi m}{h^2}\right)^{N/2}\frac{1}{N!\Gamma[N/2+1]}\int_{r_{N-1}}^{L}dr_N \cdots \int_{r_1}^{r_3}dr_2$$

$$\times \int_0^{r_2}dr_1(E-\mathscr{U})^{N/2}\theta[E-\mathscr{U}] \quad (C.58)$$

Now make a change of variables to relative positions,

$$x_1 = r_1$$
$$x_2 = r_2 - r_1$$
$$\vdots \qquad\qquad (C.59)$$
$$x_N = r_N - r_{N-1}$$

So (C.58) becomes

$$\Omega = \left(\frac{2\pi m}{h^2}\right)^{N/2} \frac{1}{N!\Gamma[N/2+1]} \int_0^{\mathscr{L}} dx_N \cdots \int_0^{\mathscr{L}} dx_1 (E - \mathscr{U})^{N/2} \theta[E - \mathscr{U}]$$

(C.60)

For any two neighboring spheres the maximum pair separation $\mathscr{L}$ is the "free volume"—the largest space available to one sphere; thus, $\mathscr{L} = L - (N-1)\sigma$, where σ is the hard-sphere diameter (i.e., the hard-rod length). Now take the Laplace transform of (C.60) with respect to the total energy E. This is an application of the transform (C.12), and the result is

$$L[\Omega] = \left(\frac{2\pi m}{h^2}\right)^{N/2} \frac{1}{N!} \int dx^N \frac{1}{s^{(N/2)+1}} \exp[-s\mathscr{U}]$$

(C.61)

Using (C.57) for the potential $\mathscr{U}$, but written in terms of x, (C.61) becomes

$$L[\Omega] = \left(\frac{2\pi m}{h^2}\right)^{N/2} \frac{1}{s^{(N/2)+1}N!} \left[\int_0^{\mathscr{L}} dx \exp[-su(x)]\right]^N$$

(C.62)

where $u(x)$ is the pair potential between one arbitrarily selected pair of neighboring spheres.

Substituting into (C.62) the hard-sphere potential for $u(x)$, we can divide the integral in two,

$$L[\Omega] = \left(\frac{2\pi m}{h^2}\right)^{N/2} \frac{1}{s^{(N/2)+1}N!} \left[\int_0^{\sigma} dx\, e^{-\infty} + \int_{\sigma}^{\mathscr{L}} dx\right]^N$$

(C.63)

Carrying out the integrations, (C.63) reduces to

$$L[\Omega] = \left(\frac{2\pi m}{h^2}\right)^{N/2} \frac{1}{s^{(N/2)+1}N!} (L - N\sigma)^N$$

(C.64)

The inverse Laplace transform of s^{-N} is [10]

$$L^{-1}\{s^{-N}\} = \frac{t^{N-1}}{(N-1)!}$$

(C.65)

so the inverse transform of (C.64) is

$$\Omega = L^{-1}\{L[\Omega]\} = \left(\frac{2\pi m}{h^2}\right)^{N/2} \frac{1}{N!} (L - N\sigma)^N \frac{E^{N/2}}{(N/2)!}$$

(C.66)

But for hard spheres the total energy is purely kinetic,

$$E = \frac{1}{2} \sum mv^2 = \frac{1}{2} NkT \qquad (C.67)$$

So (C.66) becomes

$$\Omega = \left(\frac{\pi m NkT}{h^2} \right)^{N/2} \frac{(L - N\sigma)^N}{N!(N/2)!} \qquad (C.68)$$

Using the thermal de Broglie wavelength (C.51), we finally can write

$$\boxed{\Omega = \Lambda^{-N} \left(\frac{N}{2} \right)^{N/2} \frac{(L - N\sigma)^N}{N!(N/2)!}} \qquad (C.69)$$

Equation (C.69) gives the phase-space volume accessible to an isolated one-dimensional system of hard spheres. Applying the microcanonical connection to thermodynamics, (C.5), and using Stirling's approximation, we find for the entropy,

$$S = \frac{3Nk}{2} - Nk \ln \frac{N\Lambda}{L - N\sigma} \qquad (C.70)$$

From (C.70) we can now derive all other thermodynamic properties. The results include:

$$P = \frac{NkT}{L - N\sigma} \quad \text{(pressure)} \qquad (C.71)$$

$$C_v = \tfrac{1}{2} Nk \quad \text{(isometric heat capacity)} \qquad (C.72)$$

$$\frac{\mu}{kT} = \frac{N\sigma}{L - N\sigma} + \ln \frac{N\Lambda}{L - N\sigma} \quad \text{(chemical potential)} \qquad (C.73)$$

REFERENCES

[1] J. L. Lebowitz, J. K. Percus, and L. Verlet, "Ensemble Dependence of Fluctuations with Application to Machine Computations," *Phys. Rev.*, **153**, 250 (1967).

[2] P. S. Y. Cheung, "On the Calculation of Specific Heats, Thermal Pressure Coefficients and Compressibilities in Molecular Dynamics Simulations," *Mol. Phys.*, **33**, 519 (1977).

[3] T. L. Hill, *Statistical Mechanics*, McGraw-Hill, New York, 1956; Dover edition, 1987, p. 71.

[4] E. M. Pearson, T. Halicioglu, and W. A. Tiller, "Laplace-Transform Technique for Deriving Thermodynamic Equations from the Classical Microcanonical Ensemble," *Phys. Rev. A*, **32**, 3030 (1985).

[5] R. V. Churchill, *Operational Mathematics*, 3rd. ed., McGraw-Hill, New York, 1972, pp. 33–34.

[6] R. V. Churchill, *Operational Mathematics*, 3rd. ed., McGraw-Hill, New York, 1972, transform 63, p. 462.

[7] D. Frenkel, "Free-Energy Computation and First-Order Phase Transitions" in *Molecular – Dynamics Simulation of Statistical – Mechanical Systems*, G. Ciccotti and W. G. Hoover, Eds., North-Holland, Amsterdam, 1986, Appendix A.

[8] M. Abramowitz and I. A. Stegun, *Handbook of Mathematical Functions*, Dover, New York, 1965, formula 6.1.47, p. 257.

[9] H. Takahashi, "A Simple Method for Treating the Statistical Mechanics of One-Dimensional Substances," *Proc. Phys.-Math. Soc.* (*Jpn.*), **24**, 60 (1942); reprinted in *Mathematical Physics in One Dimension*, E. H. Lieb and D. C. Mattis, Eds., Academic, New York, 1966.

[10] R. V. Churchill, *Operational Mathematics*, 3rd. ed., McGraw-Hill, New York, 1972, transform 3, p. 459.

APPENDIX D

TWO-WAY PLOTS

Most often we generate numbers to make comparisons: to test a theoretical prediction of quantity X, we generate a molecular dynamics value for X. When we have values of X from just two sources, the comparison is usually straightforward. However, if we must compare values of X from three or more sources, then the comparison can be difficult. If, in addition, the deviations among the various values of X are nonlinear, then simple $x-y$ plots or tables of numbers may fail to reveal subtle trends. These problems can sometimes be avoided by using two-way plots. Simple two-way plots are discussed in this appendix: in Section D.1 we describe how to read a two-way plot and in Section D.2 we show how to construct one.

D.1 HOW TO READ TWO-WAY PLOTS

To explore the two-way plot in a familiar situation, consider the conversion of Kelvin temperature into both Fahrenheit and Celsius. A two-way plot that serves this purpose is shown in Figure D.1. On the plot, the horizontal lines are lines of constant scale reading for both Fahrenheit and Celsius; they are level lines. The level lines fall on a simple linear, vertical scale indicated by the numbers at the left of the plot. *Two-way plots have no meaningful horizontal scale.* The two, roughly vertical lines are the two temperature scales, as indicated on the figure.

Pairs of points on the two scales are connected by generally skewed lines; these are isotherms. They are tie lines that connect the two temperature scales. To convert from one temperature scale to the other, say from 100°C to

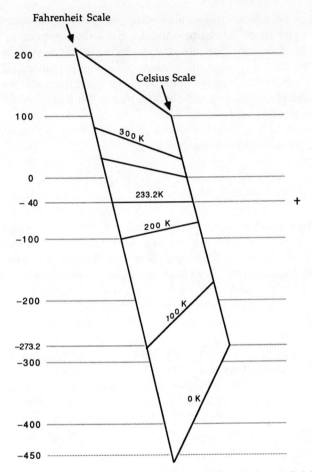

FIGURE D.1   Two-way plot for converting between Fahrenheit and Celsius temperatures.

the equivalent Fahrenheit reading, the procedure is as follows:

(a) Find the 100 degree level line.

(b) Trace the 100 degree level line to its intersection with the Celsius scale line. This locates 100°C.

(c) From 100°C, trace over the tie line, from right to left, until the Fahrenheit scale is encountered.

(d) Identify the value of the level line that coincides with the Fahrenheit scale at the end of the tie line. The value of that level line (212 degrees) is the Fahrenheit scale reading that corresponds to 100°C.

Obviously, if we wanted to use a two-way plot for temperature conversions, we would need more level lines and tie lines than appear in Figure D.1. However, in this book we are not particularly interested in converting numbers. Rather, we want to explore *relationships* among different, but presumably equivalent, representations of the same quantity. So, to continue with our example, what can the two-way plot teach us regarding the relation between Fahrenheit and Celsius?

One observation is that the slopes of the tie lines change with temperature. Thus, at low temperatures we have

$$\left(\begin{array}{c} \text{Numerical value on} \\ \text{Fahrenheit scale at given } T \end{array} \right) < \left(\begin{array}{c} \text{Numerical value on} \\ \text{Celsius scale at same } T \end{array} \right) \quad \text{(D.1)}$$

while at high temperatures we have the converse. The temperature $T = 233.2$ K is special because that tie line is parallel to and coincident with a level line (-40 degrees). The construction of the two-way plot is based on this special temperature. Recall, if we make a simple $x-y$ plot of both degrees Fahrenheit and degrees Celsius versus temperature in Kelvin, we obtain two straight lines that cross at 233.2 K. Likewise, on the two-way plot, if the tie lines are extended they will intersect at a common point (marked by a cross-hatch in Figure D.1). In contrast, a two-way plot of Fahrenheit and Rankine scales has parallel tie lines, because those temperatures scales have no common numerical value.

As a source for a two-way plot more complex than that in Figure D.1, consider the volume expansivity of liquid water. Recall that the expansivity measures the fractional change in volume caused by an isobaric change in temperature:

$$\beta = \frac{1}{V} \left(\frac{\partial V}{\partial T} \right)_P \quad \text{(D.2)}$$

Even though we anticipate that water is an anomalous liquid, we may nevertheless be struck by the behavior of its expansivity. Figure D.2 presents a two-way plot that shows how β responds to changes in pressure and temperature (data from Bridgman [1]). The figure shows the following:

Pressure Effects Along Isotherms

   (i) At 0°C, β initially increases with increasing P, goes through a maximum near 400 bars, and then decreases.

  (ii) At 20°C, β increases with P, goes through a point of inflection between 400 and 600 bars, then continues to increase beyond 600 bars.

 (iii) At 40°C, β is approximately constant, independent of pressure.

 (iv) Above 40°C, β decreases monotonically with increasing pressure.

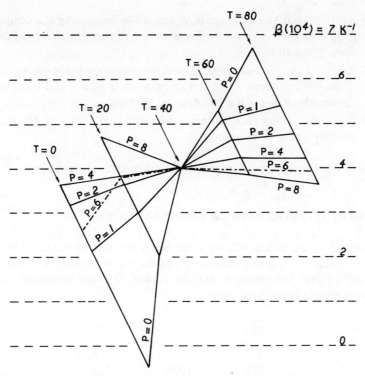

FIGURE D.2 Effects of pressure and temperature on the volume expansivity β of liquid water [1]. Temperatures are in degrees Celsius and pressures in hundreds of bars.

Temperature Effects Along Isobars

   (v) At $P = 0$, β increases quickly and monotonically with increasing T.
  (vi) For $100 \leq P \leq 400$ bars, β still increases with P, but less quickly than at $P = 0$.
 (vii) At $P = 600$ bars, β is approximately constant for $T \geq 20°C$.
(viii) At $P = 800$ bars, β decreases with increasing temperature.

These observations could likely be made almost as easily from a standard $x - y$ plot; but here are some additional observations that would be less easily extracted from other kinds of graphs:

  (ix) On increasing the pressure from 0 to 100 bars, the increase in β at 0°C is somewhat more than *twice* the increase in β at 20°C.
   (x) Likewise, on increasing the pressure from 0 to 100 bars, the decrease in β at 80°C is about *six* times the decrease at 60°C.

(xi) At $P = 0$ bar the increase in β caused by increasing the temperature from 0 to 20°C is about the *same* as the increase caused by changing T from 20 to 40°C.

(xii) At $P = 0$ the increase in β caused by increasing the temperature from 20 to 40°C is about *twice* the increase in β at $P = 100$ bars, which in turn is about *twice* the increase at 200 bars.

(xiii) Generally, along each of the isobars 100, 200, 400, and 800 bars, β is roughly linear in temperature.

You may want to explore Figure D.2 for other such observations.

D.2 HOW TO MAKE SIMPLE TWO-WAY PLOTS

Two-way plots are potentially valuable when we are confronted with two or more dependent variables tabulated against an independent variable. The structure of the temperature plot in Figure D.1 was obtained from the following simple conversion table:

$T(\mathrm{K})$	$T(°\mathrm{C})$	$T(°\mathrm{F})$
0	-273.2	-459.7
373.2	100	212

The construction procedure is as follows:

1. From each entry in a column, subtract a constant amount, for example, the median of the column. Tabulate the residuals. Taking the medians as $T_m(°\mathrm{C}) = -86.6$ and $T_m(°\mathrm{F}) = -124$, we obtain the following residuals:

	Residuals	
$T(\mathrm{K})$	(°C)	(°F)
0	-186.6	-335.7
373.2	186.6	335.7

Thus, we have

$$\text{Each original entry} = \text{median} + \text{residual} \qquad (\mathrm{D.3})$$

2. Construct a rectangular plot of $y = $ residual versus $x = $ median. The medians for each column of the table describe vertical lines; residuals locate points on the lines.

3. Draw in tie lines by connecting pairs of points on the median lines that correspond to the same value of the independent variable.
4. Construct level lines by noting that because of (D.3), a level line is a locus of points obeying

$$x + y = \text{const} \tag{D.4}$$

Thus, the level lines are a family of straight parallel lines. Their slope is determined merely by the scales used on the x- and y-axes. For example, if the scales are the same, then the slope is unity.
5. Rotate the plot so the level lines are horizontal and trace the median lines, tie lines, and level lines onto clean paper.

More sophisticated two-way plots can be generated by adapting this procedure to produce a numerical fit to the tabulated data. To learn more about two-way plots, consult the book by Tukey [2]. To delve into the details of fitting via two-way plots, consult the articles by Emerson et al. [3, 4].

REFERENCES

[1] P. W. Bridgman, *The Physics of High Pressure*, Bell and Sons, London 1958.
[2] J. W. Tukey, *Exploratory Data Analysis*, Addison-Wesley, Reading, MA, 1977, Chapters 10–11.
[3] J. D. Emerson and D. C. Hoaglin, "Analysis of Two-Way Tables by Medians" in *Understanding Robust and Exploratory Data Analysis*, D. C. Hoaglin, F. Mosteller, and J. W. Tukey, Eds., Wiley, New York, 1983.
[4] J. D. Emerson and G. Y. Wong, "Resistant Nonadditive Fits for Two-Way Tables" in *Exploring Data Tables, Trends, and Shapes*, D. C. Hoaglin, F. Mosteller, and J. W. Tukey, Eds., Wiley, New York, 1985.

APPENDIX E

ERROR PROPAGATION

At the end of a simulation we have obtained values for selected properties, say the pressure P and temperature T. Often, we then want to use those properties to compute other properties, for example, the compressibility factor $Z = P/\rho kT$. To simplify the discussion, we use *primary* quantities to mean the direct results from simulation and *secondary* quantities to mean the subsequently computed properties. Now, along with the primary quantities, we have determined their uncertainties and so the question is this: How are uncertainties in primary quantities (such as P and T) carried forward into the secondary quantities (such as Z)? This is a problem of error propagation—an important subject, but one often overlooked by simulators.

In this appendix we give the principal formulas for estimating the effects of error propagation. We also include simple examples, but we omit proofs because they can be found in any good book on error analysis, such as that by Taylor [1]. We consider two cases: the general formula (Section E.1) and a special formula (Section E.2) that applies when the quantities from the simulation are independent and adhere to Gaussian distributions.

E.1 GENERAL EXPRESSION

Consider a secondary quantity A that depends on any number of variables x_i,

$$A = A(x_1, x_2, x_3 \ldots x_n) \tag{E.1}$$

Values for the x_i have been obtained from simulation; each has an uncertainty δx_i. These uncertainties are magnitudes (> 0) that may originate from statistical errors or systematic errors or both. Thus, a value for any one x would be reported as

$$x_i \pm \delta x_i$$

The magnitude of the uncertainty δA residing in A because of the uncertainties in the x's is given by

$$\delta A = \sum_i^n \left| \frac{\partial A}{\partial x_i} \right| \delta x_i \tag{E.2}$$

Example E.1   Algebraic Sums. If A is formed by an algebraic sum of two primary quantities, x and y,

$$A = ax + by \tag{E.3}$$

then the uncertainty in A is related to the sum of the uncertainties in x and y,

$$\delta A = |a|\,\delta x + |b|\,\delta y \tag{E.4}$$

Example E.2   Algebraic Products. If A is formed by a product or quotient of two primary quantities,

$$A = axy \quad \text{or} \quad A = \frac{ax}{by} \tag{E.5}$$

then the uncertainty in A is obtained from the sum of fractional uncertainties,

$$\frac{\delta A}{|A|} = \frac{\delta x}{|x|} + \frac{\delta y}{|y|} \tag{E.6}$$

E.2   SPECIAL CASE: INDEPENDENT PRIMARY QUANTITIES

If the primary quantities x_i are known to be mutually independent and if each is distributed according to a Gaussian distribution, then the general expression (E.2) is an upper bound on the uncertainty in the secondary quantity A. A more realistic estimate of δA is provided by combining in quadrature,

$$\delta A = \sqrt{\sum_i^n \left(\frac{\partial A}{\partial x_i} \delta x_i \right)^2} \qquad (E.7)$$

This allows for the possibility that uncertainties in some of the x's will fortuitously cancel uncertainties in some of the other x's.

Example E.3 Algebraic Sums. If A is formed by an algebraic sum of two primary quantities, such as in (E.3), and if the special conditions apply, then

$$\delta A = \sqrt{(a\,\delta x)^2 + (b\,\delta y)^2} \qquad (E.8)$$

Example E.4 Algebraic Products. If A is formed by a product or quotient of two primary quantities, such as in (E.5), and if the special conditions apply, then the fractional uncertainties combine in quadrature,

$$\frac{\delta A}{|A|} = \sqrt{\left(\frac{\delta x}{x} \right)^2 + \left(\frac{\delta y}{y} \right)^2} \qquad (E.9)$$

Example E.5 Least-Squares Fits [2]. Often, several simulations are used to generate values for a quantity y as it changes with some parameter x. The y-values are then fit by least squares to some function of x. We assume x is a quantity that is constant during one run so there are no uncertainties in the x-values; for example, x might be the density or number of molecules. Each value y_i has some uncertainty δy_i, and we wish to estimate how those uncertainties propagate to the parameters obtained from the fit. Here we consider the simplest case, a straight line,

$$y = mx + b \qquad (E.10)$$

Since the precision of each y_i is inversely related to its uncertainty, we do a weighted least-squares fit in which the weights are the inverse uncertainties. Thus, for N pairs of x, y-values, we are to solve

$$\min_{\{m,b\}} \sum_i^N \left(\frac{y_i - mx_i - b}{\delta y_i} \right)^2 \qquad (E.11)$$

The solution of this minimization problem gives the slope and intercept as

$$m = \frac{1}{\Delta}\left[SS_{xy} - S_x S_y \right] \tag{E.12}$$

$$b = \frac{1}{\Delta}\left[S_{xx} S_y - S_x S_{xy} \right] \tag{E.13}$$

where

$$\Delta = SS_{xx} - \left(S_x \right)^2 \tag{E.14}$$

$$S = \sum_i^N \frac{1}{\left(\delta y_i \right)^2} \tag{E.15}$$

$$S_x = \sum_i^N \frac{x_i}{\left(\delta y_i \right)^2} \qquad S_y = \sum_i^N \frac{y_i}{\left(\delta y_i \right)^2} \tag{E.16}$$

$$S_{xx} = \sum_i^N \frac{x_i^2}{\left(\delta y_i \right)^2} \qquad S_{xy} = \sum_i^N \frac{x_i y_i}{\left(\delta y_i \right)^2} \tag{E.17}$$

Each y-value is from a different simulation, so the elements of the set $\{y\}$ are independent. If each is also a consequence of a Gaussian of instantaneous values, then (E.7) applies, and we find

$$\delta m = \sqrt{ \sum_k^N \left(\frac{\left[S x_k - S_x \right]}{\Delta(\delta y_k)^2} \delta y_k \right)^2 } \tag{E.18}$$

and

$$\delta b = \sqrt{ \sum_k^N \left(\frac{\left[S_{xx} - S_x x_k \right]}{\Delta(\delta y_k)^2} \delta y_k \right)^2 } \tag{E.19}$$

Expanding the quadratics and using (E.15)–(E.17), these expressions simplify to

$$\boxed{\delta m = \sqrt{\frac{S}{\Delta}} \qquad \delta b = \sqrt{\frac{S_{xx}}{\Delta}}} \tag{E.20}$$

Subsequently, when the fitted line (E.10) is used to estimate y at any particular x-value, the uncertainty in the computed y can be estimated by

$$\delta y = |x| \, \delta m + \delta b \qquad (E.21)$$

Press et al. [2] extend this procedure to fits of nonlinear functions.

REFERENCES

[1] J. R. Taylor, *An Introduction to Error Analysis*, University Science Books (Oxford University Press), Mill Valley, CA, 1982.

[2] W. H. Press, B. P. Flannery, S. A. Teukolsky, and W. T. Vetterling, *Numerical Recipes*, Cambridge University Press, New York, 1986, Chapter 14.

APPENDIX F

CONVENTIONS FOR NAMING VARIABLES IN FORTRAN

To ease study and verification of the programs given in Appendices I–L, we have adopted the following conventions for naming Fortran variables[†]:

1. The Fortran conventions for integer variables (names starting with letters $I-N$) and real variables (names starting with letters $A-H$ and $O-Z$) are used throughout. The only exception is that variable names starting with the letter L are reserved for logical variables.

2. The Fortran name is often identical to the actual name if it contains six or less characters, for example, ENERGY and VIRIAL.

3. If the actual name is more than six characters, a Fortran name (stem) is formed in one of the following ways:

   a. From the first syllable of the actual name if the syllable is identifiable with the quantity, for example, PRES for pressure; alternatively, from an obvious abbreviation based upon the first syllable, for example, TEMP for temperature.

   b. By contracting the actual name to six characters; usually the contraction is performed by eliminating some or all vowels. For example, we use PRTCLE for particle, DENSTY for density.

   c. By using a phonetic equivalent, for example, NABORS for neighbors.

[†]"…the art of programming is the art of organizing complexity…", E. W. Dijkstra in *Structured Programming*, O. J. Dahl, E. W. Dijkstra, and C. A. R. Hoare, Academic, London, 1972, p. 6.

    d. By analogy with the quantity's mathematical symbol or notation. For example, we use X for the Cartesian component of a position, F for force, X1 for the first time-derivative of x.

    e. By compounding two words of the actual name and applying the above conventions, especially contraction, as necessary. For example, we use ENRKIN for kinetic energy and ENRPOT for potential energy.

4. Names for functions of previously named quantities are formed by appending prefixes and/or suffixes from the following list; for example, STEP is the size of the time step, and therefore STEPSQ is the square of STEP.

5. When the addition of prefixes or suffixes extends the name beyond six characters, the name is contracted by first eliminating vowels (except the first letter if it is a vowel), then by dropping consonants starting from the end of the stem.

    Ex. 1   TEMP is the desired value of the temperature

            TMPAVE is an average temperature

    Ex. 2   ENERGY is the instantaneous potential energy

            SUMENR is the accumulated potential energy

6. Multiple functions of a variable may be indicated by appending several prefixes and/or suffixes and contracting the stem as necessary. For example,

        CUBE is the length (L) of a side of the cubic volume

        CUBEH is $\frac{1}{2}L$

        CUBESQ is L^2

        CUBHSQ is $(\frac{1}{2}L)^2$

        CUBSQH is $\frac{1}{2}L^2$

Prefixes and First Letters of a Name

FN-, FM-	Floating-point equivalent of an integer variable
I-, J-	Dummy indices such as appear in DO loops and counting operations
K-	Numerical control flags; i.e., integer parameters that dictate when or under what conditions an operation is to occur
L-	Logical control flag
MAX-	Maximum allowed value of an integer variable
N-	Total number of objects in a collection or the serial number of a particular member of a collection
R-	Function of a radial distance
SUM-	Accumulator for evaluating a sum
X-, Y-, Z-	Some function of particle positions; e.g., X(1) is the x-component of the location vector for particle 1, X1(1) is the first time derivative of X(1)

Suffixes and Last Letters of a Name

-ABS	Absolute value of a variable
-AVE	Average value of a variable
-H	Half the value of a previously evaluated variable
-IJ	Function of the distance between atoms i and j
-INV	Inverse of a previously evaluated variable
-MAX	Maximum allowed value of a real variable
-M1	Variable minus 1
-P1	Variable plus 1
-SQ	Square of a variable
-X, -Y, -Z	Cartesian components of a vector; e.g., FX is the x-component of a force

APPENDIX G

CHECKSUM
A PROGRAM TO HELP VERIFY COPIES OF FORTRAN CODES

Before attempting to execute any of the programs presented in the appendices that follow, you should verify that your copy of the code is identical to that in the listing. Program CHECKSUM can help with the verification process. CHECKSUM computes a four-digit sum for each line of Fortran code (comment lines are excluded). To obtain the sum, an integer equivalent is assigned to each Fortran character and to the ampersand and quote. Then, for each line of 72 columns, the sum of the line's contents is formed by summing its integer equivalents. The last three digits in the sum are used as the first three digits of the check sum. A fourth digit is added as a check on the previous three; that is, it is the final digit in the sum of the previous three digits. The four-digit check sum is placed in columns 77–80 of the line. A serial line number is placed in columns 74–75. An example line is shown in Figure G.1.

CHECKSUM treats a Fortran code as data, echoes each line, and appends to each the line number and the four-digit check sum. If the code already contains check sums, CHECKSUM can be used to verify them. CHECKSUM has proven valuable when code is transmitted from one machine to another and from one user to another. Note that CHECKSUM is an aid, not a foolproof tool.

368

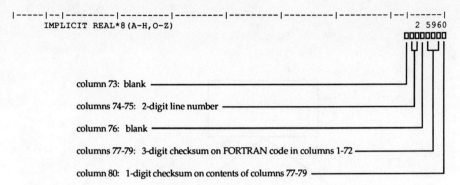

FIGURE G.1 Sample line of Fortran code showing the position of the line number and the four-digit check sum.

A hierarchy chart for CHECKSUM is given in Figure G.2. To compute or verify check sums, append to the end of CHECKSUM the code to be checked. Remove any job control lines from the code. CHECKSUM does not recognize tabs, so in the code you will need to replace any tabs with blank spaces. To conform to local conventions, you may need to change the unit numbers on input/output (i/o) statements. The following subroutines contain i/o statements:

i/o Statement	Subroutine (Number of i/o's Contained)	
READ	CKNULN	2
WRITE	main	1
	CKNULN	3
	CKNUPG	3
	CKOUT	5

In the main program you must set flags LNEW and KPRINT.

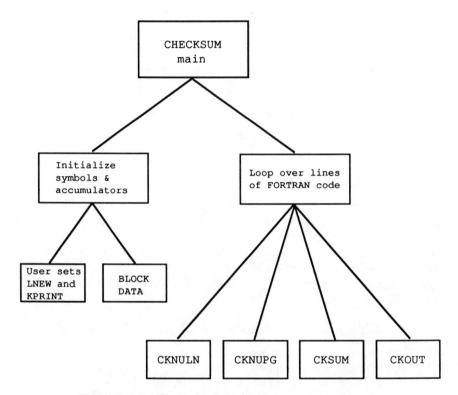

FIGURE G.2 Hierarchy chart for the program CHECKSUM

```
|-----|--|---------|---------|---------|---------|---------|---------|--|------|

C
C      filename:    checksum.f
C
C          FORTRAN program to compute checksums for each line
C                  of a FORTRAN source code.
C
C          versions:   1.1   11 April  1987
C
C-------------------------------------------------------------------------------
C
       CHARACTER*1 COLUMN(72),SYMBOL(48)                                    1 7164
       CHARACTER*32 INFILE, OTFILE
       LOGICAL LNEW, LAST                                                   2 3777
       COMMON /A/ COLUMN, SYMBOL                                            3 4699
       DATA NERROR,NLINE,NPASS/3*0/                                         4 6174
C
C...if computing new checksums, set LNEW to .TRUE.
C   if verifying existing checksums, set LNEW to .FALSE.
       LNEW = .TRUE.                                                        5 2732
C
C...printing instructions:
C       full listing? (Set KPRINT = 1)
C       or just print checksums (Set KPRINT = 0)
C       or print only if new & old sums don't match? (Set KPRINT = -1)
       KPRINT = 1                                                           6 1966
C
C...read FORTRAN line from input
   10   CALL CKNULN(KOLD,KPRINT,LNEW,LAST,NLINE,NPASS)                      7 2361
       IF (.NOT.LAST) THEN                                                  8 3586
C
C...if start of new subprogram, set line counter to zero
          CALL CKNUPG(KPRINT,NLINE)                                         9 5768
          NLINE = NLINE + 1                                               10 3261
          NPASS = NPASS + 1                                               11 2002
C
C...determine checksum for current line
          CALL CKSUM(KCHECK)                                              12 3733
          CALL CKOUT(KCHECK,KOLD,KPRINT,LNEW,NERROR,NLINE,NPASS)          13 3542
C
C...get next line of code
          GOTO 10                                                         14 1102
       ENDIF                                                              15 1292
C
       IF (.NOT.LNEW) WRITE(9,981) NERROR                                 16 7142
  981  FORMAT(//3X,'LINE CHECK COMPLETE:',I5,' UNMATCHED LINES FOUND'//)  17 4790
C
       STOP                                                               18  550
       END                                                                19  921

|-----|--|---------|---------|---------|---------|---------|---------|--|------|
```

```
|-----|--|---------|---------|---------|---------|---------|---------|--|------|
        SUBROUTINE CKNULN(KOLD,KPRINT,LNEW,LAST,NLINE,NPASS)          1 2934
        CHARACTER*1 COLUMN(72),SYMBOL(48)                             2 7164
        LOGICAL LNEW, LAST                                           3 3777
        COMMON /A/ COLUMN, SYMBOL                                     4 4699
C
C...if creating new checksums, read line
   20   IF (LNEW) THEN                                                5 3744
           READ ( 5,50,END=101) COLUMN                               6 5993
           LAST = .FALSE.                                            7 2406
C
C...if verifying existing checksums, read line & old sum
        ELSE                                                         8  932
           READ (5,55,END=101) COLUMN,KOLD                           9 7557
        ENDIF                                                       10 1292
C
C...if line is a comment, skip to next line
        IF (COLUMN(1).EQ.'C') THEN                                  11 5207
C
C...if near top of file, check for start of main source code
        IF (Npass.LT.200) THEN                                      12 4071
           DO 57 I=7,16                                             13 2530
              IF (Column(I).EQ.'M' .and. Column(I+1).EQ.'A') THEN   14 7399
                 IF (Column(I+2).EQ.'I' .and. Column(I+3).EQ.'N') THEN 15 7894
                    Nline = 0                                       16  965
                    IF (Kprint.GE.0) WRITE(6,987)                   17 5735
                 ENDIF                                              18 1292
              ENDIF                                                 19 1292
   57      CONTINUE                                                 20 1887
        ENDIF                                                       21 1292
C
        IF (Kprint.EQ.0) WRITE(6,983)                               22 5566
        IF (Kprint.EQ.1) WRITE(6,985) Column                       23 5274
        GOTO 20                                                     24 1168
     ENDIF                                                          25 1292
C
     RETURN                                                         26 1090
  101 LAST=.TRUE.                                                   27 2855
C
   50 FORMAT(72A1)                                                  28 2664
   55 FORMAT(72A1,4X,I4)                                            29 4611
  983 FORMAT(1X,'C')                                                30 3979
  985 FORMAT(72A1)                                                  31 3058
  987 FORMAT(//)                                                    32 2697
C
     RETURN                                                         33 1090
     END                                                            34  921
|-----|--|---------|---------|---------|---------|---------|---------|--|------|
```

```
|-----|--|---------|---------|---------|---------|---------|---------|--|------|
          SUBROUTINE CKNUPG(KPRINT,NLINE)                            1 6859
          CHARACTER*1 COLUMN(72), SYMBOL(48)                         2 7164
          COMMON /A/ COLUMN, SYMBOL                                  3 4699
C
C...check for new subroutine
          DO 50 I=7,10                                               4 2822
            IF (COLUMN(I).EQ.'S' .AND. COLUMN(I+1).EQ.'U') THEN      5  156
              IF (COLUMN(I+2).EQ.'B' .AND. COLUMN(I+3).EQ.'R') THEN  6  943
                NLINE = 0                                            7 2046
                IF (KPRINT.GE.0) WRITE(6,987)                        8 6747
                RETURN                                               9 1090
              ENDIF                                                 10 1292
            ENDIF                                                   11 1292
   50     CONTINUE                                                  12 1944
C
C...check for new FUNCTION subprogram
          DO 60 I=7,30                                              13 2901
            IF (COLUMN(I).EQ.'F' .AND. COLUMN(I+1).EQ.'U') THEN     14  224
              IF (COLUMN(I+2).EQ.'N' .AND. COLUMN(I+3).EQ.'C') THEN 15  673
                NLINE = 0                                           16 2046
                IF (KPRINT.GE.0) WRITE(6,987)                       17 6747
                RETURN                                              18 1090
              ENDIF                                                 19 1292
            ENDIF                                                   20 1292
   60     CONTINUE                                                  21 1887
C
C...check for BLOCK DATA
          DO 70 I=7,10                                              22 2934
            IF (COLUMN(I).EQ.'B' .AND. COLUMN(I+1).EQ.'L') THEN     23  741
              IF (COLUMN(I+2).EQ.'O' .AND. COLUMN(I+3).EQ.'C') THEN 24  460
                NLINE = 0                                           25 2046
                IF (KPRINT.GE.0) WRITE(6,987)                       26 6747
                RETURN                                              27 1090
              ENDIF                                                 28 1292
            ENDIF                                                   29 1292
   70     CONTINUE                                                  30 2057
C
  987     FORMAT(//)                                                31 2697
          RETURN                                                    32 1090
          END                                                       33  921
|-----|--|---------|---------|---------|---------|---------|---------|--|------|
```

```
|-----|--|---------|---------|---------|---------|---------|---------|--|------|

        SUBROUTINE CKSUM(KCHECK)                                 1 4824
        CHARACTER*1 COLUMN(72),SYMBOL(48)                        2 7164
        COMMON /A/ COLUMN, SYMBOL                                3 4699
C
C...loop over each column of the line
        ISUM = 0                                                 4 1405
        DO 100 I=1,72                                            5 3047
C
C...if the column is blank, move to next column
        IF (COLUMN(I).NE.' ') THEN                               6 5454
          DO 90 J=1,48                                           7 3058
            IF (COLUMN(I).EQ.SYMBOL(J)) THEN                     8 6444
C
C...once character in the column is identified, increase accumulator
C     and move to next column
              ISUM = ISUM+J                                      9 1876
              GOTO 100                                          10 1450
            ENDIF                                               11 1292
   90     CONTINUE                                              12 2237
        ENDIF                                                   13 1292
  100 CONTINUE                                                  14 2282
C
C...remove first digit from the accumulator
        ISUM  = ISUM - (ISUM/1000)*1000                         15 6309
C
C...compute checksum of the checksum
        IHUN  = ISUM/100                                        16 3047
        IRMDR = ISUM-IHUN*100                                   17 4891
        ITEN  = IRMDR/10                                        18 3126
        IONE  = IRMDR-ITEN*10                                   19 4600
        KSUM  = IHUN+ITEN+IONE                                  20 4093
        KCHECK = KSUM-(KSUM/10)*10 + 10*ISUM                    21 7939
C
        RETURN                                                  22 1090
        END                                                     23  921

|-----|--|---------|---------|---------|---------|---------|---------|--|------|
```

```
|-----|--|---------|---------|---------|---------|---------|---------|--|------|
      SUBROUTINE CKOUT(KCHECK,KOLD,KPRINT,LNEW,NERROR,NLINE,NPASS)        1 4633
      CHARACTER*1 COLUMN(72),SYMBOL(48)                                  2 7164
      LOGICAL LNEW                                                       3 2585
      COMMON /A/ COLUMN, SYMBOL                                          4 4699
C
C...if creating new checksums, print as per KPRINT
      IF (KPRINT.EQ.1) THEN                                             5 4251
         WRITE(6,989) COLUMN,NLINE,KCHECK                               6 8435
C
      ELSE IF (KPRINT.EQ.0) THEN                                        7 5364
         WRITE(6,991) NLINE,KCHECK                                      8 6725
      ENDIF                                                             9 1292
C
C...if verifying old checksums, are new & old sums the same?
      IF (.NOT.LNEW) THEN                                              10 4150
         IF (KCHECK.NE.KOLD) THEN                                      11 5678
            NERROR = NERROR + 1                                        12 2529
            IF (KPRINT.EQ.-1) WRITE(6,989) COLUMN,NLINE,KCHECK         13 1977
            WRITE(6,993) NLINE                                         14 4813
         ENDIF                                                         15 1292
C
C...if no listing requested, print periodic message to reassure user
         IF (KPRINT.EQ.-1 .AND. MOD(NPASS,500).EQ.0) WRITE(9,995) NPASS 16 3148
      ENDIF                                                            17 1292
C
  989 FORMAT(72A1,I3,1X,I4)                                            18 6174
  991 FORMAT(5X,I4,1X,I4)                                              19 5667
  993 FORMAT(/2X,'****** CURRENT & PREVIOUS CHECKSUMS DO NOT',          20 2103
     &            ' MATCH FOR LINE # ',I4//)                           21 5128
  995 FORMAT(5X,'NOW PROCESSING ',I4,'TH LINE OF CODE')                22 1472
C
      RETURN                                                           23 1090
      END                                                              24  921
|-----|--|---------|---------|---------|---------|---------|---------|--|------|
      BLOCK DATA                                                        1 2079
      CHARACTER*1 COLUMN(72),SYMBOL(48)                                 2 7164
      COMMON /A/ COLUMN, SYMBOL                                         3 4699
C
      DATA SYMBOL/'V','C','X','+','Q','J','O','A','S','U',              4 6477
     &            '&','6','R','(','/','F','1','5','P','T',              5 5061
     &            'I','.','2','G','E',')','8','N','7','Z',              6 6062
     &            '3','M','=','L','0','Y','"','H','D','-',              7 7063
     &            'W','*',1H','B','K','4','9',',','/                    8 3755
C
      END                                                              9  921
|-----|--|---------|---------|---------|---------|---------|---------|--|------|
```

APPENDIX H

ROULET
A RANDOM-NUMBER GENERATOR

The molecular dynamics programs in Appendices I, K, and L each use a random-number generator to assign initial velocities. In each, the function used is ROULET, a generator taken from Press et al.[†] ROULET uses linear congruential methods plus shuffling; it is seeded by passing any negative integer as its calling argument ISEED. The function returns a real pseudo-random value uniformly distributed on $[-1, +1]$. According to Press et al., this generator has an essentially infinite period, and it should execute on any machine. In addition to changing the value of the seed, the generator can be modified by changing the length of the shuffling vector RR and by changing the overflow bound (2^{24} in the present incarnation). The overflow bound is changed by replacing the values for parameters M1, M2, M3, IA1, IA2, IA3, IC1, IC2, and IC3 with alternatives provided by Press et al. For more on linear congruential generators, see Knuth.[‡]

[†]W. H. Press, B. P. Flannery, S. A. Teukolsky, and W. T. Vetterling, *Numerical Recipes*, Cambridge University Press, New York, 1986. © 1986 Numerical Recipes Software. Used with permission. It appears on p. 196 as FUNCTION RAN1.
[‡]D. E. Knuth, *The Art of Computer Programming*, Vol. 2, *Seminumerical Algorithms*, Addison-Wesley, Reading, MA, 1969, Chapter 3.

```
|-----|--|---------|---------|---------|---------|---------|---------|--|------|
        FUNCTION ROULET(ISEED)                                      1 4004
C
C       Create a pseudo-random number ...
C
C       © 1986 by Numerical Recipes Software. Reproduced by permission,
C          from the book Numerical Recipes: The Art of Scientific
C          Computing, published by Cambridge University Press.
C
C       Passing any negative value thru ISEED will reseed the generator.
C-------------------------------------------------------------------------
        DOUBLE PRECISION RR(97)                                    2 4464
        COMMON /RAND / RR                                          3 2529
C
C...parameter values for overflows at 2**24
        DATA M1, IA1, IC1/ 31104,   625, 6571/                     4 7074
        DATA M2, IA2, IC2/ 12960,  1741, 2731/                     5 7939
        DATA M3, IA3, IC3/ 14000,  1541, 2957/                     6 8570
        RM1 = 1./M1                                                7 1988
        RM2 = 1./M2                                                8 2103
C
C...initialize the shuffling vector RR to hold the random numbers
        IF (ISEED.LT.0) THEN                                       9 4408
          IX1 = MOD(IC1-ISEED  ,M1)                               10 4880
          IX1 = MOD(IA1*IX1+IC1,M1)                               11 4622
          IX2 = MOD(IX1, M2)                                      12 3429
          IX1 = MOD(IA1*IX1+IC1,M1)                               13 4622
          IX3 = MOD(IX1, M3)                                      14 3586
C
C...load vector RR
          DO 11 J=1,97                                            15 2608
            IX1 = MOD(IA1*IX1+IC1,M1)                             16 4622
            IX2 = MOD(IA2*IX2+IC2,M2)                             17 4925
            RR(J)= (FLOAT(IX1) + FLOAT(IX2)*RM2)*RM1              18 7018
     11   CONTINUE                                                19 1753
          ISEED=1                                                 20 1696
        ENDIF                                                     21 1292
C
C...randomly sample vector RR
        IX3 = MOD(IA3*IX3+IC3,M3)                                 22 5320
        J = 1 + (97*IX3)/M3                                       23 3519
C
        IF (J.GT.97 .OR. J.LT.1)  WRITE(6,99)                     24 7467
     99 FORMAT(//5X,'Array size for RR violated in ROULET'/)      25 5735
C
C...change interval from [0,1] to [-1,1]
        ROULET = 2.*RR(J)-1.                                      26 3801
C
C...replace this RR(J) with the next value in the sequence
        IX1 = MOD(IA1*IX1+IC1,M1)                                 27 4622
        IX2 = MOD(IA2*IX2+IC2,M2)                                 28 4925
        RR(J) = (FLOAT(IX1) + FLOAT(IX2)*RM2)*RM1                 29 7018
C
        RETURN                                                    30 1090
        END                                                       31  921
|-----|--|---------|---------|---------|---------|---------|---------|--|------|
```

APPENDIX I

MDODSS
MOLECULAR DYNAMICS CODE FOR SOFT SPHERES IN ONE DIMENSION

A one-dimensional simulation can serve as a good way to learn soft-sphere molecular dynamics. A one-dimensional simulation poses all the conceptual and practical problems that arise in molecular dynamics; nevertheless, the code is short and executes quickly. Moreover, many properties of one-dimensional systems can be computed analytically, providing unambiguous data for testing simulation results.

In the codes in Appendices I–L, global variables are passed through COMMON. COMMON blocks can be a troublesome source of errors during execution, particularly when COMMON is used to pass global scalars. To help guard access to global scalars, we pass them through arrays: RL contains the scalar reals and IN the integers. Then a subroutine can access only those global scalars that are explicitly cited in EQUIVALENCE statements. The dot chart in Figure I.1 lists the global scalars in MDODSS, cites the variable's name and its equivalent position as an element in either RL or IN, and indicates those subroutines that have access to each variable.

The program MDODSS is divided into 10 subroutines; the basic structure of the program is organized around six principal subroutines, as shown in the hierarchy chart in Figure I.2. Several of the principal subroutines, in turn, call secondary subroutines, but that more detailed structure is indicated in the notes that accompany each subroutine.

Documentary notes appear on the page facing each subroutine. Each note contains a segment of a hierarchy chart that shows how the subroutine is related to its calling and called program segments. Each note begins with

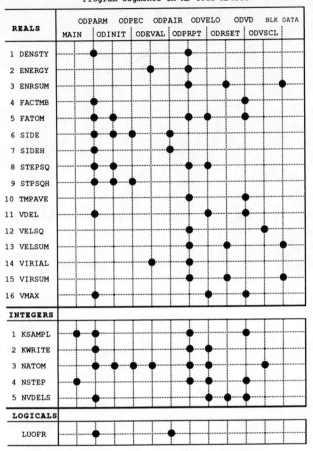

FIGURE I.1 Real, integer, and logical global scalars used in MDODSS. Arrays RL(16) and IN(5) are used to pass these scalars among the program segments shown. For example, RL(1) is equivalent to the scalar DENSTY.

lines of the form

$$\text{program segment} \Rightarrow \text{list of global scalars}$$

for example

$$\text{ODPARM} \Rightarrow \text{NATOM}$$

This notation means that the value of the global scalar NATOM is being passed from subroutine ODPARM.

To verify your copy of the code, use CHECKSUM from Appendix G. To run the program, you will need to assign values to the run parameters in

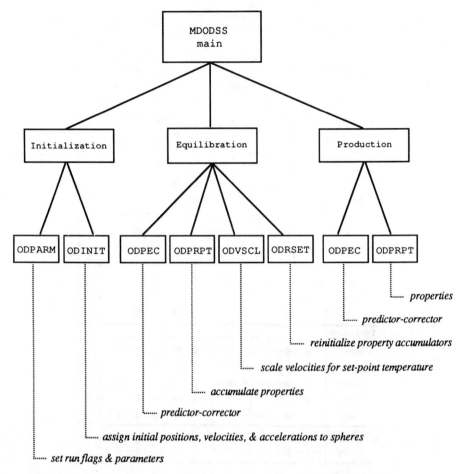

FIGURE I.2 Hierarchy chart for the one-dimensional simulation code MDODSS.

subroutine ODPARM. It is convenient to modify lines 13–23 of ODPARM to allow you to set those values interactively. You will probably have to modify the unit numbers on WRITE statements to conform to local conventions. WRITE statements appear in subroutines ODPRPT, ODVD, and ODRSET. The program contains no READ statements. Do not forget to append the random-number generator ROULET from Appendix H.

```
|-----|--|---------|---------|---------|---------|---------|---------|--|------|
C
C      filename:  mdodss.f
C
C         One-dimensional molecular dynamics for soft  spheres
C                  nearest neighbor interactions only
C
C                      version 4.4    24 Jan 1992
C-------------------------------------------------------------------------------
       IMPLICIT REAL*8(A-H,O-Z)                                      1 5702
       LOGICAL LSCALE                                                2 2428
       COMMON /INTGRS/ IN(5)                                         3 3609
       COMMON /REALS / RL(16)                                        4 3430
C
       EQUIVALENCE (IN(1),KSAMPL),  (IN(4),NSTEP)                    5 8974
C
C...initialization
       CALL ODPARM(TEMP,MAXSTP,MAXEQB,LSCALE)                        6 7962
       CALL ODINIT(HEATR,TEMP)                                       7 5027
C
C...equilibration loop
       IF (MAXEQB.GT.0) THEN                                         8 4284
         DO 3 NSTEP = 1, MAXEQB                                      9 3935
           CALL ODPEC                                               10 1708
             IF (MOD(NSTEP,KSAMPL).EQ.0) THEN                       11 7119
               CALL ODPRPT                                          12 1955
               IF (LSCALE) CALL ODVSCL(HEATR)                       13 5038
             ENDIF                                                  14 1292
    3    CONTINUE                                                   15 1720
C
C...at end of equilibration, reinitialize property accumulators
         CALL ODRSET                                                16 1911
       ENDIF                                                        17 1292
C
C...production loop
       DO 5 NSTEP = 1, MAXSTP                                       18 3542
         CALL ODPEC                                                 19 1708
         IF (MOD(NSTEP,KSAMPL).EQ.0) CALL ODPRPT                    20 7951
    5    CONTINUE                                                   21 1595
C
C...print velocity distribution
       CALL ODVD                                                    22 1641
C
       STOP                                                         23  550
       END                                                          24  921

|-----|--|---------|---------|---------|---------|---------|---------|--|------|
```

SUBROUTINE ODPARM

Run flags and parameters set by user

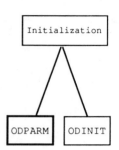

LUOFR = .true. for Lennard-Jones
       = .false. for repulsive soft-sphere

NATOM = number of spheres

STEP   = time-step, $\Delta t / \sigma (m/\varepsilon)^{1/2}$

TEMP   = set-point temperature, kT/ε

DENSTY = number density, $N\sigma^3/V$

MEQSTP = total number of time-steps to execute during equilibration

MAXSTP = total number of time steps to execute in production

KSAMPL = time-step increment at which to sample properties for accumulating averages

KWRITE = time-step increment at which to print running averages

Other flags and parameters

LSCALE = .true. scale velocities during equilibration to desired T*
       = .false. for no scaling

SIDE   = length of the one-dimensional system, L/σ

ALFA   = parameters in Gear's fifth-order predictor-corrector algorithm for second-order ODEs, from Table 4.1

VMAX   = maximum velocity to be used in sampling for velocity distribution; units are $v(m/\varepsilon)^{1/2}$

VDEL   = increments at which to sample for velocity distribution

NVDELS = number of velocity increments in distribution

In this program each sphere has unit mass.

```
|-----|--|---------|---------|---------|---------|---------|---------|--|------|
         SUBROUTINE ODPARM(TEMP,MAXSTP,MAXEQB,LSCALE)              1 9054
C
C        Here user sets run flags and parameters
C
         IMPLICIT REAL*8(A-H,O-Z)                                  2 5702
         LOGICAL LSCALE, LUOFR                                     3 3700
         COMMON /ALPHA / ALFA0,ALFA1,ALFA3,ALFA4,ALFA5             4 9144
         COMMON /INTGRS/ IN(5)                                     5 3609
         COMMON /LOGCLS/ LUOFR                                     6 3283
         COMMON /REALS / RL(16)                                    7 3430
C
         EQUIVALENCE (RL(1),DENSTY), (RL(4),FACTMB), (RL(5),FATOM)  8 2484
         EQUIVALENCE (RL(6),SIDE),   (RL(7),SIDEH),  (RL(8),STEPSQ) 9 1865
         EQUIVALENCE (RL(9),STPSQH), (RL(11),VDEL),  (RL(16),VMAX) 10 1584
         EQUIVALENCE (IN(1),KSAMPL), (IN(2),KWRITE), (IN(3),NATOM) 11 2899
         EQUIVALENCE (IN(5),NVDELS)                               12 5151
C
C...LJ (.true.) or repulsive soft spheres (.false.)?
         LUOFR = .TRUE.                                           13 2259
C
C...run parameters
         NATOM = 100                                              14 2158
         STEP  = 0.004 D0                                         15 3531
         TEMP  = 1.2D0                                            16 2653
         DENSTY= 0.5D0                                            17 3395
C
         MAXEQB =  2000                                           18 2787
         MAXSTP = 10000                                           19 2811
         KSAMPL =    20                                           20 2383
         KWRITE =    20                                           21 2563
         LSCALE = .TRUE.                                          22 2574
C
         FATOM = DFLOAT(NATOM)                                    23 3755
         SIDE  = FATOM/DENSTY                                     24 3823
         SIDEH = 0.5D0*SIDE                                       25 4509
C
         STEPSQ = STEP*STEP                                       26 3081
         STPSQH = 0.5D0*STEPSQ                                    27 4116
C
C...parameters in predictor-corrector method
         ALFA0 =   3.D0/ 16.D0                                    28 4015
         ALFA1 =251.D0/360.D0                                     29 4598
         ALFA3 = 11.D0/ 18.D0                                     30 4150
         ALFA4 =  1.D0/  6.D0                                     31 3812
         ALFA5 =  1.D0/ 60.D0                                     32 3889
C
C...parameters for velocity distribution
         FACTMB = 1.D0/DSQRT(2.D0*3.14159265D0)                   33 8952
         VMAX = 4.D0                                              34 2192
         VDEL = 0.05D0                                            35 3160
         NVDELS = 2.*VMAX/VDEL+1                                  36 4352
         RETURN                                                   37 1090
         END                                                      38  921
|-----|--|---------|---------|---------|---------|---------|---------|--|------|
```

SUBROUTINE ODINIT

ODPARM ⇒   FATOM, SIDE, STEPSQ,
STPSQH, NATOM

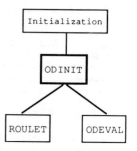

The random-number generator ROULET is given in Appendix H. Values obtained from ROULET are uniformly distributed on [−1, +1].

To obtain zero total linear momentum, the random velocities are scaled according to (5.28).

The velocity scaling to the set-point temperature includes a factor of the time-step; i.e., throughout this program velocities and their derivatives are implicitly scaled by a constant factor:

$$X1 \; = \; (dx/dt)\,\Delta t$$

$$X2 \; = \; (d^2x/dt^2)\,(\Delta t)^2/2$$

$$X3 \; = \; (d^3x/dt^3)\,(\Delta t)^3/3!$$

$$X4 \; = \; (d^4x/dt^4)\,(\Delta t)^4/4!$$

$$X5 \; = \; (d^5x/dt^5)\,(\Delta t)^5/5!$$

Otherwise, the temperature scaling is according to (5.25).

CALL ODEVAL ... We get the initial accelerations from the forces acting as a result of the initially assigned positions. Third and higher derivatives of the positions are initially set to zero in BLOCK DATA.

```
|-----|--|--------|---------|---------|---------|---------|---------|--|------|
           SUBROUTINE ODINIT(HEATR,TEMP)                            1 6118
C
C          Initialize sphere positions, velocities, & forces
C
           IMPLICIT REAL*8(A-H,O-Z)                                 2 5702
           REAL*4 ROULET                                           3 2776
           COMMON /ACCEL / X2(1000)                                 4 3979
           COMMON /FORCES/ F(1000)                                  5 3889
           COMMON /INTGRS/ IN(5)                                    6 3609
           COMMON /POSIT / X(1000)                                  7 3799
           COMMON /REALS / RL(16)                                   8 3430
           COMMON /VELO  / X1(1000)                                 9 3878
C
           EQUIVALENCE (RL(5),FATOM),  (RL(6),SIDE), (RL(8),STEPSQ) 10 1269
           EQUIVALENCE (RL(9),STPSQH), (IN(3),NATOM)               11 8570
C
C...assign uniformly spaced initial positions
           XDEL = SIDE/DFLOAT(NATOM+1)                             12 5230
           X(1) = XDEL                                             13 1944
           DO 10 I = 2,NATOM                                       14 3182
             X(I) = X(I-1) + XDEL                                  15 3238
        10 CONTINUE                                                16 1933
C
C...assign random initial velocities
           SUM = 0.                                                17 1416
           SUMSQ = 0.                                              18 1551
           IDUMMY = -23                                            19 2978
           DO 20 I = 1,NATOM                                       20 3182
             X1(I) = ROULET(IDUMMY)                                21 4330
             SUM = SUM + X1(I)                                     22 2204
        20 CONTINUE                                                23 1999
C
C...scale initial velocities so total linear momentum = zero
           DO 30 I= 1,NATOM                                        24 3261
             X1(I) = X1(I) -SUM/FATOM                              25 3845
             SUMSQ = SUMSQ + X1(I)*X1(I)                           26 3711
        30 CONTINUE                                                27 2079
C
C...scale velocities to desired temperature
           HEATR = FATOM*STEPSQ*TEMP                               28 4879
           FACTOR = DSQRT(HEATR/SUMSQ)                             29 4093
           DO 40 I = 1,NATOM                                       30 3418
             X1(I) = X1(I)*FACTOR                                  31 3036
        40 CONTINUE                                                32 2226
C
C...compute initial forces & scale to accelerations
           CALL ODEVAL                                             33 1922
           DO 50 I= 1,NATOM                                        34 3137
             X2(I) = F(I)*STPSQH                                   35 3395
        50 CONTINUE                                                36 1944
C
           RETURN                                                  37 1090
           END                                                     38  921
|-----|--|--------|---------|---------|---------|---------|---------|--|------|
```

SUBROUTINE ODPEC

ODPARM $\Rightarrow$ SIDE, STEPSQ, NATOM

Equilibration
Production

The implicit scaling of the velocities and higher derivatives of position (see Notes on ODINIT) allows us to avoid repetitive multiplications by constants at every application of the predictor.

ODPEC

In the correction loop, the computed forces F(I) are scaled by the appropriate factor of the time-step to obtain accelerations in the units of the program.

ODEVAL

Application of the boundary conditions divides into two possibilities:

CASE 1: If (Xold > SIDE) then Xnew = Xold – SIDE

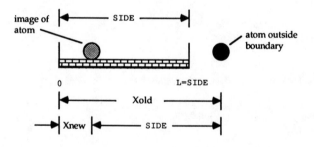

CASE 2: If (Xold < 0) then Xnew = Xold + SIDE

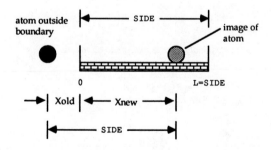

|-----|--|----------|----------|----------|----------|----------|----------|--|------|

```
      SUBROUTINE ODPEC                                            1  2798
C
C       Apply predictor-corrector algorithm
C
      IMPLICIT REAL*8(A-H,O-Z)                                    2  5702
      COMMON /ACCEL / X2(1000)                                    3  3979
      COMMON /ALPHA / ALFA0,ALFA1,ALFA3,ALFA4,ALFA5              4  9144
      COMMON /DERIV / X3(1000),X4(1000),X5(1000)                 5  9234
      COMMON /FORCES/ F(1000)                                     6  3889
      COMMON /INTGRS/ IN(5)                                       7  3609
      COMMON /POSIT / X(1000)                                     8  3799
      COMMON /REALS / RL(16)                                      9  3430
      COMMON /VELO  / X1(1000)                                   10  3878
C
      EQUIVALENCE (RL(6),SIDE), (RL(9),STPSQH), (IN(3),NATOM)    11  1865
C
C...predict
      DO 100 I=1,NATOM                                           12  3474
        X(I) = X(I)  +      X1(I) +      X2(I) +    X3(I) + X4(I) +X5(I) 13  6365
        X1(I)= X1(I) +2.D0*X2(I) + 3.D0*X3(I) + 4.D0*X4(I) + 5.D0*X5(I) 14  2552
        X2(I)= X2(I) +3.D0*X3(I) + 6.D0*X4(I) +10.D0*X5(I)      15   156
        X3(I)= X3(I) +4.D0*X4(I) +10.D0*X5(I)                   16  7973
        X4(I)= X4(I) +5.D0*X5(I)                                17  4958
C
        F(I) = 0.D0                                              18  2417
  100 CONTINUE                                                   19  2282
C
C...evaluate forces
      CALL ODEVAL                                                20  1922
C
C...correct predictions
      DO 120 I= 1,NATOM                                          21  3351
        XERROR= STPSQH*F(I) - X2(I)                              22  4532
        X (I) = X (I) + XERROR*ALFA0                             23  3823
        X1(I) = X1(I) + XERROR*ALFA1                             24  3980
        X2(I) = X2(I) + XERROR                                   25  2855
        X3(I) = X3(I) + XERROR*ALFA3                             26  4408
        X4(I) = X4(I) + XERROR*ALFA4                             27  4857
        X5(I) = X5(I) + XERROR*ALFA5                             28  4015
C
C...apply periodic boundary condition
        IF (X(I).GT.SIDE) X(I) = X(I)-SIDE                       29  6185
        IF (X(I).LT.0.D0) X(I) = X(I)+SIDE                       30  6297
  120 CONTINUE                                                   31  2169
C
      RETURN                                                     32  1090
      END                                                        33   921
```

|-----|--|----------|----------|----------|----------|----------|----------|--|------|

SUBROUTINE   ODEVAL

ODPARM ⇒ NATOM

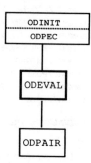

Since the forces are only between nearest
neighbors, there is only one loop over the atoms.
Consequently the program executes quickly
compared to three-dimensional simulations in
which we must consider possible interactions
among all pairs of atoms.

Whether the computed force F_{ij} is added or subtracted from F_i depends
on the sign convention chosen: we take repulsive forces as positive. For
example, if $R > 0$, the vector R points from atom $j = (i + 1)$ to atom i.

Then, if the computed force $F_{ij} > 0$, the force between i and j is repulsive: it is
tending to push the atoms apart. That is, F is pushing atom i to positions of
larger x-values, and so F_{ij} is parallel to R.

```
|-----|--|---------|---------|---------|---------|---------|---------|--|------|
      SUBROUTINE ODEVAL                                              1 3014
C
C         Evaluate forces -- nearest neighbor interactions only
C
      IMPLICIT REAL*8(A-H,O-Z)                                       2 5702
      COMMON /FORCES/ F(1000)                                        3 3889
      COMMON /INTGRS/ IN(5)                                          4 3609
      COMMON /POSIT / X(1000)                                        5 3799
      COMMON /REALS / RL(16)                                         6 3430
C
      EQUIVALENCE (RL(2),ENERGY), (RL(14),VIRIAL), (IN(3),NATOM)     7 2686
C
      ENERGY = 0.D0                                                  8 3159
      VIRIAL = 0.D0                                                  9 2620
C
C...interactions between all neighbors except first & last
      NAM1 = NATOM-1                                                10 2709
      DO 200 I=1,NAM1                                               11 3430
        R = X(I)-X(I+1)                                             12 2350
        CALL ODPAIR(R,ENR,FORCE)                                    13 4633
C
        F(I)   = F(I)   + FORCE                                     14 2541
        F(I+1) = F(I+1) - FORCE                                     15 3328
C
        ENERGY = ENERGY + ENR                                       16 4059
        VIRIAL = VIRIAL + FORCE*R                                   17 3519
  200 CONTINUE                                                      18 2349
C
C...interaction between first & last spheres
      R = X(NATOM)-X(1)                                             19 2844
      CALL ODPAIR(R,ENR,FORCE)                                      20 4633
C
      F(NATOM) = F(NATOM) + FORCE                                   21 4026
      F(1) = F(1) - FORCE                                           22 2822
C
      ENERGY = ENERGY + ENR                                         23 4059
      VIRIAL = VIRIAL + FORCE*R                                     24 3519
C
      RETURN                                                        25 1090
      END                                                           26  921
|-----|--|---------|---------|---------|---------|---------|---------|--|------|
```

SUBROUTINE ODPAIR

ODPARM $\Rightarrow$ SIDE, SIDEH

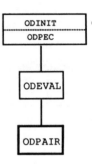

The minimum image criterion is applied here only if neighbors i and $(i + 1)$ are at opposite ends of the line of length SIDE. Initially this will occur only when computing the interaction between the first and last atoms. But during the run, the identities of the terminal atoms change because of periodic boundary conditions.

CASE 1: R > SIDEH

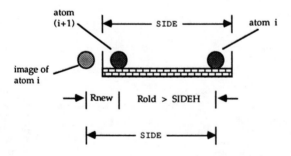

CASE 2: R < - SIDE

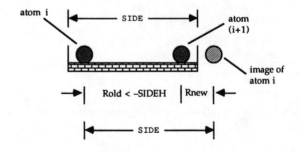

```
|-----|--|---------|---------|---------|---------|---------|---------|--|------|
        SUBROUTINE ODPAIR(R,ENR,FORCE)                                  1 5724
C
C       potential energy & force between spheres separated by R
C
        IMPLICIT REAL*8(A-H,O-Z)                                        2 5702
        LOGICAL LUOFR                                                   3 2103
        COMMON /REALS / RL(16)                                          4 3430
        COMMON /LOGCLS/ LUOFR                                           5 3283
C
        EQUIVALENCE (RL(6),SIDE), (RL(7),SIDEH)                         6 8491
C
C...if I & I+1 are at opposite ends of the system, use an image distance
        IF (R.GT. SIDEH) R = R-SIDE                                     7 5038
        IF (R.LT.-SIDEH) R = R+SIDE                                     8 5173
        R6 = 1.D0/R**6                                                  9 2956
C
C...check whether interaction is LJ or repulsive soft-sphere
        IF (LUOFR) THEN                                                10 2686
C
C...Lennard-Jones
        ENR   = 4.D0*R6*(R6-1.D0)                                      11 5689
        FORCE = 24.D0*R6*(2.D0*R6-1.D0)/R                              12 7771
        ELSE                                                           13  932
C
C...repulsive soft-sphere
        ENR   = R6*R6                                                  14 1911
        FORCE = 12.D0*ENR/R                                            15 3687
        ENDIF                                                          16 1292
C
        RETURN                                                        17 1090
        END                                                           18  921
|-----|--|---------|---------|---------|---------|---------|---------|--|------|
```

SUBROUTINE ODPRPT

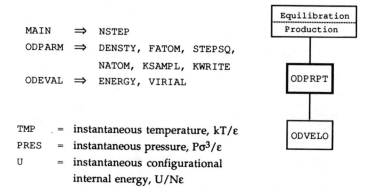

MAIN $\Rightarrow$ NSTEP

ODPARM $\Rightarrow$ DENSTY, FATOM, STEPSQ,

NATOM, KSAMPL, KWRITE

ODEVAL $\Rightarrow$ ENERGY, VIRIAL

TMP   . = instantaneous temperature, kT/ε

PRES = instantaneous pressure, $P\sigma^3/\varepsilon$

U   = instantaneous configurational
              internal energy, $U/N\varepsilon$

Since the H-function is not a time average, but rather is computed from the instantaneous velocity distribution, we only compute the distribution when H is to be printed.

These one-dimensional results for configurational internal energy, density, and total energy can be compared with analytic values computed from eqs. (4.61)-(4.64) at the same temperature and pressure as obtained from the simulation.

```
|-----|--|---------|---------|---------|---------|---------|---------|--|------|

      SUBROUTINE ODPRPT                                                1 3047
C
C        Accumulate properties at intervals of KSAMPL time-steps
C
      IMPLICIT REAL*8(A-H,O-Z)                                         2 5702
      COMMON /INTGRS/ IN(5)                                            3 3609
      COMMON /REALS / RL(16)                                           4 3430
      COMMON /VELO  / X1(1000)                                         5 3878
C
      EQUIVALENCE (RL(1),DENSTY),   (RL(2),ENERGY),   (RL(3),ENRSUM)   6 3014
      EQUIVALENCE (RL(5),FATOM),    (RL(8),STEPSQ),   (RL(10),TMPAVE)  7 1775
      EQUIVALENCE (RL(12),VELSQ),   (RL(13),VELSUM),  (RL(14),VIRIAL)  8 2394
      EQUIVALENCE (RL(15),VIRSUM),  (IN(1),KSAMPL),   (IN(2),KWRITE)   9 2822
      EQUIVALENCE (IN(3),NATOM),    (IN(4),NSTEP)                     10 8592
C
C...sum the squares of velocities to get temperature
      VELSQ = 0.D0                                                    11 2383
      DO 300 I=1,NATOM                                                12 3610
        VELSQ = VELSQ + X1(I)*X1(I)                                   13 3890
  300 CONTINUE                                                        14 2428
C
      VELSUM = VELSUM + VELSQ                                         15 3339
      ENRSUM = ENRSUM + ENERGY                                        16 4228
      VIRSUM = VIRSUM + VIRIAL                                        17 3070
C
C...instantaneous properties
      TMP  = VELSQ/(FATOM*STEPSQ)                                     18 4453
      PRES = DENSTY*(TMP + VIRIAL/FATOM)                              19 6095
      U = ENERGY/FATOM                                                20 2923
      ETOTAL = 0.5D0*TMP + U                                          21 4239
C
C...running averages
      SAMPLS = NSTEP/KSAMPL                                           22 4071
      TMPAVE = VELSUM/(FATOM*SAMPLS*STEPSQ)                           23 6691
      ENRAVE = ENRSUM/(FATOM*SAMPLS)                                  24 5410
      PREAVE = DENSTY*(TMPAVE + VIRSUM/(FATOM*SAMPLS))                25 8491
C
C...sample for velocity distribution
      CALL ODVELO(H)                                                  26 2697
C
C...print at KWRITE intervals
      IF (MOD(NSTEP,KWRITE).EQ.0) THEN                                27 7298
        WRITE(6,91) NSTEP,TMP,TMPAVE,U,ENRAVE,PRES,PREAVE,H,ETOTAL    28 3643
   91   FORMAT(I7,8F10.4)                                             29 4611
      ENDIF                                                           30 1292
C
      RETURN                                                          31 1090
      END                                                             32  921

|-----|--|---------|---------|---------|---------|---------|---------|--|------|
```

SUBROUTINE ODVELO

ODPARM $\Rightarrow$ FACTMB, FATOM, STEPSQ,
                          VDEL, VMAX, KWRITE,
                          NATOM, NVDELS

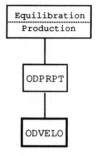

To sample for the velocity distribution, we divide the range of sampled velocities into an odd number of shells, NVDELS,

$$\text{NVDELS} = 2(v_{max}/\Delta v) + 1$$

Then the central shell is centered on $v = 0$; this central shell is numbered

$$N_h = v_{max}/\Delta v + 1 + 0.5$$

The factor of 0.5 appears because the shell extends from $-\Delta v/2$ to $+\Delta v/2$ and in Fortran the integer value of N_h is obtained by truncation. Thus,

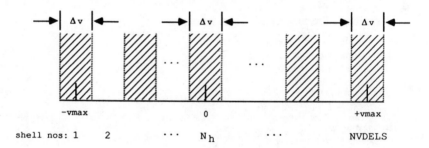

Then the shell number for any positive or negative velocity $v(t)$ is

$$N_{shell} = v(t)/\Delta v + N_h$$

The H-function is computed by (3.45).

```
|-----|--|----------|----------|----------|----------|----------|----------|--|------|

        SUBROUTINE ODVELO(H)                                         1 3788
C
C       Accumulate velocity distribution function
C
        IMPLICIT REAL*8(A-H,O-Z)                                     2 5702
        COMMON /INTGRS/ IN(5)                                        3 3609
        COMMON /REALS / RL(16)                                       4 3430
        COMMON /VELO  / X1(1000)                                     5 3878
        COMMON /VDISTR/ NVEL(200),NVIN(200)                          6 7210
C
        EQUIVALENCE (RL(5),FATOM),  (RL(8),STEPSQ)                   7 7973
        EQUIVALENCE (RL(11),VDEL),  (RL(16),VMAX),   (IN(2),KWRITE)  8 2013
        EQUIVALENCE (IN(3),NATOM),  (IN(4),NSTEP),   (IN(5),NVDELS)  9 2383
C
C...initialize accumulators for instantaneous distribution
        DO 400 I=1,NVDELS                                           10 4172
          NVIN(I)=0                                                 11 2079
  400   CONTINUE                                                    12 2574
C
        STEP = DSQRT(STEPSQ)                                        13 3193
C
        DO 420 I=1,NATOM                                            14 3643
C
C...remove factor of time-step from velocity
          X1DIM = X1(I)/STEP                                        15 3148
C
        IF (DABS(X1DIM) .LT. VMAX) THEN                             16 5825
          NSHELL = X1DIM/VDEL + 1.5 + VMAX/VDEL                     17 6501
C
C...contribution to running average distribution
          NVEL(NSHELL) = NVEL(NSHELL)+1                             18 6466
C
C...contribution to instantaneous distribution
          NVIN(NSHELL) = NVIN(NSHELL)+1                             19 6264
        ENDIF                                                       20 1292
  420   CONTINUE                                                    21 2451
C
C...at KWRITE intervals, compute H function from instantaneous distribution
        IF (MOD(NSTEP,KWRITE).EQ.0) THEN                            22 7298
          H = 0.D0                                                  23 2024
C
        DO 450 K=1,NVDELS                                           24 4240
          FOFV = DFLOAT(NVIN(K))/FATOM                              25 4981
          IF (FOFV.GT.0.) H = H + FOFV*DLOG(FOFV)                   26 6411
  450   CONTINUE                                                    27 2406
C
        H = H*VDEL                                                  28 2507
        ENDIF                                                       29 1292
C
        RETURN                                                      30 1090
        END                                                         31  921

|-----|--|----------|----------|----------|----------|----------|----------|--|------|
```

SUBROUTINE ODVD

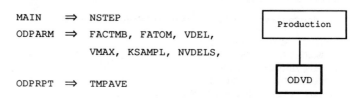

```
MAIN    ⟹  NSTEP
ODPARM  ⟹  FACTMB, FATOM, VDEL,
           VMAX, KSAMPL, NVDELS,

ODPRPT  ⟹  TMPAVE
```

$$FACT = \frac{1}{\sqrt{2\pi \langle T \rangle}}$$

SAMPLS = number of times the velocity distribution has been sampled during the run.

DO 600 ... loop over the velocity shells, each of thickness Δv.

SUBROUTINE ODVSCL

```
ODINIT  ⟹  MAIN  ⟹  HEATR
ODPARM  ⟹  NATOM
ODPRPT  ⟹  VELSQ
```

$$FACTOR = \sqrt{\frac{N(\Delta t)^2\ TEMP}{\sum v_i^2}}$$

where TEMP is the set-point temperature. This is the scale factor that appears in (5.25), except here the square of the time-step is included to convert the velocity sum from program units to pure reduced units.

```
|-----|--|---------|---------|---------|---------|---------|---------|--|------|

       SUBROUTINE ODVD                                          1 2732
C
C          Print velocity distribution
C
       IMPLICIT REAL*8(A-H,O-Z)                                 2 5702
       COMMON /INTGRS/ IN(5)                                    3 3609
       COMMON /REALS / RL(16)                                   4 3430
       COMMON /VDISTR/ NVEL(200),NVIN(200)                      5 7210
       EQUIVALENCE (RL(4),FACTMB), (RL(5),FATOM), (RL(11),VDEL) 6 2079
       EQUIVALENCE (RL(10),TMPAVE),(RL(16),VMAX), (IN(1),KSAMPL)7 2013
       EQUIVALENCE (IN(4),NSTEP), (IN(5),NVDELS)                8 8873
C
       WRITE(6,95)                                              9 2855
   95  FORMAT(///)                                             10 2462
C
       FACT   = FACTMB/DSQRT(TMPAVE)                            11 4475
       SAMPLS = NSTEP/KSAMPL                                    12 4071
C
       DO 600 K=1,NVDELS                                        13 4071
          V = -VMAX + VDEL*(K-1)                                14 4059
C
C...computed average distribution
          FVAVE = DFLOAT(NVEL(K))/(FATOM*SAMPLS)                15 7120
C
C...analytic Maxwell distribution
          FANAL = FACT*DEXP(-V*V/(2.D0*TMPAVE))*VDEL            16 8873
C
          WRITE(6,97) V, FVAVE, FANAL                           17 5386
   97     FORMAT(5X,F7.3,2(5X,F8.4))                            18 6703
  600  CONTINUE                                                 19 2237
C
       WRITE(6,95)                                              20 2855
       RETURN                                                   21 1090
       END                                                      22  921

|-----|--|---------|---------|---------|---------|---------|---------|--|------|

       SUBROUTINE ODVSCL(HEATR)                                  1 4239
C
C      During equilibration, scale velocities to set-point temperature
C
       IMPLICIT REAL*8(A-H,O-Z)                                  2 5702
       COMMON /INTGRS/ IN(5)                                     3 3609
       COMMON /REALS / RL(16)                                    4 3430
       COMMON /VELO  / X1(1000)                                  5 3878
       EQUIVALENCE (RL(12),VELSQ), (IN(3),NATOM)                 6 8244
C
       FACTOR = DSQRT(HEATR/VELSQ)                               7 4183
       DO 700 I=1,NATOM                                          8 3597
          X1(I) = X1(I)*FACTOR                                   9 3036
  700  CONTINUE                                                 10 2406
C
       RETURN                                                   11 1090
       END                                                      12  921
|-----|--|---------|---------|---------|---------|---------|---------|--|------|
```

```
|-----|--|---------|---------|---------|---------|---------|---------|--|------|
      SUBROUTINE ODRSET                                                 1 3003
C
C      At end of equilibration, reinitialize property accumulators
C
      IMPLICIT REAL*8(A-H,O-Z)                                         2 5702
      COMMON /INTGRS/ IN(5)                                            3 3609
      COMMON /REALS / RL(16)                                          4 3430
      COMMON /VDISTR/ NVEL(200),NVIN(200)                             5 7210
C
      EQUIVALENCE (RL(3),ENRSUM), (RL(13),VELSUM), (RL(15),VIRSUM)     6 2338
      EQUIVALENCE (IN(5),NVDELS)                                       7 5151
C
      ENRSUM = 0.D0                                                    8 2811
      VELSUM = 0.D0                                                    9 2754
      VIRSUM = 0.D0                                                   10 2507
C
      DO 500 I = 1,NVDELS                                             11 3890
        NVEL(I) = 0                                                   12 2170
  500 CONTINUE                                                        13 2293
C
      WRITE(6,93)                                                     14 2989
   93 FORMAT(/10X,'*** Equilibration ended ***'/)                     15 7108
      RETURN                                                          16 1090
      END                                                             17  921
|-----|--|---------|---------|---------|---------|---------|---------|--|------|


      BLOCK DATA                                                       1 2079
      IMPLICIT REAL*8(A-H,O-Z)                                         2 5702
C
      COMMON /ACCEL / X2(1000)                                         3 3979
      COMMON /DERIV / X3(1000),X4(1000),X5(1000)                       4 9234
      COMMON /FORCES/ F(1000)                                          5 3889
      COMMON /REALS / RL(16)                                          6 3430
      COMMON /VDISTR/ NVEL(200),NVIN(200)                             7 7210
      EQUIVALENCE (RL(3),ENRSUM), (RL(13),VELSUM), (RL(15),VIRSUM)     8 2338
C
      DATA F   /1000*0.D0/                                            9 4161
      DATA X2 /1000*0.D0/                                            10 4262
      DATA X3 /1000*0.D0/                                            11 4341
      DATA X4 /1000*0.D0/                                            12 4497
      DATA X5 /1000*0.D0/                                            13 4217
      DATA NVEL /200*0/                                              14 3632
      DATA VELSUM, ENRSUM, VIRSUM/3*0.D0/                            15 7197
C
      END                                                            16  921
|-----|--|---------|---------|---------|---------|---------|---------|--|------|
```

APPENDIX J

GENTCF
A GENERIC CODE FOR AUTOCORRELATION FUNCTIONS

The code that follows, GENTCF, is a skeleton program for computing generic autocorrelation functions. Thus, from time-dependent values of a dynamic variable $X(t)$, GENTCF computes the normalized autocorrelation function

$$\hat{\Psi}(t) = \frac{\langle X(t_0) X(t_0 + t) \rangle}{\langle X(t_0) X(t_0) \rangle}$$

GENTCF takes values of $X(t)$ from a disk file that holds the data in the form of a two-column table. Column 1 contains an integer counter that numbers the values of X, while column 2 contains the $X(t)$-values. GENTCF expects the table to be organized in the form FORMAT (I4, D12.5). The calculation is an implementation of Algorithm III given in Section 7.1.2, so GENTCF makes only one pass through the disk file of X-values.

The documentation of the code follows the format described in Appendix I. Figure J.1 contains the dot chart of global scalars and the subroutines that access them. The hierarchy chart for GENTCF is in Figure J.2. To conform to local conventions, you may need to change the unit numbers on WRITE statements that appear in subroutines TCLOOP and TCNORM.

To run GENTCF, assign in the main program the name for the disk file that holds the X-values. Run flags and parameters are set in subroutine TCPARM. If you want the correlation function printed at intervals of time in picoseconds, then additional parameters must be entered in subroutine TCONVT. A first test of GENTCF can be done by computing the autocorrelation of $X(t) = \sin 2\pi t$. The normalized result should be $\cos 2\pi t$.

399

Program segments in code GENTCF

REALS	MAIN	TCPARM	TCONVT	TCLOOP	TCSHFT	TCNORM
1 TSTEP		●	●			●
2 XAVE				●		●
3 XSQAV				●		●
INTEGERS						
1 NDELS		●	●	●	●	●
2 NLEFT		●		●		
3 NORGNS		●		●		
4 KINTVL		●	●			●

FIGURE J.1 Real and integer global scalars used in GENTCF. Arrays RL(3) and IN(4) are used to pass these scalars among the program segments shown.

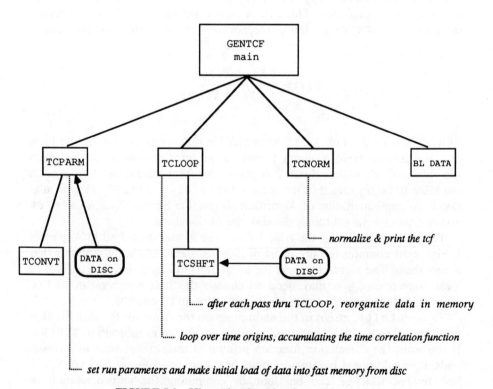

FIGURE J.2 Hierarchy chart for the code GENTCF.

```
|-----|--|---------|---------|---------|---------|---------|---------|--|------|
C
C         filename:  gentcf.f
C
C         Compute a generic autocorrelation function
C
C              version 2.2     26 Jan 1992
C
C  --------------------------------------------------------------------------
C
          IMPLICIT REAL*8(A-H,O-Z)                                   1 5702
          LOGICAL LTIME                                             2 2620
          CHARACTER*32 Dfile                                        3 2642
C...enter filename that has data
          DFILE = 'filename'                                        4 2541
          OPEN (10,FILE=DFILE,ACCESS='sequential',FORM='formatted') 5 8446
C...set parameters
          CALL TCPARM(LTIME, NSKIP)                                 6 5140
C...read down disc thru number of initial time-steps to skip
          IF (NSKIP.GT.0) THEN                                      7 4330
             DO 20 I = 1, NSKIP                                     8 3452
                READ (10,91) NDUMMY                                 9 4666
   91           FORMAT(I6)                                         10 2338
   20        CONTINUE                                              11 1999
          ENDIF                                                    12 1292
C...loop over time-origins
          CALL TCLOOP                                              13 1674
          CLOSE (10)                                               14 1696
C...normalize & print time correlation function
          CALL TCNORM(LTIME)                                       15 3520
C
          STOP                                                     16  550
          END                                                      17  921

|-----|--|---------|---------|---------|---------|---------|---------|--|------|
```

SUBROUTINE TCPARM

Run flags and parameters set by user:

NMAX   = total number of time-steps on disc
file

NSKIP   = number of initial time-steps to
omit from the calculation

NORGNS = number of time origins to use;

NORGNS < NMAX

LTIME   = .true. to print the computed tcf
at intervals of psec

        = .false. to print at unit intervals
at which data is stored on disc

CALL TCONVT ... if the program is to print the computed tcf in psecs, then
we convert the dimensionless time interval to psec.

NDELS   = number of time-steps that span the delay time over which the tcf
is to be computed; NDELS << NMAX

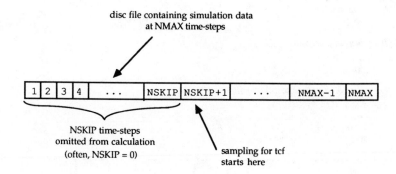

disc file containing simulation data
at NMAX time-steps

NSKIP time-steps
omitted from calculation
(often, NSKIP = 0)

sampling for tcf
starts here

The range of delay times over which tcf is to be computed is $0 \le t \le (\text{NDELS})\,\Delta t$

```
|-----|--|---------|---------|---------|---------|---------|---------|--|------|

      SUBROUTINE TCPARM(LTIME, NSKIP)                                 1 6231
C
C          calculational parameters
C
      IMPLICIT REAL*8(A-H,O-Z)                                        2 5702
      LOGICAL LTIME                                                  3 2620
      COMMON /DATER/ X(1000),NN(1000)                                4 6747
      COMMON /SCALRS/ RL(3), IN(4)                                   5 5140
      EQUIVALENCE (RL(1), TSTEP), (IN(1), NDELS)                     6 8468
      EQUIVALENCE (IN(2), NLEFT), (IN(3), NORGNS), (IN(4), KINTVL)   7 2923
C
C...number of steps on disc & number to skip
      NMAX  = 1000                                                   8 2260
      NSKIP = 0                                                      9 1900
C
C...number of time-origins to use in calculation
      NORGNS = 900                                                  10 2596
C
C...print delay-time in picosec (.true.) or in time-steps (.false.)?
      LTIME = .TRUE.                                                11 2776
C
C...set dimensional time-step, based on argon parameters
      IF (LTIME) THEN                                               12 3205
         CALL TCONVT(TSTEP,KINTVL,NDELS)                            13 6691
      ELSE                                                          14  932
C
C...maximum delay-time (in time-steps) over which to compute TCF
         NDELS = 50                                                 15 2215
      ENDIF                                                         16 1292
C
C...check for sufficient time-steps on disc to get to requested time-delay
      NLEFT = NMAX-NSKIP                                            17 3890
      IF (NDELS.GT.NLEFT) NDELS = NLEFT-1                           18 7715
C
C...move (NDELS+1) sets of data from disc to fast memory
      NDELP1 = NDELS+1                                              19 3519
      DO 100 I=1, NDELP1                                            20 4149
         READ(10,93)  NN(I), X(I)                                   21 5320
   93    FORMAT(I4,D12.5)                                           22 4486
  100 CONTINUE                                                      23 2282
      RETURN                                                        24 1090
      END                                                           25  921

|-----|--|---------|---------|---------|---------|---------|---------|--|------|
```

SUBROUTINE TCLOOP

TCPARM ⇒ NDELS, NLEFT, NORGNS

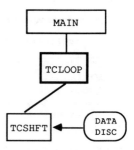

NCHECK ... If the time origin gets close to the end of the stored data, there may not be enough remaining data for a complete sampling of the desired delay time NDELS. If not, we sample the available time-steps.

NORM(J) ... Here we accumulate the number of samples drawn for each value of the delay time. The number of samples may not be the same for all delays if time-steps near the end of the disc file are used as time origins.

CALL TCSHFT ... In this program, once a time-step has been used as an origin, it is not used in any subsequent calculation, so we remove that data from memory to make room for a new time-step.

```
|-----|--|---------|---------|---------|---------|---------|---------|--|------|

        SUBROUTINE TCLOOP                                        1 2765
C
C           Loop over time-origins for autocorrelation
C
        IMPLICIT REAL*8(A-H,O-Z)                                 2 5702
        COMMON /ANS  / TCF(1000), NORM(1000)                     3 6736
        COMMON /DATER/ X(1000),   NN(1000)                       4 6747
        COMMON /SCALRS/ RL(3), IN(4)                             5 5140
        EQUIVALENCE (RL(2), XAVE), (RL(3), XSQAV)                6 6994
        EQUIVALENCE (IN(1), NDELS), (IN(2), NLEFT), (IN(3), NORGNS)  7 2495
C
        SUM = 0.                                                 8 1416
        SUMSQ = 0.                                               9 1551
C
C...loop over time-origins
        DO 300 K = 1,NORGNS                                     10 3991
          XX = X(1)                                             11  998
          SUM = SUM + XX                                        12 1450
          SUMSQ = SUMSQ + XX*XX                                 13 2215
C
C...at intervals, print reassurance to user
          IF (MOD(NN(1),1000) .EQ. 0) WRITE(6,95) NN(1)         14 9852
   95     FORMAT(5X,'gentcf at time-origin ',I5)                15 4835
C
C...check number of time-steps remaining on disc
          NCHECK = NLEFT-K                                      16 3812
          IF (NDELS.LE.NCHECK) NDELAY = NDELS                   17 7939
          IF (NDELS.GT.NCHECK) NDELAY = NCHECK                  18 7838
          IF (NDELAY .GT. 0) THEN                               19 4813
C
C..loop over delay time from origin
            DO 200 J=1,NDELAY                                   20 4138
              JP1 = J+1                                         21 1023
              NORM(J) = NORM(J) + 1                             22 3069
              TCF(J)  = TCF(J) + XX*X(JP1)                      23 3384
  200       CONTINUE                                           24 2349
C
C...realign data in fast memory
            CALL TCSHFT(NDELS,NDELAY,NCHECK)                    25 7647
          ENDIF                                                 26 1292
  300   CONTINUE                                               27 2428
C
        XAVE = SUM/DFLOAT(NORGNS)                               28 4093
        XSQAV = SUMSQ/DFLOAT(NORGNS)                            29 4127
C
        RETURN                                                 30 1090
        END                                                    31  921

|-----|--|---------|---------|---------|---------|---------|---------|--|------|
```

SUBROUTINE TCSHFT

By shifting the data in fast memory we are able to compute the complete tcf by making a single pass through the data, while not having to store all the data in fast memory. This strategy succeeds because typically the amount of data to be sampled is much greater that the total delay time over which the tcf is to be computed; i.e., NMAX >> NDELS .

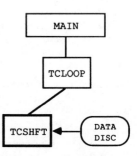

The shift is simply

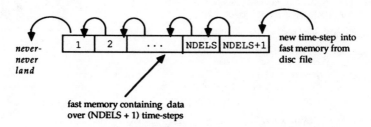

never-never land

fast memory containing data over (NDELS + 1) time-steps

```
|-----|--|---------|---------|---------|---------|---------|--|------|
       SUBROUTINE TCSHFT(NDELS,NDELAY,NCHECK)                    1 8738
C
C          realign data in fast memory
C
       IMPLICIT REAL*8(A-H,O-Z)                                  2 5702
       COMMON /DATER/ X(1000),NN(1000)                           3 6747
C
C...shift contents of fast memory up one slot
       DO 400 J = 1,NDELAY                                       4 4363
         JP1 = J+1                                               5 1023
         NN(J) = NN(JP1)                                         6 2732
         X(J)  = X(JP1)                                          7 1674
  400  CONTINUE                                                  8 2574
C
C...if there are still time-steps on disc, read in the next one
       IF (NDELS.LT.NCHECK) THEN                                 9 5612
         NP1 = NDELAY+1                                         10 2888
         READ(10,93) NN(NP1),X(NP1)                             11 6185
  93     FORMAT(I4,D12.5)                                       12 4486
       ENDIF                                                    13 1292
C
       RETURN                                                   14 1090
       END                                                      15  921
|-----|--|---------|---------|---------|---------|---------|--|------|
```

SUBROUTINE   TCNORM

$$\text{TCPARM} \Rightarrow \text{NDELS, KINTVL, TSTEP}$$
$$\text{TCLOOP} \Rightarrow \text{XAVE, XSQAV}$$

The function printed here is the normalized autocorrelation function:

$$\hat{\psi}(t) \;=\; \frac{\langle X(t_0)\, X(t_0 + t)\rangle}{\langle X(t_0)\, X(t_0)\rangle}$$

so we expect it to be unity at $t = 0$. The first computed value, however, is not at $t = 0$; rather, it is at $t = \Delta t$, so we expect the first printed value to be

$$\hat{\psi}(\Delta t) \;<\; 1 \quad \text{but} \;\approx\; 1$$

To ensure the acf decays to zero at large delay times, compute

$$\frac{\langle X(t_0)\, X(t_0 + t)\rangle \;-\; \langle X(t_0)\rangle^2}{\langle X(t_0)\, X(t_0)\rangle \;-\; \langle X(t_0)\rangle^2}$$

by

```
XAVESQ   =   XAVE*XAVE
TCFZIP(J) = (TCF(J)/DFLOAT(NORM(J))-XAVESQ)/(XSQAV-XAVESQ)
```

```
|-----|--|---------|---------|---------|---------|---------|---------|--|------|
          SUBROUTINE TCNORM(LTIME)                                    1 4611
C
C         Normalize and print results
C
          IMPLICIT REAL*8(A-H,O-Z)                                    2 5702
          LOGICAL LTIME                                              3 2620
          COMMON /ANS/ TCF(1000), NORM(1000)                         4 6736
          COMMON /SCALRS/ RL(3), IN(4)                               5 5140
          EQUIVALENCE (RL(1), TSTEP), (RL(2), XAVE), (RL(3), XSQAV)  6  325
          EQUIVALENCE (IN(1), NDELS), (IN(4), KINTVL)                7 9335
C
          WRITE(6, 97) XAVE                                          8 3339
    97    FORMAT(/5X,"Average = ", F8.4/)                            9 5858
C
C...loop over delay times
          DO 500 J=1,NDELS                                          10 3733
            TCFNM = TCF(J)/(XSQAV*DFLOAT(NORM(J)))                  11 6286
C
            IF (LTIME) THEN                                         12 3205
              PSECS = TSTEP*J*1.D12*KINTVL                          13 5892
              WRITE(6,98)  PSECS, TCFNM                             14 5049
            ELSE                                                    15  932
              WRITE(6,99) J, TCFNM                                  16 4666
            ENDIF                                                   17 1292
C
   500    CONTINUE                                                  18 2293
C
    98    FORMAT(10X,F9.3,2X,1PE15.3)                               19 7007
    99    FORMAT(10X,I3,3X,1PE13.3)                                 20 6770
C
          RETURN                                                    21 1090
          END                                                       22  921
|-----|--|---------|---------|---------|---------|---------|---------|--|------|
```

SUBROUTINE   TCONVT

DELTA   = Δt, the dimensionless value of
the time-step used in the
simulation to integrate the
equations of motion. For
Lennard-Jones, it is typically
about 0.004.

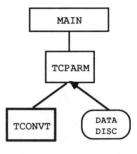

KINTVL = Usually a tcf is not computed at
every time-step, but, say, every
ten time-steps.

```
|-----|--|---------|---------|---------|---------|---------|---------|--|------|
        SUBROUTINE TCONVT (TSTEP,KINTVL,NDELS)                          1 7782
        IMPLICIT REAL*8 (A-H,O-Z)                                       2 5702
C
C
C         Convert time-step to picoseconds using argon parameters
C
        WTMOL = 40.D0                                                   3 3441
        AVONO = 0.60225D24                                             4 3609
        BOLTZ = 1.382D-16                                              5 3968
        SIGMA = 3.405D0                                                6 3531
        EPSIL = 120.D0                                                 7 3126
        TUNIT = SIGMA*DSQRT(WTMOL/(EPSIL*BOLTZ*AVONO))/1.D8            8  819
C
C...value of dimensionless integration time-step
        DELTA = 0.004D0                                                9 4060
C
C...enter time-step interval at which data stored on disc
        KINTVL = 10                                                   10 2349
C
C...maximum delay-time (picosec) over which to compute TCF
        TMAX = 3.D0                                                   11 2237
C
        TSTEP = DELTA*TUNIT                                           12 3935
        NDELS = TMAX*1.D-12/(TSTEP*KINTVL)                           13 7704
C
        RETURN                                                       14 1090
        END                                                          15  921

|-----|--|---------|---------|---------|---------|---------|---------|--|------|
        BLOCK DATA                                                     1 2079
        IMPLICIT REAL*8 (A-H,O-Z)                                      2 5702
        COMMON /ANS/ TCF(1000), NORM(1000)                            3 6736
        DATA TCF/1000*0.D0/                                           4 4385
        DATA NORM/1000*0/                                             5 3845
        END                                                           6  921
|-----|--|---------|---------|---------|---------|---------|---------|--|------|
```

APPENDIX K

MDHS
MOLECULAR DYNAMICS CODE FOR HARD SPHERES

The following code, MDHS, performs hard-sphere molecular dynamics using the algorithm described in Chapter 3. Documentary notes are provided for selected subroutines; the documentation follows the format described in Appendix I. Figure K.1 contains the dot chart of global integer scalars and the subroutines that access them. Figure K.2 does the same for global real scalars. The hierarchy chart for MDHS appears in Figure K.3. To conform to local conventions, you may need to change the unit numbers on WRITE statements that appear in subroutines HSPRTY, HSMOVE, HSPRNT, HSGOFR, HSRSET, and HSVD; the code contains no READ statements.

The program is written in a system of units based on the mass m of one sphere and the length L of the edge of the cubic container. Since the hard-sphere potential is purely repulsive, the units of energy (and therefore of time and velocity) are arbitrary.

To verify your copy of the code, use CHECKSUM from Appendix G. To run MDHS, append the random-number generator ROULET from Appendix H, then set run flags and parameters in subroutine HSFLAG.

Program segments in MD code MDHS

INTEGERS	MAIN	HSFLAG	HSFCC	HSIVEL	HSTABL	HSCTIM	HSNXPR	HSMOVE	HSCRSH	HSUPDT	HSPRTY	HSGSAM	HSGOFR	HSPRNT	HSRSET	HSVELO	HSVD	HSORDR	BL DATA
1 KSAMPL	●	●							●			●	●			●			
2 KWRITE		●											●						
3 NATOM		●	●	●	●			●	●				●	●	●	●		●	
4 NATOMA									●	●	●								
5 NATOMB									●	●	●	●							
6 NCLSNS	●	●							●			●				●			
7 NVDELS		●													●		●		

FIGURE K.1 Integer global scalars used in MDHS. Array IN(7) is used to pass these scalars among the program segments shown.

Program segments in MD code MDHS

REALS	MAIN	HSFLAG	HSFCC	HSIVEL	HSTABL	HSCTIM	HSNXPR	HSMOVE	HSCRSH	HSUPDT	HSPRTY	HSGSAM	HSGOFR	HSPRNT	HSRSET	HSVELO	HSVD	HSORDR	BL DATA
1 BZERO										●									
2 DIST			●														●		
3 ETA			●									●	●						
4 RDEL			●							●		●	●				●		
5 SAMPLS										●									●
6 SIGSQ			●				●		●							●			●
7 SUMB									●		●					●			●
8 SUMKE									●										
9 TIME	●																		●
10 TNEXT	●						●	●											●
11 TOTKE											●				●		●		●
12 TZERO											●								
13 VDEL			●												●	●			
14 VMAX			●													●	●		

FIGURE K.2 Real global scalars used in MDHS. Array RL(14) is used to pass these scalars among the program segments shown.

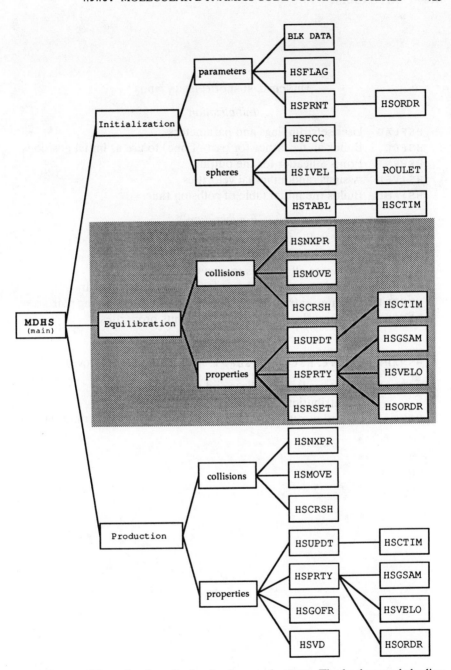

FIGURE K.3   Hierarchy chart for hard-sphere code MDHS. The background shading helps visually separate the three segments: initialization, equilibration, and production.

PRINCIPAL SUBROUTINES IN MDHS

Initialization

HSFLAG   User sets run flags and parameters
HSFCC   Builds an fcc lattice (or part of one) to use as initial positions
HSPRNT   Prints a header for the output
HSIVEL   Assigns random initial velocities
HSTABL   Builds the initial table of collision times

Equilibration

HSNXPR   Reads table to identify the two spheres that will next
          collide and to find their time of collision
HSMOVE   Advances all spheres to next collision
HSCRSH   Computes postcollision velocities for colliding pair
HSUPDT   Updates entries in collision-time table for colliding pair
HSPRTY   At intervals, accumulates and prints properties

Production

HSNXPR, HSMOVE, HSCRSH, HSUPDT, HSPRTY (same as above)

HSGOFR   At intervals, normalizes and prints the radial
          distribution function
HSVD   At end of run, normalizes and prints velocity distribution

```
|-----|--|---------|---------|---------|---------|---------|---------|--|------|
C
C              Molecular dynamics for hard spheres
C
C          filename:  mdhs.f       version 3.3      9 Dec 1990
C   ------------------------------------------------------------------------
      IMPLICIT REAL*8(A-H,O-Z)                                    1 5702
      COMMON /INTGRS/ IN(7)                                       2 3711
      COMMON /REALS / RL(14)                                      3 3777
      EQUIVALENCE (RL(9),TIME),   (RL(10),TNEXT)                  4 8750
      EQUIVALENCE (IN(1),KSAMPL),  (IN(6),NCLSNS)                 5 8727
C
C...initialization
      CALL HSFLAG(KRDF, MAXEQB, MAXCLN)                           6 6804
      CALL HSFCC                                                  7 1450
      CALL HSPRNT                                                 8 2057
      CALL HSIVEL                                                 9 2068
      CALL HSTABL                                                10 2316
C
C...equilibration
      IF (MAXEQB.GT.0) THEN                                      11 4284
        DO 5 NCLSNS = 1, MAXEQB                                  12 3890
          CALL HSNXPR                                            13 1887
          TIME = TIME + TNEXT                                    14 3294
          CALL HSMOVE                                            15 1900
          CALL HSCRSH                                            16 1876
          CALL HSUPDT                                            17 2136
          IF (MOD(NCLSNS,KSAMPL).EQ.0) CALL HSPRTY               18 8222
    5     CONTINUE                                               19 1595
        CALL HSRSET                                              20 1922
      ENDIF                                                      21 1292
C
C...production
      DO 7 NCLSNS = 1, MAXCLN                                    22 3902
        CALL HSNXPR                                              23 1887
        TIME = TIME + TNEXT                                      24 3294
        CALL HSMOVE                                              25 1900
        CALL HSCRSH                                              26 1876
        CALL HSUPDT                                              27 2136
        IF (MOD(NCLSNS,KSAMPL).EQ.0) CALL HSPRTY                 28 8222
        IF (MOD(NCLSNS,KRDF  ).EQ.0) CALL HSGOFR                 29 7603
    7   CONTINUE                                                 30 1708
C
      CALL HSVD                                                  31 1652
      STOP                                                       32  550
      END                                                        33  921
|-----|--|---------|---------|---------|---------|---------|---------|--|------|
```

SUBROUTINE   HSFLAG

Run flags and parameters set by user:

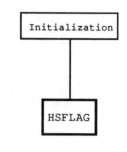

NATOM   = number of spheres; this code can
handle any number, not just the
numbers that completely fill an fcc
lattice.

ETA   = packing fraction; η *must* be < 0.74
and for a fluid, η < 0.49. See phase
diagram in Figure 3.11.

MAXCLN = total number of collisions to compute during production

KWRITE = collision interval at which to print properties that
monitor the run. KWRITE should be an integer multiple
of KSAMPL.

MAXEQB = total number of collisions to compute during
equilibration.

KRDF   = collision interval at which to print radial distribution
function.

SIGMA = sphere diameter in units of the box length.

VMAX   = maximum sphere velocity to be sampled in computing
the velocity distribution. The units are arbitrary, but since
the initial velocities are assigned on [–1, 1], VMAX = 5 is a
conservative choice.

KSAMPL = collision interval at which we accumulate values for the
pressure, velocity distribution, and g(r).

RDEL   = thickness of spherical shells used in sampling for g(r).
RDEL is initially assigned in units of a sphere diameter, but
it is immediately scaled to program units.

```
|-----|--|---------|---------|---------|---------|---------|---------|--|------|
         SUBROUTINE HSFLAG(KRDF, MAXEQB, MAXCLN)                       1 7894
C
C         Parameters and flags typically set by user
C
         IMPLICIT REAL*8(A-H,O-Z)                                      2 5702
         COMMON /INTGRS/ IN(7)                                         3 3711
         COMMON /REALS / RL(14)                                        4 3777
         EQUIVALENCE (RL(3),ETA),    (RL(4),RDEL),   (RL(6),SIGSQ)     5 1269
         EQUIVALENCE (RL(13),VDEL), (RL(14),VMAX),  (IN(1),KSAMPL)     6 2259
         EQUIVALENCE (IN(2),KWRITE),(IN(3),NATOM),  (IN(7),NVDELS)     7 2901
C
         PI = 3.14159265D0                                            8 3980
C
C...assign number of spheres & packing fraction
         NATOM  = 108                                                 9 2079
         ETA    = 0.3D0                                              10 2484
         DENSTY = 6.D0*ETA/PI                                        11 4486
C
         MAXCLN = 200                                                12 2338
         KWRITE = 10                                                 13 2507
         MAXEQB = 200                                                14 2439
C
         KRDF = 1000                                                 15 2686
C
C...scale sphere diameter to units of box length (L = 1)
         SIDE = (DFLOAT(NATOM)/DENSTY)**(1.D0/3.D0)                  16 9773
         SIGMA= 1.D0/SIDE                                            17 3496
         SIGSQ= SIGMA*SIGMA                                          18 3317
C
C...sampling increment and delta R for g(r)
         KSAMPL= 10                                                  19 2327
         RDEL  = 0.025D0                                             20 3519
C
C...scale RDEL to units of box length
         RDEL  = RDEL*SIGMA                                          21 3913
C
C...parameters for sampling the velocity distribution
         VDEL = 0.05D0                                               22 3160
         VMAX = 5.D0                                                 23 1911
         NVDELS = 2.*VMAX/VDEL+1                                     24 4352
C
         RETURN                                                      25 1090
         END                                                         26  921
|-----|--|---------|---------|---------|---------|---------|---------|--|------|
```

```
|-----|--|---------|----------|----------|----------|----------|----------|--|------|
      SUBROUTINE HSFCC                                                  1 2541
C
C
C            Assign  initial  positions  based  on  fcc  lattice
C
      IMPLICIT REAL*8(A-H,O-Z)                                          2 5702
      COMMON /INTGRS/ IN(7)                                             3 3711
      COMMON /POSIT / X(1000),Y(1000),Z(1000)                          4 8659
      COMMON /REALS / RL(14)                                           5 3777
      EQUIVALENCE (RL(2),DIST), (IN(3),NATOM)                          6 8222
C
C...number of unit cells
      NUNIT = (DFLOAT(NATOM)/4.)**(1./3.)+ 0.1                         7 8312
C
C...check whether user asked for non-fcc number of atoms
   10 NCHECK = 4*(NUNIT**3)                                            8 5757
      IF (NCHECK.LT.NATOM) THEN                                        9 5218
        NUNIT = NUNIT + 1                                             10 2686
        GOTO 10                                                      11 1102
      ENDIF                                                          12 1292
C
C...length of cell edge based on side = 1
      DIST = 1.D0/DFLOAT(NUNIT)                                      13 5218
C
C...assign positions to the four spheres of the unit cell
      X(1)  = 0.D0                                                  14 2248
      Y(1)  = 0.D0                                                  15 2574
      Z(1)  = 0.D0                                                  16 2518
      X(2)  = 0.D0                                                  17 2305
      Y(2)  = 0.5D0*DIST                                            18 4127
      Z(2)  = 0.5D0*DIST                                            19 4060
      X(3)  = 0.5D0*DIST                                            20 3878
      Y(3)  = 0.D0                                                  21 2710
      Z(3)  = 0.5D0*DIST                                            22 4149
      X(4)  = 0.5D0*DIST                                            23 4026
      Y(4)  = 0.5D0*DIST                                            24 4352
      Z(4)  = 0.D0                                                  25 2800
C
C...replicate unit cell NUNIT times in each direction
      M = 0                                                         26 1001
      KCT = 0                                                       27 1359
      DO 14 I = 1,NUNIT                                             28 3351
      DO 14 J = 1,NUNIT                                             29 3205
      DO 14 K = 1,NUNIT                                             30 3597
        DO 12 IJ = 1,4                                              31 2574
C
C...if number of positions assigned equals number of spheres, get out
        IF (KCT.GE.NATOM) RETURN                                    32 4419
          X(IJ+M) = X(IJ)+DIST*(K-1)                                33 4868
          Y(IJ+M) = Y(IJ)+DIST*(J-1)                                34 5139
          Z(IJ+M) = Z(IJ)+DIST*(I-1)                                35 5162
          KCT = KCT + 1                                             36 1887
   12   CONTINUE                                                    37 1810
        M = M + 4                                                   38 1472
   14 CONTINUE                                                      39 2046
      RETURN                                                        40 1090
      END                                                           41  921
|-----|--|---------|----------|----------|----------|----------|----------|--|------|
```

|-----|--|---------|----------|----------|---------|----------|----------|--|------|

```
      SUBROUTINE HSIVEL                                          1 3159
C
C        Assign random initial velocities to spheres
C
      IMPLICIT REAL*8(A-H,O-Z)                                   2 5702
      REAL*4 ROULET                                             3 2776
      COMMON /INTGRS/ IN(7)                                     4 3711
      COMMON /REALS / RL(14)                                    5 3777
      COMMON /VEL    / X1(1000),Y1(1000),Z1(1000)              6 9009
C
      EQUIVALENCE (RL(8),SUMKE), (IN(3),NATOM)                  7 8581
      DATA SUMX,SUMY,SUMZ/3*0.D0/                               8 6275
C
C...assign random initial velocities on [-1,+1]; ROULET is given in Appendix H.
      MSEED = -29                                               9 2732
      DO 100 I = 1,NATOM                                       10 3474
         XX = ROULET(MSEED)                                    11 3182
         YY = ROULET(MSEED)                                    12 3845
         ZZ = ROULET(MSEED)                                    13 3722
         XYZ = 1./DSQRT(XX*XX +YY*YY +ZZ*ZZ)                   14 6927
C
         X1(I) = XX/XYZ                                        15 2046
         Y1(I) = YY/XYZ                                        16 3036
         Z1(I) = ZZ/XYZ                                        17 2855
C
         SUMX = SUMX + X1(I)                                   18 2260
         SUMY = SUMY + Y1(I)                                   19 3250
         SUMZ = SUMZ + Z1(I)                                   20 3070
  100 CONTINUE                                                 21 2282
C
      SUMKE = 0.D0                                             22 2855
      FATOM = DFLOAT(NATOM)                                    23 3755
      DO 120 I = 1,NATOM                                       24 3351
C
C...adjust initial velocities so total linear momentum is zero; see Eq. (3.40)
         X1(I) = X1(I) - SUMX/FATOM                            25 3878
         Y1(I) = Y1(I) - SUMY/FATOM                            26 4868
         Z1(I) = Z1(I) - SUMZ/FATOM                            27 4688
C
C...accumulate kinetic energy
         SUMKE = SUMKE + X1(I)*X1(I) + Y1(I)*Y1(I) + Z1(I)*Z1(I)  28  190
  120 CONTINUE                                                 29 2169
C
      SUMKE = 0.50D0*SUMKE                                     30 5016
C
      RETURN                                                   31 1090
      END                                                      32  921
```

|-----|--|---------|----------|----------|---------|----------|----------|--|------|

```
|-----|--|----------|----------|----------|----------|----------|----------|--|------|
         SUBROUTINE HSNXPR                                                   1 2978
C
C            Identify next pair of colliding spheres
C
         IMPLICIT REAL*8(A-H,O-Z)                                           2 5702
         COMMON /INTGRS/ IN(7)                                              3 3711
         COMMON /REALS / RL(14)                                             4 3777
         COMMON /TIMCOL/ TC(1000),NPAIR(1000)                               5 7377
         EQUIVALENCE (RL(10),TNEXT)                                         6 5072
         EQUIVALENCE (IN(3),NATOM), (IN(4),NATOMA), (IN(5),NATOMB)          7 2439
C
         TNEXT = 1.D8                                                       8 2349
         DO 400 II = 1,NATOM                                                9 3979
           IF (TC(II).LT.TNEXT) THEN                                       10 4868
              TNEXT = TC(II)                                               11 2338
              NATOMA= II                                                   12 1786
           ENDIF                                                           13 1292
   400   CONTINUE                                                          14 2574
C
         NATOMB = NPAIR(NATOMA)                                            15 4048
         RETURN                                                            16 1090
         END                                                               17  921
|-----|--|----------|----------|----------|----------|----------|----------|--|------|

         SUBROUTINE HSRSET                                                  1 3014
C
C...Reinitialize accumulators at end of equilibration
C
         IMPLICIT REAL*8(A-H,O-Z)                                           2 5702
         COMMON /INTGRS/ IN(7)                                             3 3711
         COMMON /MAXWEL/ NVELX(300),NVELY(300),NVELZ(300)                   4 1337
         COMMON /RDF   / NGOFR(300)                                         5 4352
         COMMON /REALS / RL(14)                                             6 3777
         EQUIVALENCE (RL(4),RDEL),     (RL(5),SAMPLS), (RL(7),SUMB)         7 2158
         EQUIVALENCE (RL(11),TOTKE), (IN(7),NVDELS)                         8 9009
C
         SUMB   = 0.D0                                                      9 2596
         SAMPLS = 0.D0                                                     10 2754
         TOTKE  = 0.D0                                                     11 2811
C
         DO 800 I=1,NVDELS                                                 12 3980
            NVELX(I)=0                                                     13 2204
            NVELY(I)=0                                                     14 2530
            NVELZ(I)=0                                                     15 2473
   800   CONTINUE                                                          16 2383
C
         NSHELL = 1./RDEL                                                  17 3665
         DO 820 I = 1, NSHELL                                             18 4183
            NGOFR(I) = 0                                                   19 2170
   820   CONTINUE                                                          20 2260
C
         WRITE(6,981)                                                      21 3115
   981   FORMAT(/)                                                         22 2428
         RETURN                                                            23 1090
         END                                                               24  921
|-----|--|----------|----------|----------|----------|----------|----------|--|------|
```

SUBROUTINE HSMOVE

HSFLAG $\Rightarrow$ NATOM
HSNXPR $\Rightarrow$ TNEXT

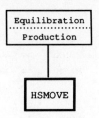

The application of pbc is the 3-D extension of the
1-D situation used in routine ODPEC in Appendix I.
Note figure in notes for ODPEC. The check for negative
collision times is merely an indulgence in paranoia.

```
|-----|--|---------|---------|---------|---------|---------|---------|--|------|
      SUBROUTINE HSMOVE                                                 1 2990
C
C         Move all spheres thru duration since last collision
C
      IMPLICIT REAL*8(A-H,O-Z)                                          2 5702
      COMMON /INTGRS/ IN(7)                                             3 3711
      COMMON /POSIT / X(1000),Y(1000),Z(1000)                           4 8659
      COMMON /REALS / RL(14)                                            5 3777
      COMMON /TIMCOL/ TC(1000),NPAIR(1000)                              6 7377
      COMMON /VEL   / X1(1000),Y1(1000),Z1(1000)                        7 9009
      EQUIVALENCE (RL(10),TNEXT), (IN(3),NATOM)                         8 8581
C
      DO 400 I = 1,NATOM                                                9 3766
C
C...shorten all collision times by the current increment
          TC(I) = TC(I) - TNEXT                                        10 3351
          IF (TC(I) .LT. 0.D0) THEN                                    11 5005
             WRITE(6,941) I, NPAIR(I), TC(I)                           12 6804
  941        FORMAT(//10X,2I5,5X,"Collision time negative = ",D12.3//) 13 8772
             STOP                                                      14  550
          ENDIF                                                        15 1292
C
C...advance all spheres
          X(I) = X(I) + X1(I)*TNEXT                                    16 3845
          Y(I) = Y(I) + Y1(I)*TNEXT                                    17 4835
          Z(I) = Z(I) + Z1(I)*TNEXT                                    18 4655
C
C...apply periodic boundary conditions
          IF (X(I).GT.1.D0) X(I)=X(I)-1.D0                             19 6567
          IF (Y(I).GT.1.D0) Y(I)=Y(I)-1.D0                             20 7557
          IF (Z(I).GT.1.D0) Z(I)=Z(I)-1.D0                             21 7377
          IF (X(I).LT.0.D0) X(I)=X(I)+1.D0                             22 6488
          IF (Y(I).LT.0.D0) Y(I)=Y(I)+1.D0                             23 7478
          IF (Z(I).LT.0.D0) Z(I)=Z(I)+1.D0                             24 7298
  400 CONTINUE                                                         25 2574
      RETURN                                                           26 1090
      END                                                              27  921
|-----|--|---------|---------|---------|---------|---------|---------|--|------|
```

SUBROUTINE HSCRSH

HSFLAG $\Rightarrow$ KSAMPL, SIGSQ

HSNXPR $\Rightarrow$ NATOMA, NATOMB

MAIN $\Rightarrow$ NCLSNS

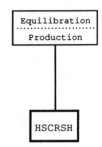

Here we determine the postcollision velocities of the colliding pair by solving (3.17) and (3.18).

When the collision partners were identified in subroutine HSCTIM, we checked for possible collisions between sphere i and j and between i and all images of j. If it is an image that is involved in the collision, we need not store that fact; instead, we merely compute here the minimum image distance. Either j or that image of j closest to sphere i *must necessarily* be the one colliding with i.

B $= \Delta\underline{v} \cdot \Delta\underline{r}$ is the momentum transfer on collision and is used in subroutine HSPRTY to compute the virial contribution to the pressure; see (3.52).

SUMKE ...   All sphere velocities are constant, except for those of the colliding pair; hence, the kinetic energy could possibly change only as a result of changes in the velocities of the colliding spheres.

```
|-----|--|----------|----------|----------|----------|----------|----------|--|------|
          SUBROUTINE HSCRSH                                                 1 2967
C
C              Determine change in velocities of colliding pair
C
          IMPLICIT REAL*8(A-H,O-Z)                                         2 5702
          COMMON /INTGRS/ IN(7)                                            3 3711
          COMMON /POSIT / X(1000),Y(1000),Z(1000)                          4 8659
          COMMON /REALS / RL(14)                                           5 3777
          COMMON /VEL   / X1(1000),Y1(1000),Z1(1000)                       6 9009
          EQUIVALENCE (IN(1),KSAMPL), (IN(4),NATOMA), (IN(5),NATOMB)       7 2811
          EQUIVALENCE (IN(6),NCLSNS)                                       8 4835
          EQUIVALENCE (RL(6),SIGSQ) , (RL(7),SUMB), (RL(8),SUMKE)          9 1573
C
          RX = X(NATOMA) - X(NATOMB)                                      10 4172
          RY = Y(NATOMA) - Y(NATOMB)                                      11 5162
          RZ = Z(NATOMA) - Z(NATOMB)                                      12 4981
C
C...closest image distance is that for collision
          IF (RX.GT.0.5D0) RX = RX -1.D0                                  13 5487
          IF (RY.GT.0.5D0) RY = RY -1.D0                                  14 6477
          IF (RZ.GT.0.5D0) RZ = RZ -1.D0                                  15 6297
          IF (RX.LT.-.5D0) RX = RX +1.D0                                  16 5274
          IF (RY.LT.-.5D0) RY = RY +1.D0                                  17 6264
          IF (RZ.LT.-.5D0) RZ = RZ +1.D0                                  18 6084
          RR = RX*RX + RY*RY + RZ*RZ                                      19 4093
C
C...difference in sphere velocities
          VX = X1(NATOMA) - X1(NATOMB)                                    20 4396
          VY = Y1(NATOMA) - Y1(NATOMB)                                    21 5386
          VZ = Z1(NATOMA) - Z1(NATOMB)                                    22 5207
C
C...project velocity difference onto line of centers
          B = (RX*VX + RY*VY + RZ*VZ)/SIGSQ                               23 5140
          IF (MOD(NCLSNS,KSAMPL).EQ.0) SUMB = SUMB + B                    24 8806
C
C...velocity change due to collision
          DELVX = RX*B                                                    25 2372
          DELVY = RY*B                                                    26 3036
          DELVZ = RZ*B                                                    27 2912
C
C...update kinetic energy
          SUMKE=SUMKE-0.5D0*(X1(NATOMA)**2 + Y1(NATOMA)**2 + Z1(NATOMA)**2) 28 4240
          SUMKE=SUMKE-0.5D0*(X1(NATOMB)**2 + Y1(NATOMB)**2 + Z1(NATOMB)**2) 29 5320
C
          X1(NATOMA) = X1(NATOMA) - DELVX                                 30 5016
          Y1(NATOMA) = Y1(NATOMA) - DELVY                                 31 6006
          Z1(NATOMA) = Z1(NATOMA) - DELVZ                                 32 5825
          X1(NATOMB) = X1(NATOMB) + DELVX                                 33 5375
          Y1(NATOMB) = Y1(NATOMB) + DELVY                                 34 6365
          Z1(NATOMB) = Z1(NATOMB) + DELVZ                                 35 6185
C
          SUMKE=SUMKE+0.5D0*(X1(NATOMA)**2 + Y1(NATOMA)**2 + Z1(NATOMA)**2) 36 3889
          SUMKE=SUMKE+0.5D0*(X1(NATOMB)**2 + Y1(NATOMB)**2 + Z1(NATOMB)**2) 37 4969
          RETURN                                                         38 1090
          END                                                            39  921
|-----|--|----------|----------|----------|----------|----------|----------|--|------|
```

```
|-----|--|---------|---------|---------|---------|---------|---------|--|------|
      SUBROUTINE HSUPDT                                             1 3227
C
C    Update table for all subsequent collisions of colliding pair
C
      IMPLICIT REAL*8 (A-H,O-Z)                                     2 5702
      COMMON /INTGRS/ IN(7)                                         3 3711
      COMMON /ITHSPH/ XI, YI, ZI, VXI, VYI, VZI                     4 7906
      COMMON /POSIT / X(1000),Y(1000),Z(1000)                       5 8659
      COMMON /REALS / RL(14)                                        6 3777
      COMMON /TIMCOL/ TC(1000),NPAIR(1000)                          7 7377
      COMMON /VEL   / X1(1000),Y1(1000)   ,Z1(1000)                 8 9009
      EQUIVALENCE (IN(3),NATOM), (IN(4),NATOMA), (IN(5),NATOMB)     9 2439
C
      DO 540 I = 1,NATOM                                           10 3597
C
C...set flag for spheres that need collision-time update
      NFLAG = 0                                                    11 1786
      IF (I.EQ.NATOMA .OR. NPAIR(I).EQ.NATOMA) NFLAG = 1           12 8266
      IF (I.EQ.NATOMB .OR. NPAIR(I).EQ.NATOMB) NFLAG = 1           13 8985
C
C...find collision partners for spheres being updated
      IF (NFLAG.EQ.1) THEN                                         14 3890
         XI  = X(I)                                                15 1214
         YI  = Y(I)                                                16 1876
         ZI  = Z(I)                                                17 1753
         VXI = X1(I)                                               18 1393
         VYI = Y1(I)                                               19 2057
         VZI = Z1(I)                                               20 1933
         TC(I) = 1.D8                                              21 2215
C
C...inner loop over spheres J
         DO 520 J = 1,NATOM                                        22 3216
            IF (I.NE.J) THEN                                       23 3126
               CALL HSCTIM(I,J,CTIME)                              24 4633
C
C...replace entry in collision-time table if CTIME < current entry
               IF (CTIME.LT.TC(I)) THEN                            25 4699
                  TC(I) = CTIME                                    26 2169
                  NPAIR(I) = J                                     27 1898
               ENDIF                                               28 1292
C
               IF (CTIME.LT.TC(J)) THEN                            29 4543
                  TC(J) = CTIME                                    30 2013
                  NPAIR(J) = I                                     31 1898
               ENDIF                                               32 1292
            ENDIF                                                  33 1292
  520    CONTINUE                                                  34 2170
      ENDIF                                                        35 1292
  540 CONTINUE                                                     36 2406
      RETURN                                                       37 1090
      END                                                          38  921
|-----|--|---------|---------|---------|---------|---------|---------|--|------|
```

```
|-----|--|---------|---------|---------|---------|---------|---------|--|------|
        SUBROUTINE HSTABL                                            1 3407
C
C          Build table of collision times
C
        IMPLICIT REAL*8(A-H,O-Z)                                     2 5702
        COMMON /INTGRS/ IN(7)                                        3 3711
        COMMON /ITHSPH/ XI, YI, ZI, VXI, VYI, VZI                    4 7906
        COMMON /POSIT / X(1000),Y(1000),Z(1000)                      5 8659
        COMMON /TIMCOL/ TC(1000),NPAIR(1000)                         6 7377
        COMMON /VEL   / X1(1000),Y1(1000),Z1(1000)                   7 9009
        EQUIVALENCE (IN(3),NATOM)                                    8 4879
C
C...initially set elements of table to arbitrarily high values
        DO 200 I = 1,NATOM                                           9 3531
            TC(I) = 1.D10                                           10 2462
  200   CONTINUE                                                    11 2349
C
C...loop over spheres I
        NM1 = NATOM - 1                                             12 2620
        DO 240 I = 1,NM1                                            13 3463
          IP1 = I + 1                                               14 1326
          XI  = X(I)                                                15 1214
          YI  = Y(I)                                                16 1876
          ZI  = Z(I)                                                17 1753
          VXI = X1(I)                                               18 1393
          VYI = Y1(I)                                               19 2057
          VZI = Z1(I)                                               20 1933
C
C...inner loop over spheres J
          DO 220 J = IP1,NATOM                                      21 3665
                CALL HSCTIM(I,J,CTIME)                              22 4633
C
C...place entry in table if CTIME < current entry;  also save name of collision partner
            IF (CTIME.LT.TC(I)) THEN                                23 4699
                TC(I) = CTIME                                       24 2169
                NPAIR(I) = J                                        25 1898
            ENDIF                                                   26 1292
C
            IF (CTIME.LT.TC(J)) THEN                                27 4543
                TC(J) = CTIME                                       28 2013
                NPAIR(J) = I                                        29 1898
            ENDIF                                                   30 1292
  220     CONTINUE                                                  31 2226
  240   CONTINUE                                                    32 2451
C
        RETURN                                                      33 1090
        END                                                         34  921
|-----|--|---------|---------|---------|---------|---------|---------|--|------|
```

SUBROUTINE   HSCTIM

HSFLAG $\Rightarrow$ SIGSQ

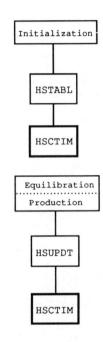

Given the positions and velocities of any two spheres, this routine determines whether the spheres will collide and, if so, when. As discussed in Section 3.2.1, there are three possibilities: (i) the spheres are moving away from one another, (ii) the spheres are approaching but will not collide, and (iii) the spheres will collide.

We consider possible collisions not only between spheres i and j, but also between i and periodic images of j. In spite of claims in the literature, we find it possible that i could collide with some image of j that is not necessarily the image closest to i; for example, see Figure 3.4. Such situations are common at low densities, but they may also occur at high densities.

RRX   ... This is a device to simplify the determination of the periodic images of sphere j; such a device is part of the motivation for using the box length L as the unit of length rather than use the sphere diameter.

CTIME   ... To deter round-off error, we use this method, rather than the standard formula, for solving the quadratic (3.24). The method is taken from Press, et al., *Numerical Recipes*, Section 5.5.

```
|-----|--|----------|----------|----------|----------|----------|----------|--|------|
         SUBROUTINE HSCTIM(I,J,TMIN)                                           1 5735
C
C           Determine collision times for spheres I & J
C
         IMPLICIT REAL*8(A-H,O-Z)                                              2 5702
         COMMON /ITHSPH/ XI, YI, ZI, VXI, VYI, VZI                            3 7906
         COMMON /POSIT / X(1000),Y(1000),Z(1000)                              4 8659
         COMMON /REALS / RL(14)                                                5 3777
         COMMON /VEL    / X1(1000),Y1(1000),Z1(1000)                          6 9009
         EQUIVALENCE (RL(6),SIGSQ)                                            7 4396
         TMIN = 1.D8                                                           8 2394
C
C...difference in velocities of the I-J pair
         VX = VXI - X1(J)                                                      9 1685
         VY = VYI - Y1(J)                                                     10 2675
         VZ = VZI - Z1(J)                                                     11 2495
         A  = VX*VX + VY*VY + VZ*VZ                                           12 3193
         RRX = XI - (X(J) - 2.D0)                                             13 3744
         RRY = YI - (Y(J) - 2.D0)                                             14 4734
         RRZ = ZI - (Z(J) - 2.D0)                                            15 4554
C
C...loop over sphere J and its 26 images
         DO 300 JX = 1,3                                                      16 2855
            RX = RRX - DFLOAT(JX)                                             17 2912
            BX = RX*VX                                                        18 1427
            RXSQ = RX*RX                                                      19 1371
C
            DO 300 JY = 1,3                                                   20 3182
               RY = RRY - DFLOAT(JY)                                         21 3902
               BY = RY*VY                                                     22 2417
               RYSQ = RY*RY                                                   23 2361
C
               DO 300 JZ = 1,3                                                24 3126
                  RZ = RRZ - DFLOAT(JZ)                                       25 3722
                  B  = BX + BY + RZ*VZ                                        26 3283
C
C...eliminate I-J pairs moving away from one another
                  IF (B.LT.0.D0) THEN                                         27 4611
                     C  = RXSQ + RYSQ + RZ*RZ - SIGSQ                         28 3722
                     AC = A*C                                                 29  954
                     BSQ= B*B                                                 30 2215
C
C...eliminate I-J pairs that do not collide
                     IF (AC.LT.BSQ) THEN                                      31 3542
C
C...use less positive root of quadratic for collision time
                        CTIME = -C/(B - DSQRT(BSQ - AC))                      32 5487
C
C...store time of soonest collision with I
                        IF (CTIME.LT.TMIN) TMIN = CTIME                       33 6107
                     ENDIF                                                    34 1292
                  ENDIF                                                       35 1292
  300    CONTINUE                                                             36 2428
         RETURN                                                               37 1090
         END                                                                  38  921
|-----|--|----------|----------|----------|----------|----------|----------|--|------|
```

SUBROUTINE HSPRTY

```
MAIN    ⟹  NCLSNS, TIME
HSFLAG  ⟹  SIGSQ, KSAMPL, KWRITE
HSNXPR  ⟹  NATOMA, NATOMB
HSCRSH  ⟹  SUMB, SUMKE
HSPRTY  ⟹  BZERO, TZERO, SAMPLS
```

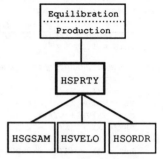

BZERO ... initial value (either at start of equilibration or start of production) for momentum transfer on collision.

TZERO ... initial time, either at start of equilibration or start of production.

BZERO and TZERO are stored in COMMON so that they will be available on subsequent calls of HSPRTY. They are used in obtaining the virial contribution to the pressure; see (3.52).

HINST is the instantaneous H-function.

Note that since HSPRTY is called only every KSAMPL collisions, the print interval KWRITE should be an integer multiple of KSAMPL.

```
|-----|--|---------|---------|---------|---------|---------|---------|--|------|

         SUBROUTINE HSPRTY                                              1 3227
C
C          Accumulate & print properties at intervals
C
         IMPLICIT REAL*8(A-H,O-Z)                                       2 5702
         COMMON /INTGRS/ IN(7)                                          3 3711
         COMMON /REALS / RL(14)                                         4 3777
         COMMON /VEL   / X1(1000),Y1(1000)   ,Z1(1000)                  5 9009
C
         EQUIVALENCE (RL(1),BZERO),   (RL(5),SAMPLS), (RL(6),SIGSQ)     6 1506
         EQUIVALENCE (RL(7),SUMB),    (RL(8),SUMKE),  (RL(9),TIME)      7 2226
         EQUIVALENCE (RL(11),TOTKE),  (RL(12),TZERO)                    8 8682
         EQUIVALENCE (IN(1),KSAMPL),  (IN(2),KWRITE)                    9 9380
         EQUIVALENCE (IN(4),NATOMA),  (IN(5),NATOMB),  (IN(6),NCLSNS)  10 2394
C
         IF (SAMPLS.LT.0.9) THEN                                       11 5016
           TZERO = TIME                                                12 2260
           BZERO = SUMB                                                13 2473
         ENDIF                                                         14 1292
         SAMPLS = SAMPLS + 1.D0                                        15 3722
C
C...accumulate running average kinetic energy
         TOTKE = TOTKE + SUMKE                                         16 3924
         AVEKE = TOTKE/DFLOAT(NCLSNS/KSAMPL)                           17 7052
C
C...radial distribution function g(r)
         CALL HSGSAM                                                   18 1988
C
C...velocity distribution & H-funciton
         CALL HSVELO(HINST)                                            19 3485
C
C...virial from momentum transfer on collision
         TELASP = (TIME - TZERO)/DFLOAT(KSAMPL)                        20 7478
         BELASP = SUMB - BZERO                                         21 4262
         IF (TELASP.GT.0.D0) VIRIAL = -(SIGSQ*BELASP)/(2.D0*AVEKE*TELASP) 22 3485
C
C...print at KWRITE intervals
         IF (MOD(NCLSNS,KWRITE).EQ.0)   THEN                           23 7388
           CALL HSORDR(ALAMDA,ALAM1)                                   24 5139
           WRITE(6,99) NCLSNS,TIME,HINST,ALAM1,NATOMA,NATOMB,SUMKE,VIRIAL 25 5342
      99   FORMAT(1X,I6,D13.4,1X,2F8.4,2I5,2F10.5)                     26  730
         ENDIF                                                         27 1292
C
         RETURN                                                        28 1090
         END                                                           29  921

|-----|--|---------|---------|---------|---------|---------|---------|--|------|
```

SUBROUTINE HSGSAM

HSFLAG $\Rightarrow$ RDEL, NATOM

This routine accumulates the numerator $N(r, \Delta r)$ in (6.102) for $g(r)$ by counting the number of pairs of spheres i and j whose centers are separated by distances ($r \pm \Delta r$). The accumulation locates the smallest image distance.

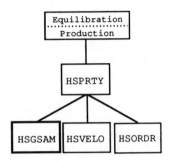

RDEL $= \Delta r$; typically 0.025 σ, where σ is the sphere diameter

NGOFR $= N(r, \Delta r)$ is the numerator

NSHELL . . .identifies a spherical shell of thickness RDEL centered on atom i. A shell having NSHELL $= n$ encompasses a shell volume of radius $r = n\Delta r \pm \Delta r/2$, where n is an integer. Solving for n gives

$$n = int (r/\Delta r + 1/2)$$

where the sign on 1/2 is positive because the int-operation is realized by truncation.

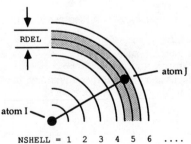

```
|-----|--|---------|---------|---------|---------|---------|---------|--|------|
      SUBROUTINE HSGSAM                                        1 3070
C
C          Sample for radial distribution function
C
      IMPLICIT REAL*8(A-H,O-Z)                                 2 5702
      COMMON /INTGRS/ IN(7)                                    3 3711
      COMMON /POSIT / X(1000),Y(1000),Z(1000)                  4 8659
      COMMON /RDF   / NGOFR(300)                               5 4352
      COMMON /REALS / RL(14)                                   6 3777
      EQUIVALENCE (RL(4),RDEL), (IN(3),NATOM)                  7 8671
C
C...outer loop over spheres I
      NM1 = NATOM-1                                            8 2620
      DO 600 I=1,NM1                                           9 3249
        XI = X(I)                                             10 1214
        YI = Y(I)                                             11 1876
        ZI = Z(I)                                             12 1753
        IP1 = I+1                                             13 1326
C
C...inner loop over spheres J > I
        DO 600 J=IP1,NATOM                                    14 3676
          RX = XI-X(J)                                        15 1629
          RY = YI-Y(J)                                        16 2619
          RZ = ZI-Z(J)                                        17 2439
C
C...apply minimum image criteria
          IF (RX.GT.0.5D0) RX = RX - 1.D0                     18 5487
          IF (RY.GT.0.5D0) RY = RY - 1.D0                     19 6477
          IF (RZ.GT.0.5D0) RZ = RZ - 1.D0                     20 6297
          IF (RX.LT.-.5D0) RX = RX + 1.D0                     21 5274
          IF (RY.LT.-.5D0) RY = RY + 1.D0                     22 6264
          IF (RZ.LT.-.5D0) RZ = RZ + 1.D0                     23 6084
          RIJ = DSQRT(RX*RX + RY*RY + RZ*RZ)                  24 5498
C
          IF (RIJ.LE.0.5D0) THEN                              25 4802
             ISHELL = RIJ/RDEL + 0.5D0                        26 5139
             NGOFR(ISHELL) = NGOFR(ISHELL)+1                  27 6321
          ENDIF                                               28 1292
  600 CONTINUE                                                29 2237
C
      RETURN                                                  30 1090
      END                                                     31  921
|-----|--|---------|---------|---------|---------|---------|---------|--|------|
```

SUBROUTINE HSGOFR

HSFLAG $\Rightarrow$ ETA, RDEL, SIGSQ, KSAMPL,
                       NATOM
MAIN $\Rightarrow$ NCLSNS

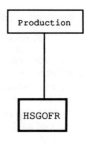

NRDELS = number of spherical shells
sampled; unless one is devious (or looking for heart
aches), the number should be no more than that
provided by half the box length

ORGINS = the number of spheres used as centers-of-origins in
sampling for g(r) in HSGSAM. To avoid sampling each pair of spheres twice,
we sample only N(N–1)/2 unique pairs; see Section 6.4.1

VOLSHL   = volume of the spherical shell of radius r and thickness Δr

$$\text{VOLSHL} \;=\; \int_{r-\Delta r/2}^{r+\Delta r/2} 4\pi\, r^2\, dr \;\;=\;\; 4\pi\, r^2\, \Delta r \;\;+\;\; \frac{\pi}{3}(\Delta r)^3$$

```
|-----|--|---------|---------|---------|---------|---------|---------|--|------|

      SUBROUTINE HSGOFR                                        1 2945
C
C         Normalize counters for radial distribution function
C
      IMPLICIT REAL*8 (A-H,O-Z)                                2 5702
      COMMON /INTGRS/ IN(7)                                    3 3711
      COMMON /RDF   / NGOFR(300)                               4 4352
      COMMON /REALS / RL(14)                                   5 3777
      EQUIVALENCE (IN(1),KSAMPL), (IN(3),NATOM), (IN(6),NCLSNS) 6 2237
      EQUIVALENCE (RL(3),ETA),    (RL(4),RDEL),   (RL(6),SIGSQ) 7 1269
C
      PI = 3.14159265D0                                        8 3980
      SIGMA  = DSQRT(SIGSQ)                                    9 3216
      DENSTY = 6.D0*ETA/PI                                    10 4486
C
C...print heading
      WRITE(6,971) NCLSNS                                     11 4239
      WRITE(6,973) NATOM,DENSTY                               12 6275
      WRITE(6,975)                                            13 3148
C
      NRDELS= 0.5D0/RDEL-1                                    14 5139
      ORIGNS = (NCLSNS/KSAMPL)*(NATOM/2)                      15 6624
C
C...loop over radial shells
      RDELSI = RDEL/SIGMA                                     16 3946
      DO 700 J=1,NRDELS                                       17 3979
        IF (NGOFR(J).GT.0) THEN                               18 4453
          RADIUS= RDELSI*DFLOAT(J)                            19 4868
          VOLSHL= 4.D0*PI*RDELSI*RADIUS*RADIUS + (PI*RDELSI**3)/3.D0 20 3711
          GR   = DFLOAT(NGOFR(J))/(DENSTY*ORIGNS*VOLSHL)      21 8895
C
C...correct first value of g(r); the first sampling interval  is only half of RDEL
C       because the spheres are impenetrable
          IF (RADIUS.LT.(1.+RDELSI)) THEN                     22 6107
            RADIUS = 1.+0.5*RDELSI                            23 4341
            GR = 2.*GR                                        24 1944
          ENDIF                                               25 1292
          WRITE(6,977) J,RADIUS,NGOFR(J),GR                  26 7467
        ENDIF                                                 27 1292
  700 CONTINUE                                                28 2406
C
      WRITE(6,979)                                            29 3430
C
  971 FORMAT('1'////T20,'G(R) AT COLLISION NUMBER ',I6/)      30 1203
  973 FORMAT(T25,'NATOM = ',I4,3X,'DENSITY = ',F6.3,//)       31 3159
  975 FORMAT(29X,'I',6X,'R',6X,'NUMERATR',3X,'G(R)'/)         32 3227
  977 FORMAT(28X,I3,3X,F7.4,3X,I6,F8.3)                       33 9447
  979 FORMAT('1'///)                                          34 4071
C
      RETURN                                                  35 1090
      END                                                     36  921

|-----|--|---------|---------|---------|---------|---------|---------|--|------|
```

```
|-----|--|---------|---------|---------|---------|---------|---------|--|------|
      SUBROUTINE HSVELO(HINST)                                       1 4576
C
C        Accumulate velocity distribution & H-function
C
      IMPLICIT REAL*8(A-H,O-Z)                                       2 5702
      COMMON /INTGRS/ IN(7)                                          3 3711
      COMMON /MAXWEL/ NVELX(300),NVELY(300),NVELZ(300)              4 1337
      COMMON /REALS / RL(14)                                         5 3777
      COMMON /VEL   / X1(1000),Y1(1000),Z1(1000)                    6 9009
      DIMENSION NVINX(300), NVINY(300), NVINZ(300)                  7  325
      EQUIVALENCE (RL(13),VDEL), (RL(14),VMAX)                      8 8367
      EQUIVALENCE (IN(3),NATOM), (IN(7),NVDELS)                     9 8772
C
      DO 870 I=1,NVDELS                                            10 3924
        NVINX(I) = 0                                               11 2103
        NVINY(I) = 0                                               12 2439
        NVINZ(I) = 0                                               13 2372
  870 CONTINUE                                                     14 2327
C
C...see routine ODVELO in Appendix I;  NVIN_ are instantaneous, NVEL_ are averages
      OFFSET = 1.5 + VMAX/VDEL                                      15 3452
      DO 880 I=1,NATOM                                             16 3496
        NSHELL = X1(I)/VDEL + OFFSET                               17 4936
        NVELX(NSHELL) = NVELX(NSHELL)+1                            18 6523
        NVINX(NSHELL) = NVINX(NSHELL)+1                            19 6321
          NSHELL = Y1(I)/VDEL + OFFSET                             20 5263
          NVELY(NSHELL) = NVELY(NSHELL)+1                          21 7186
          NVINY(NSHELL) = NVINY(NSHELL)+1                          22 6983
        NSHELL = Z1(I)/VDEL + OFFSET                               23 5207
        NVELZ(NSHELL) = NVELZ(NSHELL)+1                            24 7063
        NVINZ(NSHELL) = NVINZ(NSHELL)+1                            25 6860
  880 CONTINUE                                                     26 2305
C
C...compute instantaneous H-function
      FATOM = DFLOAT(NATOM)                                         27 3755
      HH = 0.D0                                                     28 2406
      DO 890 K=1,NVDELS                                            29 4341
        IF (NVINX(K) .GT. 0) THEN                                  30 4778
          FOFV = DFLOAT(NVINX(K))/FATOM                            31 5016
          HH   = HH + FOFV*DLOG(FOFV)                              32 4554
        ENDIF                                                      33 1292
          IF (NVINY(K) .GT. 0) THEN                                34 5106
            FOFV = DFLOAT(NVINY(K))/FATOM                          35 5342
            HH =  HH + FOFV*DLOG(FOFV)                             36 4554
          ENDIF                                                    37 1292
        IF (NVINZ(K) .GT. 0) THEN                                  38 5049
          FOFV = DFLOAT(NVINZ(K))/FATOM                            39 5285
          HH   = HH + FOFV*DLOG(FOFV)                              40 4554
        ENDIF                                                      41 1292
  890 CONTINUE                                                     42 2507
      HINST = HH*VDEL/3.                                           43 4341
      RETURN                                                       44 1090
      END                                                          45  921
|-----|--|---------|---------|---------|---------|---------|---------|--|------|
```

|-----|--|---------|---------|---------|---------|---------|---------|--|------|

```
        SUBROUTINE HSVD                                               1 2743
C
C              Print velocity distribution
C
        IMPLICIT REAL*8 (A-H,O-Z)                                     2 5702
        COMMON /INTGRS/ IN(7)                                         3 3711
        COMMON /MAXWEL/ NVELX(300),NVELY(300),NVELZ(300)             4 1337
        COMMON /REALS / RL(14)                                        5 3777
C
        EQUIVALENCE (RL(11),TOTKE), (RL(13),VDEL), (RL(14),VMAX)     6 2103
        EQUIVALENCE (IN(1),KSAMPL), (IN(3),NATOM), (IN(6),NCLSNS)    7 2237
        EQUIVALENCE (IN(7),NVDELS)                                    8 5263
C
        WRITE(6,911)                                                  9 3014
        PI = 3.14159265D0                                           10 3980
C
C...NSAMPL is number of times velocity distribution was sampled in run
        NSAMPL = NCLSNS/KSAMPL                                       11 4352
        FNORM  = DFLOAT(NATOM*NSAMPL)                                12 5601
        TMPAVE = 2.D0*TOTKE/(3.*FNORM)                               13 6624
        FACT   = 1.D0/DSQRT(2.D0*PI*TMPAVE)                          14 6815
        HMB = 0.D0                                                   15 2787
C
C...loop over velocity shells, each of thickness VDEL
        DO 900 K=1,NVDELS                                            16 4420
            V = -VMAX + VDEL*(K-1)                                   17 4059
C
C...sampled, time-average distribution
            FOFV = DFLOAT(NVELX(K)+NVELY(K)+NVELZ(K))/(3.*FNORM)    18  796
C
C...analytic Maxwell distribution
            FANAL = FACT*DEXP(-V*V/(2.D0*TMPAVE))*VDEL              19 8873
            HMB   = HMB + FANAL*DLOG(FANAL)                         20 6398
            IF (FOFV.GT.0.D0) WRITE(6,913) V, FOFV, FANAL           21 8828
  900   CONTINUE                                                    22 2585
C
C...HMB is H-function from Maxwell; the run's H should be close to HMB; see § 3.5.2
        HMB = HMB*VDEL                                               23 4026
C
        WRITE(6,911)                                                24 3014
        WRITE(6,915) HMB                                            25 4161
C
  911   FORMAT(///)                                                 26 2620
  913   FORMAT(5X,F7.3,2(5X,F8.4))                                  27 6893
  915   FORMAT(10X,'Analytic H-function = ',F8.3/)                  28 6859
        RETURN                                                      29 1090
        END                                                         30  921
```

|-----|--|---------|---------|---------|---------|---------|---------|--|------|

SUBROUTINE HSORDR

HSFLAG ⟹ NATOM

HSFCC   ⟹ DIST

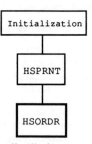

The positional order parameter λ is defined in (3.42). ALAMDA is the value used on the call from HSPRNT during initialization and merely tests that the fcc lattice has been properly constructed.

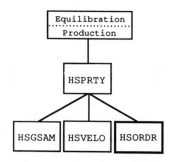

ALAM1 is used on all calls from HSPRTY; it is computed using distances relative to a body-fixed frame (i.e., one point on the lattice) rather than a space-fixed frame (the origin used for the position vectors). Otherwise, the entire lattice could be displaced relative to a space-fixed axis, so the order parameter decreases, but the lattice itself remains intact.

```
|-----|--|---------|---------|---------|---------|---------|--|------|
      SUBROUTINE HSORDR(ALAMDA,ALAM1)                          1 6220
C
C         Compute Verlet's translational order parameter
C
      IMPLICIT REAL*8(A-H,O-Z)                                 2 5702
      COMMON /INTGRS/ IN(7)                                    3 3711
      COMMON /POSIT / X(1000), Y(1000), Z(1000)               4 8659
      COMMON /REALS / RL(14)                                   5 3777
      EQUIVALENCE (RL(2),DIST), (IN(3),NATOM)                  6 8222
      PI4D = 4.D0*3.14159265D0/DIST                            7 7715
      ALAMDA = DCOS(PI4D*X(1)) + DCOS(PI4D*Y(1)) + DCOS(PI4D*Z(1))  8 2024
      ALAM1  = 0.D0                                            9 2631
      FATOM  = DFLOAT(NATOM)                                  10 3755
      DO 1000 I = 2,NATOM                                     11 3889
         ALAMDA = ALAMDA + DCOS(PI4D*X(I))                    12 6231
     &                   + DCOS(PI4D*Y(I))                    13 3766
     &                   + DCOS(PI4D*Z(I))                    14 3700
         ALAM1 = ALAM1   + DCOS(PI4D*(X(I)-X(1)))             15 7030
     &                   + DCOS(PI4D*(Y(I)-Y(1)))             16 5498
     &                   + DCOS(PI4D*(Z(I)-Z(1)))             17 5375
1000     CONTINUE                                             18 2631
      ALAMDA = ALAMDA/(3.D0*FATOM)                            19 5982
      ALAM1  = ALAM1/(3.D0*(FATOM-1.))                        20 6578
      RETURN                                                  21 1090
      END                                                     22  921
|-----|--|---------|---------|---------|---------|---------|--|------|
```

```
|-----|--|---------|---------|---------|---------|---------|---------|--|------|
      SUBROUTINE HSPRNT                                                 1 3148
C
C         Print initial output heading
C
      IMPLICIT REAL*8(A-H,O-Z)                                          2 5702
      COMMON /INTGRS/ IN(7)                                             3 3711
      COMMON /REALS / RL(14)                                            4 3777
      EQUIVALENCE (RL(3),ETA),  (IN(3),NATOM)                           5 7940
C... virial (Z – 1) from Carnahan-Starling equation of state
      CSVIR = (1.D0+ETA*(1.D0+ETA*(1.D0-ETA)))/(1.D0-ETA)**3 - 1.D0     6 3586
C
      DENSTY = 6.D0*ETA/3.14159265D0                                    7 7333
      WRITE(6,990)                                                      8 3496
      WRITE(6,991) NATOM,DENSTY                                         9 6310
      WRITE(6,992) ETA,CSVIR                                           10 4846
      WRITE(6,994)                                                     11 3609
      WRITE(6,995)                                                     12 3328
C... check that translational order parameter is unity
      CALL HSORDR(ALAMDA,ALAM1)                                        13 5139
      WRITE(6,997) ALAMDA                                             14 4723
C
  990 FORMAT('1',////,3X,67('-'))                                      15 8132
  991 FORMAT(//15X,'MOLECULAR DYNAMICS FOR ',I4,' HARD SPHERES',       16 3328
     &       //25X,'RHO*SIGMA**3 = ',F6.4,7X/)                         17 8266
C
  992 FORMAT(12X,'PACKING FRACTION = ',F6.4,4X,'C-S VIRIAL = ',F6.3/)   18 4262
  994 FORMAT(/3X,67('-')/)                                             19 5959
C
  995 FORMAT(3X,'COLL NO.',2X,' TIME',5X,'H-FUNC',                      20 2158
     &       3X,'ORDER',2X,'COLLIDING',1X,'K ENERGY',3X,'VIRIAL'       21 4060
     &       /,T33,'PARM',T42,'PAIR')                                  22 7209
C
  997 FORMAT(6X,'0',22X,F8.4)                                          23 6994
      RETURN                                                          24 1090
      END                                                             25  921

|-----|--|---------|---------|---------|---------|---------|---------|--|------|

      BLOCK DATA                                                       1 2079
      IMPLICIT REAL*8(A-H,O-Z)                                         2 5702
      COMMON /MAXWEL/ NVELX(300),NVELY(300),NVELZ(300)                 3 1337
      COMMON /RDF   / NGOFR(300)                                       4 4352
      COMMON /REALS / RL(14)                                           5 3777
      COMMON /TIMCOL/ TC(1000), NPAIR(1000)                            6 7377
      EQUIVALENCE (RL(5),SAMPLS), (RL(7),SUMB)                         7 8356
      EQUIVALENCE (RL(9),TIME),   (RL(11),TOTKE)                       8 8783
C
      DATA SAMPLS,SUMB,TIME,TOTKE/4*0.D0/                              9 8895
      DATA NPAIR/1000*0/                                              10 3935
      DATA NGOFR/ 300*0/                                             11 3711
      DATA NVELX,NVELY,NVELZ/300*0, 300*0, 300*0/                     12 1641
      END                                                             13  921
|-----|--|---------|---------|---------|---------|---------|---------|--|------|
```

APPENDIX L

MDSS
MOLECULAR DYNAMICS CODE FOR SOFT SPHERES

The following code, MDSS, performs soft-sphere molecular dynamics using the predictor–corrector algorithm described in Chapter 4. Documentary notes are provided for selected subroutines; the documentation follows the format described in Appendix I. Figures L.1 and L.2 contain dot charts of global scalars and the subroutines that access them. The hierarchy chart for MDSS is given in Figure L.3. To conform to local conventions, you may need to change the unit numbers on WRITE statements in subroutines SSMNTR, SSRSET, SSPRNT, and SSGOFR; the code contains no READ statements. The program uses the following system of fundamental units:

Mass m = mass of one atom

Length σ = distance to zero in Lennard-Jones potential

Energy ε = depth of minimum in Lennard-Jones potential

Time $\sigma(m/\varepsilon)^{1/2}$

Units of derived quantities are given in Table 5.2.

To verify your copy of the code, use CHECKSUM from Appendix G. To run MDSS, append the random-number generator ROULET from Appendix H, then set run flags and parameters in subroutine SSFLAG.

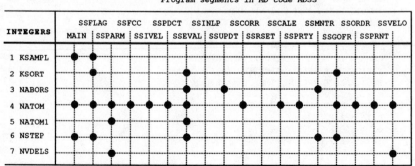

Program segments in MD code MDSS

FIGURE L.1 Integer global scalars used in MDSS. Array IN(7) passes these scalars among the program segments shown.

438

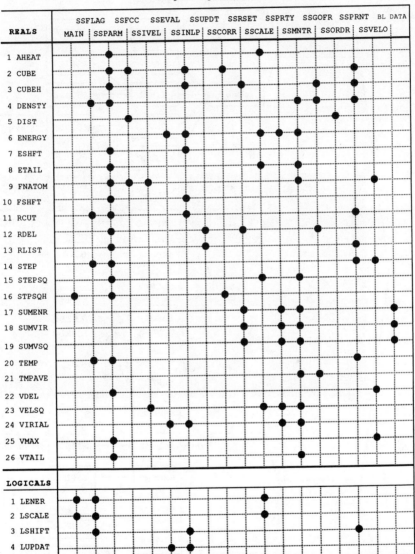

Program segments in MD code MDSS

FIGURE L.2 Real and logical scalars used in MDSS. Arrays RL(26) and LG(4) pass these scalars among the program segments shown. (Subroutines SSPDCT, SSPBC, and SSX2SC do not access arrays RL and LG, and so they are omitted from this table.)

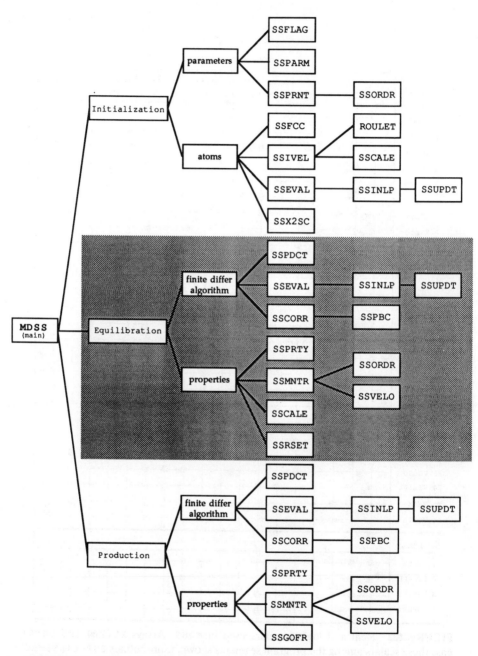

FIGURE L.3 Hierarchy chart for hard-sphere code M D S S. The background shading helps visually separate the three segments: initialization, equilibration, and production.

```
|-----|--|---------|---------|---------|---------|---------|---------|--|------|
C
C                Molecular  Dynamics  for  Soft  Spheres
C
C            filename:  mdss.f           version 5.5          13 Dec 1991
C-------------------------------------------------------------------------------
C
      IMPLICIT REAL*8 (A-H,O-Z)                                        1 5702
      LOGICAL LG(4),LENER,LSCALE                                      2 6073
      COMMON /INTGRS/ IN(7)                                            3 3711
      COMMON /LGCLS / LG                                               4 2990
      COMMON /REALS / RL(26)                                           5 3496
C
      EQUIVALENCE (IN(1),KSAMPL), (IN(4),NATOM), (IN(6),NSTEP)        6 2293
      EQUIVALENCE (LG(1),LENER),  (LG(2),LSCALE)                      7 8817
      EQUIVALENCE (RL(16),STPSQH)                                      8 4880
C
C...initialization
      CALL SSFLAG(KRDF,KWRITE,MAXSTP,MAXEQB,NDEAD,DENER)              9 2136
      CALL SSPARM                                                    10 1685
      CALL SSFCC                                                     11 1168
      CALL SSPRNT                                                    12 1764
      CALL SSIVEL                                                    13 1775
      CALL SSEVAL                                                    14 1641
      CALL SSX2SC(STPSQH,NATOM)                                      15 4161
C
C...equilibration
      DO 5 NSTEP = 1, MAXEQB                                         16 3801
        CALL SSPDCT                                                  17 1764
        CALL SSEVAL                                                  18 1641
        CALL SSCORR                                                  19 1315
        CALL SSPRTY                                                  20 1843
        IF (MOD(NSTEP,KWRITE).EQ.0)   CALL SSMNTR(1)                21 8648
        IF (NSTEP.LT.(MAXEQB-NDEAD)) THEN                            22 7232
           IF (LSCALE .OR. LENER) CALL SSCALE(DENER)                23 7131
        ENDIF                                                        24 1292
    5 CONTINUE                                                       25 1595
C
      CALL SSRSET                                                    26 1630
C
C...production
      DO 7 NSTEP = 1, MAXSTP                                         27 3654
        CALL SSPDCT                                                  28 1764
        CALL SSEVAL                                                  29 1641
        CALL SSCORR                                                  30 1315
        IF (MOD(NSTEP,KSAMPL).EQ.0)   CALL SSPRTY                   31 7849
        IF (MOD(NSTEP,KWRITE).EQ.0)   CALL SSMNTR(KSAMPL)          32 9942
        IF (MOD(NSTEP,KRDF).EQ.0)     CALL SSGOFR                   33 7221
    7 CONTINUE                                                       34 1708
C
      STOP                                                           35  550
      END                                                            36  921

|-----|--|---------|---------|---------|---------|---------|---------|--|------|
```

SUBROUTINE SSFLAG

Run Parameters

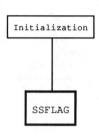

NATOM   = number of atoms

DENSTY = density $N\sigma^3/V$

TEMP   = desired value of the temperature in units of ε/k

STEP   = integration time-step, in units of $\sigma(m/\varepsilon)^{1/2}$

RCUT   = distance at which pair potential is truncated, in units of σ

Run Flags

MAXEQB = number of steps to execute during equilibration

MAXSTP = number of steps to execute during production

KWRITE = time-step interval at which monitor properties are printed

KRDF   = time-step interval at which g(r) is printed

KSORT   = time-step interval at which neighbor-list is updated and sorting for g(r) is performed

NSTEP   = time-step counter

NDEAD   = number of time-steps between equilibration and start of production

Each atom has unit mass.

```
|-----|--|---------|---------|---------|---------|---------|---------|--|------|

      SUBROUTINE SSFLAG(KRDF,KWRITE,MAXSTP,MAXEQB,NDEAD,DENER)      1 3227
C
C         Run parameters and flags typically set by user
C
      IMPLICIT REAL*8 (A-H,O-Z)                                     2 5702
      LOGICAL LG(4),LENER,LSCALE,LSHIFT                            3 7939
      COMMON /INTGRS/ IN(7)                                        4 3711
      COMMON /LGCLS / LG                                           5 2990
      COMMON /REALS / RL(26)                                       6 3496
C
      EQUIVALENCE (IN(1),KSAMPL), (IN(2),KSORT),   (IN(4),NATOM)   7 2338
      EQUIVALENCE (IN(6),NSTEP)                                    8 4745
      EQUIVALENCE (LG(1),LENER),  (LG(2),LSCALE), (LG(3),LSHIFT)   9 2844
      EQUIVALENCE (RL(4),DENSTY), (RL(11),RCUT),  (RL(14),STEP)   10 2237
      EQUIVALENCE (RL(20),TEMP)                                   11 5139
C
C...run parameters
      NATOM  = 108                                                12 2079
      DENSTY = 0.7D0                                              13 3508
      TEMP   = 1.2D0                                              14 2653
      STEP = 0.004D0                                              15 3531
      RCUT =    2.5D0                                             16 2158
C
C...run flags
      MAXEQB = 1000                                               17 2721
      MAXSTP = 5000                                               18 2473
      KWRITE =    20                                              19 2563
      KRDF   = 2500                                               20 2574
      KSORT  =   10                                               21 1797
      KSAMPL =   10                                               22 2327
      NSTEP  =    0                                               23 1696
      NDEAD  =  100                                               24 2596
C
C...to scale velocities for temperature during equilibration, set LSCALE = .TRUE.
      LSCALE = .TRUE.                                             25 2574
C
C...to scale velocities for set-point total energy, set LENER = .TRUE.
      IF (.NOT.LSCALE) THEN                                       26 3991
        LENER = .FALSE.                                           27 2945
C
C...if scaling energy, load the set-point value
        IF (LENER) DENER = -2.0D0                                 28 5599
      ENDIF                                                       29 1292
C
C...if using shifted-force potential, set LSHIFT = .TRUE.
C...for potential merely truncated, set LSHIFT = .FALSE.
      LSHIFT = .FALSE.                                            30 3070
C
      RETURN                                                      31 1090
      END                                                         32  921

|-----|--|---------|---------|---------|---------|---------|---------|--|------|
```

SUBROUTINE SSPARM

SSFLAG ⇒ NATOM, STEP, DENSTY, RCUT, TEMP

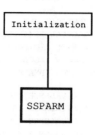

RLIST ... Sphere radius (in units of σ) about each atom, to which the neighbor-list is maintained. The list of neighbors for atom i contains more than just the atoms currently interacting with i, to allow for the possibility that some atoms j having r_{ij} > RCUT will cross RCUT and interact with i *before* the list is next updated. We don't want the neighbor-list too large, else we lose the speed advantage.

AHEAT ... contains a factor of $(\Delta t)^2$ because the velocities are implicitly scaled by Δt.

RDEL ... the increment in radial distance at which the histogram for g(r) is constructed. We want RDEL small, to obtain a well-defined curve for g(r). But if it is too small, the sampling population in each Δr is inadequate for reliable statistics. RDEL is in units of σ.

ALFA0 ... are the parameters in Gear's fifth-order predictor-corrector algorithm for second-order ODEs. See Table 9.1 in C. W. Gear, *Numerical Initial Value Problems in Ordinary Differential Equations*, Prentice-Hall, Englewood Cliffs, 1971, p. 154. But 3/16 for ALFA0 seems to be somewhat better than 3/20 (C. W. Gear, Argonne National Lab Report, ANL-7126, 1966).

The shifted force $f_s(r)$ is given by (5.8) for the (12, 6) Lennard-Jones potential. The corresponding "shifted-force" potential $u_s(r)$ is given in (5.12).

The expressions for ETAIL and VTAIL are from (6.9) and (6.15), respectively.

```
|-----|--|----------|----------|----------|----------|----------|----------|--|------|
      SUBROUTINE SSPARM                                             1 2776
C
C          Calculate parameters from run parameters set in SSFLAG
C
      IMPLICIT REAL*8 (A-H,O-Z)                                     2 5702
      COMMON /ALPHA / ALFA0 ,ALFA1 ,ALFA3 ,ALFA4 ,ALFA5            3 9144
      COMMON /INTGRS/ IN(7)                                         4 3711
      COMMON /REALS / RL(26)                                        5 3496
      EQUIVALENCE (IN(4),NATOM) , (IN(5),NATOM1), (IN(7),NVDELS)    6 2473
      EQUIVALENCE (RL(1),AHEAT) , (RL(2),CUBE)   , (RL(3),CUBEH),   7 2237
     &   (RL(4),DENSTY) , (RL(7),ESHFT) , (RL(8),ETAIL),           8 1551
     &   (RL(9),FNATOM), (RL(10),FSHFT), (RL(11),RCUT), (RL(12),RDEL),  9 4420
     &   (RL(13),RLIST), (RL(14),STEP) , (RL(15),STEPSQ),(RL(16),STPSQH), 10 4352
     &   (RL(20),TEMP) , (RL(22),VDEL) , (RL(25),VMAX), (RL(26),VTAIL)  11 3586
C
C...functions of particle number and time-step
      FNATOM = NATOM                                               12 2394
      NATOM1 = NATOM-1                                             13 2978
      STEPSQ = STEP*STEP                                           14 3081
      STPSQH = 0.5D0*STEPSQ                                        15 4116
C
C...size of cube, cut-off & list distances, scale factor for velocities, & sampling increment for g(r)
      VOL   = FNATOM/DENSTY                                        16 3586
      CUBE  = VOL**(1.D0/3.D0)                                     17 5353
      CUBEH = 0.5D0*CUBE                                           18 4240
      RLIST = RCUT+0.3D0                                           19 3418
      IF (RLIST.GT.CUBEH) RLIST= CUBEH                            20 6309
      IF (RCUT.GT.CUBEH)  RCUT = CUBEH-0.35D0                     21 7467
      AHEAT= 3.D0*FNATOM*STEPSQ*TEMP                               22 6792
      RDEL = 0.025D0                                               23 3519
C
C...parameters in predictor-corrector method
      ALFA0 =   3.D0/ 16.D0                                        24 4015
      ALFA1 =251.D0/360.D0                                         25 4598
      ALFA3 = 11.D0/ 18.D0                                         26 4150
      ALFA4 =  1.D0/  6.D0                                         27 3812
      ALFA5 =  1.D0/ 60.D0                                         28 3889
C
C...shifted-force constants
      RCINV = 1.D0/RCUT                                            29 2710
      RC6INV= RCINV**6                                             30 2710
      ESHFT = RC6INV*(28.D0-52.D0*RC6INV)                         31 7423
      FSHFT = 48.D0*RCINV*RC6INV*(RC6INV-.5D0)                     32 8402
C
C...long-range corrections to pressure and internal energy & parameters for velocity distribution
      PI   = 3.14159265D0                                          33 3980
      RC3  = RCUT**3                                               34 2394
      RC9  = RC3**3                                                35 2563
      ETAIL= 8.D0*PI*DENSTY*(1.0D0/(9.D0*RC9)-1.D0/(3.D0*RC3))     36 5005
      VTAIL= 96.D0*PI*DENSTY*(0.5D0/(3.D0*RC3)-1.D0/(9.D0*RC9))    37 5094
C
      VMAX = 5.                                                    38 1179
      VDEL = 0.05                                                  39 2428
      NVDELS = 2.*VMAX/VDEL + 1.01                                 40 5094
      RETURN                                                       41 1090
      END                                                          42  921
|-----|--|----------|----------|----------|----------|----------|----------|--|------|
```

```
|-----|--|---------|---------|---------|---------|---------|---------|--|------|
      SUBROUTINE SSFCC                                             1 2259
C
C        Assign initial positions based on face-centered cubic lattice
C
      IMPLICIT REAL*8 (A-H,O-Z)                                    2 5702
      COMMON /INTGRS/ IN(7)                                        3 3711
      COMMON /POSIT / X(256),Y(256),Z(256)                         4 6589
      COMMON /REALS / RL(26)                                       5 3496
      EQUIVALENCE (IN(4),NATOM)                                    6 5027
      EQUIVALENCE (RL(2),CUBE), (RL(5),DIST), (RL(9),FNATOM)       7 1742
C
C...number of basic lattice units, then check whether user asked for non-fcc number of atoms
      NUNIT = (FNATOM/4.)**(1./3.)+.1                              8 6488
   5  NCHECK = 4*(NUNIT**3)                                        9 5410
      IF (NCHECK.LT.NATOM) THEN                                   10 5218
         NUNIT = NUNIT + 1                                        11 2686
         GOTO 5                                                   12  763
      ENDIF                                                       13 1292
C
C...set lattice distance based on cube of side = CUBE
      DIST  = 0.5D0*CUBE/DFLOAT(NUNIT)                            14 6804
C
C...assign positions of first four atoms
      X(1) = 0.D0                                                 15 2248
      Y(1) = 0.D0                                                 16 2574
      Z(1) = 0.D0                                                 17 2518
      X(2) = 0.D0                                                 18 2305
      Y(2) = DIST                                                 19 2215
      Z(2) = DIST                                                 20 2158
      X(3) = DIST                                                 21 1966
      Y(3) = 0.D0                                                 22 2710
      Z(3) = DIST                                                 23 2237
      X(4) = DIST                                                 24 2114
      Y(4) = DIST                                                 25 2440
      Z(4) = 0.D0                                                 26 2800
C
C...replicate first four positions over NUNITS
      M=0                                                         27 1001
      KCT = 0                                                     28 1359
      DO 12 I=1,NUNIT                                             29 3126
      DO 12 J=1,NUNIT                                             30 2978
      DO 12 K=1,NUNIT                                             31 3362
         DO 10 IJ=1,4                                             32 2697
C
C...if number of positions assigned exceeds number atoms, get out
         IF (KCT.LT.NATOM) THEN                                   33 4486
            X(IJ+M) = X(IJ) +2.D0*DIST*(K-1)                      34 6477
            Y(IJ+M) = Y(IJ) +2.D0*DIST*(J-1)                      35 6747
            Z(IJ+M) = Z(IJ) +2.D0*DIST*(I-1)                      36 6770
         ENDIF                                                    37 1292
         KCT = KCT + 1                                            38 1887
  10     CONTINUE                                                 39 1933
         M=M+4                                                    40 1472
  12  CONTINUE                                                    41 1810
      RETURN                                                      42 1090
      END                                                         43  921
|-----|--|---------|---------|---------|---------|---------|---------|--|------|
```

|-----|--|---------|---------|---------|---------|---------|---------|--|------|

```
      SUBROUTINE SSIVEL                                          1 2866
C
C        Assign random initial velocities
C
      IMPLICIT REAL*8 (A-H,O-Z)                                  2 5702
      REAL*4 ROULET                                             3 2776
      COMMON /INTGRS/ IN(7)                                     4 3711
      COMMON /REALS / RL(26)                                    5 3496
      COMMON /VEL   / X1(256), Y1(256), Z1(256)               6 6938
C
      EQUIVALENCE (IN(4),NATOM)                                 7 5027
      EQUIVALENCE (RL(9),FNATOM), (RL(23),VELSQ)              8 8682
C
      DATA SUMX,SUMY,SUMZ,VELSQ/4*0.0D0/                        9 7995
      MSEED = -30509                                           10 3698
C
```
C. . . *assign random velocity components on (-1, +1); ROULET is given in Appendix H*
```
      DO 100 I=1,NATOM                                         11 3474
        X1(I)=ROULET(MSEED)                                    12 3935
        Y1(I)=ROULET(MSEED)                                    13 4262
        Z1(I)=ROULET(MSEED)                                    14 4206
        SUMX =SUMX+X1(I)                                       15 2260
        SUMY =SUMY+Y1(I)                                       16 3250
        SUMZ =SUMZ+Z1(I)                                       17 3070
  100 CONTINUE                                                 18 2282
C
```
C. . . *scale velocities so that total linear momentum is zero; see Eq. (5.28)*
```
      DO 120 I=1,NATOM                                         19 3351
        X1(I) = X1(I) - SUMX/FNATOM                           20 4150
        Y1(I) = Y1(I) - SUMY/FNATOM                           21 5140
        Z1(I) = Z1(I) - SUMZ/FNATOM                           22 4969
        VELSQ = VELSQ + X1(I)**2 +Y1(I)**2 +Z1(I)**2          23 8176
  120 CONTINUE                                                 24 2169
C
```
C. . . *scale velocities to set-point temperature or to set-point total energy*
```
      CALL SSCALE(0.)                                          25 2620
C
      RETURN                                                   26 1090
      END                                                      27  921
```

|-----|--|---------|---------|---------|---------|---------|---------|--|------|

```
|-----|--|---------|---------|---------|---------|---------|---------|--|------|

      SUBROUTINE SSPDCT                                              1 2855
C
C         Use fifth-order Taylor series to predict positions
C         and their derivatives at the next time-step
C
      IMPLICIT REAL*8 (A-H,O-Z)                                      2 5702
      COMMON /ACCEL / X2(256)      ,Y2(256)      ,Z2(256)            3 7221
      COMMON /DERIV3/ X3(256)      ,Y3(256)      ,Z3(256)            4 8053
      COMMON /DERIV4/ X4(256)      ,Y4(256)      ,Z4(256)            5 8659
      COMMON /DERIV5/ X5(256)      ,Y5(256)      ,Z5(256)            6 7535
      COMMON /FORCE / FX(256)      ,FY(256)      ,FZ(256)            7 6938
      COMMON /INTGRS/ IN(7)                                         8 3711
      COMMON /POSIT / X(256),Y(256),Z(256)                          9 6589
      COMMON /VEL   / X1(256)      ,Y1(256)      ,Z1(256)          10 6938
      EQUIVALENCE (IN(4),NATOM)                                    11 5027
C
C...velocities and higher derivatives are implicitly scaled by appropriate factors of the time-step,
C       see routine ODINIT in Appendix I.
C
C...Note that the numerical coefficients in the Taylor series form a Pascal triangle.
C
      DO 200 I=1,NATOM                                             12 3531
       X(I) = X(I) + X1(I)+X2(I)+X3(I)+X4(I)+X5(I)                 13 6365
       Y(I) = Y(I) + Y1(I)+Y2(I)+Y3(I)+Y4(I)+Y5(I)                 14 8671
       Z(I) = Z(I) + Z1(I)+Z2(I)+Z3(I)+Z4(I)+Z5(I)                 15 8255
C
       X1(I) = X1(I)+2.D0*X2(I)+3.D0*X3(I)+4.D0*X4(I)+5.D0*X5(I)   16 2552
       Y1(I) = Y1(I)+2.D0*Y2(I)+3.D0*Y3(I)+4.D0*Y4(I)+5.D0*Y5(I)   17 4532
       Z1(I) = Z1(I)+2.D0*Z2(I)+3.D0*Z3(I)+4.D0*Z4(I)+5.D0*Z5(I)   18 4172
C
       X2(I) = X2(I)+3.D0*X3(I)+6.D0*X4(I)+10.D0*X5(I)             19  156
       Y2(I) = Y2(I)+3.D0*Y3(I)+6.D0*Y4(I)+10.D0*Y5(I)            20 1809
       Z2(I) = Z2(I)+3.D0*Z3(I)+6.D0*Z4(I)+10.D0*Z5(I)            21 1506
C
       X3(I) = X3(I)+4.D0*X4(I)+10.D0*X5(I)                       22 7973
       Y3(I) = Y3(I)+4.D0*Y4(I)+10.D0*Y5(I)                       23 9290
       Z3(I) = Z3(I)+4.D0*Z4(I)+10.D0*Z5(I)                       24 9054
C
       X4(I) = X4(I)+5.D0*X5(I)                                   25 4958
       Y4(I) = Y4(I)+5.D0*Y5(I)                                   26 5948
       Z4(I) = Z4(I)+5.D0*Z5(I)                                   27 5768
C
C...initialize force arrays
       FX(I) = 0.0D0                                              28 2798
       FY(I) = 0.0D0                                              29 3126
       FZ(I) = 0.0D0                                              30 3069
  200 CONTINUE                                                    31 2349
C
      RETURN                                                      32 1090
      END                                                         33  921

|-----|--|---------|---------|---------|---------|---------|---------|--|------|
```

```
|-----|--|---------|---------|---------|---------|---------|---------|--|------|

      SUBROUTINE SSEVAL                                            1 2732
C
C        Evaluate forces on each atom using predicted positions
C
      IMPLICIT REAL*8 (A-H,O-Z)                                    2 5702
      LOGICAL LG(4),LUPDAT                                         3 4521
      COMMON /INTGRS/ IN(7)                                        4 3711
      COMMON /LGCLS / LG                                           5 2990
      COMMON /NABLST/ LIST(15000)   ,NPOINT(256)                  6 8097
      COMMON /POSIT / X(256),Y(256),Z(256)                        7 6589
      COMMON /REALS / RL(26)                                       8 3496
      COMMON /ZLOOP / XI, YI,ZI,RIJ,I , J ,JBEGIN,JEND            9  167
C
      EQUIVALENCE (IN(2),KSORT) , (IN(3),NABORS) , (IN(4),NATOM)  10 2091
      EQUIVALENCE (IN(5),NATOM1), (IN(6),NSTEP)                   11 8299
      EQUIVALENCE (RL(6),ENERGY), (RL(24),VIRIAL), (LG(4),LUPDAT) 12 3227
C
C...initialize instantaneous potential energy and virial
      ENERGY=0.D0                                                 13 3159
      VIRIAL=0.D0                                                 14 2620
C
C...set counter & flag for neighbor list update at intervals
      NABORS=0                                                    15 1775
      LUPDAT= .FALSE.                                             16 2990
      IF (MOD(NSTEP,KSORT).EQ.0) LUPDAT= .TRUE.                   17 8222
C
C...begin outer loop over atoms
      DO 300 I=1,NATOM1                                           18 3788
C
C...use neighbor list to find neighbors of atom I; for the relation between arrays LIST and
C   NPOINT, see Figure 5.5
         JBEGIN=NPOINT(I)                                         19 3654
         JEND  =NPOINT(I+1)-1                                     20 3935
C
C...at KSORT intervals, use all (NATOM – 1) atoms as neighbors of I
         IF (LUPDAT) THEN                                         21 3182
            NPOINT(I)= NABORS+1                                   22 3474
            JBEGIN   = I+1                                        23 2237
            JEND     = NATOM                                      24 2260
         ENDIF                                                    25 1292
C
C...store position of atom I
         XI=X(I)                                                  26 1214
         YI=Y(I)                                                  27 1876
         ZI=Z(I)                                                  28 1753
         CALL SSINLP                                              29 1988
  300 CONTINUE                                                    30 2428
C
      IF (LUPDAT) NPOINT(NATOM)=NABORS+1                          31 6286
      RETURN                                                      32 1090
      END                                                         33  921

|-----|--|---------|---------|---------|---------|---------|---------|--|------|
```

SUBROUTINE SSINLP

SSPARM	⇒	CUBEH, ESHFT, FSHFT
SSEVAL	⇒	LUPDAT, ENERGY, VIRIAL
SSEVAL	⇒	JBEGIN, JEND, X1, Y1, Z1, I
SSFLAG	⇒	RCUT, LSHIFT

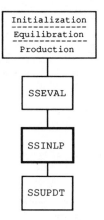

DO 400 ... The atoms j possibly interacting with i are taken from the neighbor-list (LUPDAT = .FALSE.) except at KSORT intervals (LUPDAT = .TRUE.), when all $(N - 1)$ atoms are sampled for possible interactions with i.

IF (XIJ ... This applies the minimum image criterion, i.e., for a spherically cut-off potential, only one of the 26 images of j (that closest to i) interacts with i. See W. W. Wood, in *Physics of Simple Liquids*, North-Holland, Amsterdam, 1968, Chapter 5. Recall the atom positions here are on [0, CUBE]. To find the image of j closest to i we consider i centered at a cube of edge = CUBE and apply the periodic boundary condition to r_{ij}; see subroutine SSPBC. To invoke the minimum image criteria, we could call subroutine SSPBC rather than use these six IF statements; however, we opt for speed here since the overhead incurred in the subroutine call inside the double loop can slow the execution of each time-step by 30%.

CALL SSUPDT ... This subroutine is called only at intervals of KSORT, not every time-step. See subroutine SSEVAL.

IF LSHIFT = .TRUE. ... then the simulation is for a shifted-force potential rather than the truncated potential. The shifted-force potential is useful for verifying the code because the energy conservation should be excellent.

```
|-----|--|----------|----------|----------|----------|----------|----------|--|------|
      SUBROUTINE SSINLP                                                   1 3070
C
C        Perform inner loop over atoms to get force on I
C
      IMPLICIT REAL*8 (A-H,O-Z)                                           2 5702
      LOGICAL LG(4),LSHIFT,LUPDAT                                         3 6387
      COMMON /FORCE / FX(256)       ,FY(256)       ,FZ(256)              4 6938
      COMMON /LGCLS / LG                                                  5 2990
      COMMON /NABLST/ LIST(15000)   ,NPOINT(256)                         6 8097
      COMMON /POSIT / X(256),Y(256),Z(256)                               7 6589
      COMMON /REALS / RL(26)                                              8 3496
      COMMON /ZLOOP / XI, YI, ZI, RIJ, I, J, JBEGIN, JEND                9  167
      EQUIVALENCE (RL(2),CUBE) ,   (RL(3),CUBEH) ,  (RL(6),ENERGY)      10 2226
      EQUIVALENCE (RL(7),ESHFT),   (RL(10),FSHFT), (RL(11),RCUT)        11 1720
      EQUIVALENCE (RL(24),VIRIAL), (LG(3),LSHIFT), (LG(4),LUPDAT)       12 3395
C
      DO 400 JX=JBEGIN,JEND                                              13 4981
        J=JX                                                            14  482
        IF (.NOT.LUPDAT) J=LIST(JX)                                     15 4789
C
C...components of vector between atoms I and J
        XIJ = XI-X(J)                                                   16 1764
        YIJ = YI-Y(J)                                                   17 2754
        ZIJ = ZI-Z(J)                                                   18 2574
C
C...find image of J closest to I
        IF (XIJ.LT.-CUBEH) XIJ = XIJ + CUBE                             19 5421
        IF (XIJ.GT. CUBEH) XIJ = XIJ - CUBE                             20 5285
        IF (YIJ.LT.-CUBEH) YIJ = YIJ + CUBE                             21 6411
        IF (YIJ.GT. CUBEH) YIJ = YIJ - CUBE                             22 6275
        IF (ZIJ.LT.-CUBEH) ZIJ = ZIJ + CUBE                            23 6231
        IF (ZIJ.GT. CUBEH) ZIJ = ZIJ - CUBE                            24 6095
      RSQ = XIJ*XIJ + YIJ*YIJ + ZIJ*ZIJ                                25 4947
      RIJ = DSQRT(RSQ)                                                  26 2260
      IF (LUPDAT.AND.RIJ.LE.CUBEH) CALL SSUPDT                          27 7726
C
      IF (RIJ.LE.RCUT) THEN                                             28 3766
        RSQINV=1.D0/RSQ                                                 29 2653
        R6INV =RSQINV*RSQINV*RSQINV                                     30 4239
C
C...ENR is u(r), FOR is -(1/r)(du/dr)
        ENR= 4.D0*R6INV*(R6INV-1.D0)                                    31 6680
        FOR= RSQINV*48.D0*R6INV*(R6INV-0.5D0)                           32 8200
        IF (LSHIFT) ENR = ENR + ESHFT + RIJ*FSHFT                       33 6770
        IF (LSHIFT) FOR = FOR - RSQINV*RIJ*FSHFT                        34 6602
        FX(I) = FX(I)+FOR*XIJ                                           35 3058
        FX(J) = FX(J)-FOR*XIJ                                           36 3115
        FY(I) = FY(I)+FOR*YIJ                                           37 4048
        FY(J) = FY(J)-FOR*YIJ                                           38 4105
        FZ(I) = FZ(I)+FOR*ZIJ                                           39 3867
        FZ(J) = FZ(J)-FOR*ZIJ                                           40 3924
        ENERGY = ENERGY+ENR                                             41 4059
        VIRIAL = VIRIAL-FOR*RSQ                                         42 3744
      ENDIF                                                             43 1292
  400 CONTINUE                                                          44 2574
      RETURN                                                            45 1090
      END                                                               46  921
|-----|--|----------|----------|----------|----------|----------|----------|--|------|
```

```
|-----|--|---------|---------|---------|---------|---------|---------|--|------|

      SUBROUTINE SSCORR                                               1 2406
C
C         Correct predicted positions and their derivatives
C
      IMPLICIT REAL*8 (A-H,O-Z)                                       2 5702
      COMMON /ACCEL / X2(256)       ,Y2(256)       ,Z2(256)          3 7221
      COMMON /ALPHA / ALFA0 ,ALFA1 ,ALFA3 ,ALFA4 ,ALFA5              4 9144
      COMMON /DERIV3/ X3(256)       ,Y3(256)       ,Z3(256)          5 8053
      COMMON /DERIV4/ X4(256)       ,Y4(256)       ,Z4(256)          6 8659
      COMMON /DERIV5/ X5(256)       ,Y5(256)       ,Z5(256)          7 7535
      COMMON /FORCE / FX(256)       ,FY(256)       ,FZ(256)          8 6938
      COMMON /INTGRS/ IN(7)                                          9 3711
      COMMON /POSIT / X(256),Y(256),Z(256)                          10 6589
      COMMON /REALS / RL(26)                                        11 3496
      COMMON /VEL   / X1(256)       ,Y1(256)       ,Z1(256)         12 6938
C
      EQUIVALENCE (IN(4),NATOM), (RL(2),CUBE), (RL(16),STPSQH)      13 1810
C
      DO 500 I=1,NATOM                                              14 3485
       XERROR = STPSQH*FX(I)-X2(I)                                  15 4565
       YERROR = STPSQH*FY(I)-Y2(I)                                  16 5555
       ZERROR = STPSQH*FZ(I)-Z2(I)                                  17 5375
C
       X (I)=X (I)+XERROR*ALFA0                                     18 3823
       X1(I)=X1(I)+XERROR*ALFA1                                     19 3980
       X2(I)=X2(I)+XERROR                                           20 2855
       X3(I)=X3(I)+XERROR*ALFA3                                     21 4408
       X4(I)=X4(I)+XERROR*ALFA4                                     22 4857
       X5(I)=X5(I)+XERROR*ALFA5                                     23 4015
C
       Y (I)=Y (I)+YERROR*ALFA0                                     24 4813
       Y1(I)=Y1(I)+YERROR*ALFA1                                     25 4970
       Y2(I)=Y2(I)+YERROR                                           26 3845
       Y3(I)=Y3(I)+YERROR*ALFA3                                     27 5397
       Y4(I)=Y4(I)+YERROR*ALFA4                                     28 5847
       Y5(I)=Y5(I)+YERROR*ALFA5                                     29 5005
C
       Z (I)=Z (I)+ZERROR*ALFA0                                     30 4633
       Z1(I)=Z1(I)+ZERROR*ALFA1                                     31 4790
       Z2(I)=Z2(I)+ZERROR                                           32 3665
       Z3(I)=Z3(I)+ZERROR*ALFA3                                     33 5218
       Z4(I)=Z4(I)+ZERROR*ALFA4                                     34 5667
       Z5(I)=Z5(I)+ZERROR*ALFA5                                     35 4824
C
C...Apply periodic boundary conditions
          CALL SSPBC(X(I),Y(I),Z(I),0.D0,CUBE,CUBE)                 36 9863
  500 CONTINUE                                                      37 2293
      RETURN                                                        38 1090
      END                                                           39  921

|-----|--|---------|---------|---------|---------|---------|---------|--|------|
```

```
|-----|--|---------|----------|---------|----------|----------|---------|--|------|
      SUBROUTINE SSPBC(X,Y,Z,BLOW,BHI,EDGE)                              1 9616
C
C         Apply periodic boundary conditions to point (X, Y, Z) contained in a cube of
C         side of length EDGE .  Coordinates for side = 0 correspond to BLOW and those
C         for side = EDGE to BHI.  For a picture, see notes to routine ODPEC in Appendix I.
C
      IMPLICIT REAL*8 (A-H,O-Z)                                          2 5702
C
      IF (X.LT.BLOW) THEN                                                3 4150
        X = X + EDGE                                                     4 1562
      ELSEIF (X.GT.BHI) THEN                                             5 4756
        X = X - EDGE                                                     6 1922
      ENDIF                                                              7 1292
C
      IF (Y.LT.BLOW) THEN                                                8 4486
        Y = Y + EDGE                                                     9 2226
      ELSEIF (Y.GT.BHI) THEN                                            10 5083
        Y = Y - EDGE                                                    11 2585
      ENDIF                                                             12 1292
C
      IF (Z.LT.BLOW) THEN                                               13 4420
        Z = Z + EDGE                                                    14 2103
      ELSEIF (Z.GT.BHI) THEN                                            15 5027
        Z = Z - EDGE                                                    16 2462
      ENDIF                                                             17 1292
C
      RETURN                                                           18 1090
      END                                                              19  921

|-----|--|---------|----------|---------|----------|----------|---------|--|------|

      SUBROUTINE SSPRTY                                                  1 2934
C
C         Accumulate property values every KSAMPL time-steps
C
      IMPLICIT REAL*8 (A-H,O-Z)                                          2 5702
      COMMON /INTGRS/ IN(7)                                             3 3711
      COMMON /REALS / RL(26)                                            4 3496
      COMMON /VEL   / X1(256), Y1(256), Z1(256)                         5 6938
      EQUIVALENCE (IN(4),NATOM)                                         6 5027
      EQUIVALENCE (RL(6),ENERGY),  (RL(17),SUMENR), (RL(18),SUMVIR)     7 2619
      EQUIVALENCE (RL(19),SUMVSQ), (RL(23),VELSQ) , (RL(24),VIRIAL)     8 2305
C
C...sum square of velocities
      VELSQ =0.D0                                                        9 2383
      DO 700 I=1,NATOM                                                  10 3597
        VELSQ =VELSQ +X1(I)**2   +Y1(I)**2   +Z1(I)**2                  11 8176
  700 CONTINUE                                                          12 2406
C
C...accumulate sums for running averages
      SUMENR=SUMENR+ENERGY                                              13 4228
      SUMVIR=SUMVIR+VIRIAL                                              14 3070
      SUMVSQ=SUMVSQ+VELSQ                                               15 2439
      RETURN                                                            16 1090
      END                                                               17  921
|-----|--|---------|----------|---------|----------|----------|---------|--|------|
```

SUBROUTINE SSMNTR

MAIN   ⟹   NSTEP

SSFLAG   ⟹   DENSTY

SSPARM   ⟹   ETAIL, VTAIL, FNATOM, STEPSQ

SSINLP   ⟹   ENERGY, VIRIAL

SSUPDT   ⟹   NABORS

SSPRPT   ⟹   SUMENR, SUMVIR, SUMVSQ, VELSQ

EPOT   = potential energy at current time-step

ETOTAL = total energy at current time-step

```
|-----|--|---------|----------|----------|----------|----------|----------|--|------|
      SUBROUTINE SSMNTR(KSAMPL)                                              1 4857
C
C         Print properties at intervals
C
      IMPLICIT REAL*8 (A-H,O-Z)                                             2 5702
      COMMON /INTGRS/ IN(7)                                                 3 3711
      COMMON /REALS / RL(26)                                               4 3496
      EQUIVALENCE (IN(3),NABORS) ,  (IN(6),NSTEP)                          5 8390
      EQUIVALENCE (RL(4),DENSTY),   (RL(6),ENERGY) ,  (RL(8),ETAIL)        6 3069
      EQUIVALENCE (RL(9),FNATOM),   (RL(15),STEPSQ), (RL(17),SUMENR)       7 2484
      EQUIVALENCE (RL(18),SUMVIR),  (RL(19),SUMVSQ), (RL(21),TMPAVE)       8 2103
      EQUIVALENCE (RL(23),VELSQ) ,  (RL(24),VIRIAL), (RL(26),VTAIL)        9 2192
C
C...instantaneous property values at current time-step
      EKIN  =VELSQ/(2.D0*FNATOM*STEPSQ)                                    10 6826
      EPOT  =ENERGY/FNATOM+ETAIL                                           11 4936
      ETOTAL=EKIN+EPOT                                                     12 3418
      TMP   =2.D0*EKIN/3.D0                                                13 5263
      PRES  =DENSTY*(TMP - VIRIAL/(3.D0*FNATOM) - VTAIL/3.D0)              14 1483
C
C...average property values over duration of run
      DENOM =FNATOM*DFLOAT(NSTEP/KSAMPL)                                   15 7445
      EKAVE =SUMVSQ/(2.D0*DENOM*STEPSQ)                                    16 6792
      EPAVE =SUMENR/DENOM+ETAIL                                            17 4868
      TMPAVE=2.D0*EKAVE/3.D0                                               18 5454
      PREAVE=DENSTY*(TMPAVE - SUMVIR/(3.D0*DENOM) - VTAIL/3.D0)            19 2158
C
C...get the order parameter & H-function
      CALL SSORDR(ALAMDA,ALAM1)                                            20 4846
      CALL SSVELO(HINST)                                                   21 3193
      WRITE(6,940) NSTEP,TMP,TMPAVE,EPOT,EPAVE,PRES,PREAVE,ALAM1,HINST,    22 5780
     &             NABORS,ETOTAL                                           23 2822
  940 FORMAT(2X,I6,6F9.4,2(1X,F7.3),1X,I7,1X,F8.5)                         24 2046
      RETURN                                                              25 1090
      END                                                                 26  921
|-----|--|---------|----------|----------|----------|----------|----------|--|------|
```

```
|-----|--|---------|----------|----------|----------|----------|----------|--|------|
      SUBROUTINE SSUPDT                                                      1 2934
C
C     At KSORT intervals, sample for g(r) & update the neighbor list
C
      IMPLICIT REAL*8 (A-H,O-Z)                                             2 5702
      COMMON /INTGRS/ IN(7)                                                 3 3711
      COMMON /NABLST/ LIST(15000)   ,NPOINT(256)                           4 8097
      COMMON /RDF   / NGOFR(300), GR(300)                                  5 6613
      COMMON /REALS / RL(26)                                               6 3496
      COMMON /ZLOOP / XI, YI, ZI, RIJ, I, J, JBEGIN, JEND                  7  167
      EQUIVALENCE (IN(3),NABORS), (RL(12),RDEL), (RL(13),RLIST)            8 2439
C
C...sampling for g(r) explained in notes to routine HSGSAM, Appendix K
      NSHELL = RIJ/RDEL+0.5D0                                              9 5207
      NGOFR(NSHELL) = NGOFR(NSHELL)+1                                     10 6466
C
C...see Figure 5.5 for description of this neighbor list
      IF (RIJ.LE.RLIST) THEN                                             11 4284
         NABORS = NABORS+1                                                12 2721
         LIST(NABORS)=J                                                   13 2721
      ENDIF                                                               14 1292
      RETURN                                                              15 1090
      END                                                                 16  921

|-----|--|---------|----------|----------|----------|----------|----------|--|------|

      SUBROUTINE SSORDR(ALAMDA,ALAM1)                                      1 5937
C
C           Compute Verlet's translational order parameter
C
      IMPLICIT REAL*8(A-H,O-Z)                                            2 5702
      COMMON /INTGRS/ IN(7)                                               3 3711
      COMMON /POSIT / X(256),Y(256),Z(256)                               4 6589
      COMMON /REALS / RL(26)                                              5 3496
      EQUIVALENCE (IN(4),NATOM), (RL(5),DIST)                            6 8323
         PI4D = 4.D0*3.14159265D0/DIST                                    7 7715
         ALAMDA = DCOS(PI4D*X(1)) + DCOS(PI4D*Y(1)) + DCOS(PI4D*Z(1))     8 2024
         ALAM1  = 0.D0                                                    9 2631
         FATOM  = DFLOAT(NATOM)                                          10 3755
C
C...for distinction between ALAMDA & ALAM1, see notes to routine HSORDR, Appendix K
      DO 1000 I = 2,NATOM                                                11 3889
         ALAMDA = ALAMDA + DCOS(PI4D*X(I))                               12 6231
     &                   + DCOS(PI4D*Y(I))                               13 3766
     &                   + DCOS(PI4D*Z(I))                               14 3700
         ALAM1 = ALAM1  + DCOS(PI4D*(X(I)-X(1)))                         15 7030
     &                   + DCOS(PI4D*(Y(I)-Y(1)))                         16 5498
     &                   + DCOS(PI4D*(Z(I)-Z(1)))                         17 5375
 1000    CONTINUE                                                         18 2631
         ALAMDA = ALAMDA/(3.D0*FATOM)                                    19 5982
         ALAM1  = ALAM1/(3.D0*(FATOM-1.))                                20 6578
      RETURN                                                              21 1090
      END                                                                 22  921
|-----|--|---------|----------|----------|----------|----------|----------|--|------|
```

```
|-----|--|---------|----------|---------|---------|---------|---------|--|------|
         SUBROUTINE SSVELO(HINST)                                    1  4284
C
C          Accumulate H-function
C
         IMPLICIT REAL*8(A-H,O-Z)                                    2  5702
         COMMON /INTGRS/ IN(7)                                       3  3711
         COMMON /REALS / RL(26)                                      4  3496
         COMMON /VEL   / X1(256),Y1(256),Z1(256)                     5  6938
         DIMENSION NVINX(300), NVINY(300), NVINZ(300)                6   325
         EQUIVALENCE (IN(4),NATOM), (IN(7),NVDELS), (RL(9),FNATOM)    7  2732
         EQUIVALENCE (RL(14),STEP), (RL(22),VDEL),  (RL(25),VMAX)     8  1719
C
         DO 870 I=1,NVDELS                                           9  3924
           NVINX(I) = 0                                             10  2103
           NVINY(I) = 0                                             11  2439
           NVINZ(I) = 0                                             12  2372
  870    CONTINUE                                                   13  2327
C
C...sample for instantaneous velocity distribution; see routine ODVELO in Appendix I
         VFACT = VDEL*STEP                                          14  2945
         OFFSET = 1.5 + VMAX/VDEL                                   15  3452
         DO 880 I=1,NATOM                                           16  3496
           IF (DABS(X1(I)) .LT. VMAX) THEN                          17  5511
             NSHELL = X1(I)/VFACT + OFFSET                          18  4419
             NVINX(NSHELL) = NVINX(NSHELL)+1                        19  6321
               NSHELL = Y1(I)/VFACT + OFFSET                        20  4745
               NVINY(NSHELL) = NVINY(NSHELL)+1                      21  6983
             NSHELL = Z1(I)/VFACT + OFFSET                          22  4688
             NVINZ(NSHELL) = NVINZ(NSHELL)+1                        23  6860
           ENDIF                                                    24  1292
  880    CONTINUE                                                   25  2305
C
C...compute instantaneous H-function
         HH = 0.D0                                                  26  2406
         DO 890 K=1,NVDELS                                          27  4341
           IF (NVINX(K) .GT. 0) THEN                                28  4778
             FOFV = DFLOAT(NVINX(K))/FNATOM                         29  5296
             HH   = HH + FOFV*DLOG(FOFV)                            30  4554
           ENDIF                                                    31  1292
             IF (NVINY(K) .GT. 0) THEN                              32  5106
               FOFV = DFLOAT(NVINY(K))/FNATOM                       33  5623
               HH = HH + FOFV*DLOG(FOFV)                            34  4554
             ENDIF                                                  35  1292
           IF (NVINZ(K) .GT. 0) THEN                                36  5049
             FOFV = DFLOAT(NVINZ(K))/FNATOM                         37  5566
             HH   = HH + FOFV*DLOG(FOFV)                            38  4554
           ENDIF                                                    39  1292
  890    CONTINUE                                                   40  2507
         HINST = HH*VDEL/3.                                         41  4341
         RETURN                                                     42  1090
         END                                                        43   921
|-----|--|---------|----------|---------|---------|---------|---------|--|------|
```

```
|-----|--|----------|----------|----------|----------|----------|----------|--|------|

      SUBROUTINE SSGOFR                                               1 2653
C
C         Normalize counters for radial distribution function
C
      IMPLICIT REAL*8 (A-H,O-Z)                                       2 5702
      COMMON /INTGRS/ IN(7)                                           3 3711
      COMMON /RDF   / NGOFR(300), GR(300)                             4 6613
      COMMON /REALS / RL(26)                                          5 3496
C
      EQUIVALENCE (IN(2),KSORT), (IN(4),NATOM) , (IN(6),NSTEP)        6 1821
      EQUIVALENCE (RL(3),CUBEH), (RL(4),DENSTY), (RL(12),RDEL)        7 3092
      EQUIVALENCE (RL(21),TMPAVE)                                     8 5049
C
      PI = 3.14159265D0                                              9 3980
C
C...print heading
      WRITE(6,963) NSTEP                                             10 4116
      WRITE(6,965) NATOM,DENSTY,TMPAVE                               11 7502
      WRITE(6,967)                                                   12 3081
C
C...see notes to routine HSGOFR, Appendix K
      NRDELS =CUBEH/RDEL-1                                           13 4835
      ORIGNS =(NSTEP/KSORT)*(NATOM/2)                                14 6006
C
C...loop over radial shells
      DO 800 J=1,NRDELS                                             15 3957
         GR(J) = 0.D0                                                16 2473
         IF (NGOFR(J).GT.0) THEN                                     17 4453
            RADIUS=RDEL*DFLOAT(J)                                    18 4565
            VOLSHL=4.D0*PI*RDEL*RADIUS*RADIUS + (PI*RDEL**3)/3.D0    19 3115
            GR(J) =DFLOAT(NGOFR(J))/(DENSTY*ORIGNS*VOLSHL)           20 9357
            WRITE(6,969) J,RADIUS,NGOFR(J),GR(J)                     21 7939
         ENDIF                                                       22 1292
  800 CONTINUE                                                       23 2383
C
      WRITE(6,971)                                                   24 3137
C
  963 FORMAT('1'////T20,'RADIAL DISTRIBUTION FUNCTION AT TIME-'      25 2495
     &        ,'STEP',I6/)                                           26 3407
  965 FORMAT(T18,'NATOM = ',I4,3X,'DENSITY = ',F6.3,3X,              27 3148
     &        'AVE TEMP = ',F6.3//)                                  28 4453
  967 FORMAT(29X,'I',6X,'R',6X,'NUMERATR',3X,'G(R)'/)                29 3160
  969 FORMAT(28X,I3,3X,F6.3,3X,I6,F8.3)                              30 9133
  971 FORMAT('1'///)                                                 31 3777
C
      RETURN                                                         32 1090
      END                                                            33  921

|-----|--|----------|----------|----------|----------|----------|----------|--|------|
```

SUBROUTINE   SSCALE

SSFLAG ⟹ NATOM, LENER, LSCALE
SSPARM ⟹ AHEAT, STEPSQ, ETAIL
SSINLP ⟹ ENERGY
SSPRTY ⟹ VELSQ

VELSQ = current sum of squares of
          velocities

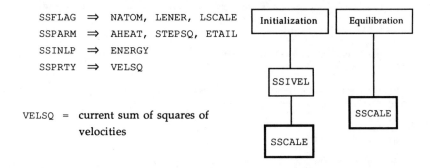

The scaling to the set-point temperature T is according to (5.25):

$$\left(v_i^{new}\right)^2 \; = \; \left(v_i^{old}\right)^2\left[\frac{3\,NkT}{\sum\left(v_i^{old}\right)^2}\right] \; = \; \left(v_i^{old}\right)^2\left[\frac{AHEAT}{SUMSQ}\right]$$

where v_i and T are in units of ε, σ, and m (the unit of mass) = 1. The factor of $(\Delta t)^2$ in AHEAT cancels with the factor of Δt in each v_i.

The scaling to the set-point energy is according to (5.22).

SUBROUTINE   SSX2SC

This routine is only called once during a run. At the start of initialization it is used to scale the inital forces to the program units for acceleration. The scale factor is STPSQH = $(\Delta t)^2/2$.

```
|-----|--|---------|---------|---------|---------|---------|---------|--|------|

      SUBROUTINE SSCALE(DENER)                                    1 4442
C
C         Scale velocities during equilibration for set-point
C
      IMPLICIT REAL*8 (A-H,O-Z)                                   2 5702
      LOGICAL LG(4),LENER,LSCALE                                  3 6073
      COMMON /INTGRS/ IN(7)                                       4 3711
      COMMON /LGCLS / LG                                          5 2990
      COMMON /REALS / RL(26)                                      6 3496
      COMMON /VEL   / X1(256), Y1(256), Z1(256)                   7 6938
C
      EQUIVALENCE (IN(4),NATOM),   (LG(1),LENER),   (LG(2),LSCALE) 8 2473
      EQUIVALENCE (RL(1),AHEAT),   (RL(6),ENERGY),  (RL(8),ETAIL)  9 2192
      EQUIVALENCE (RL(15),STEPSQ),(RL(23),VELSQ)                 10 8323
C
C...scale factor for temperature
      IF (LSCALE) FACTOR = DSQRT(AHEAT/VELSQ)                    11 6028
C
C...scale factor for total energy
      IF (LENER) THEN                                            12 3137
        EPOT = ENERGY/NATOM + ETAIL                              13 4778
        DEK  = DENER - EPOT                                      14 3834
        FACTOR=DSQRT(2.*NATOM*DEK*STEPSQ/VELSQ)                  15 7760
      ENDIF                                                      16 1292
C
C...apply the scaling
      DO 600 I=1,NATOM                                           17 3429
        X1(I)=X1(I)*FACTOR                                       18 3036
        Y1(I)=Y1(I)*FACTOR                                       19 3698
        Z1(I)=Z1(I)*FACTOR                                       20 3575
  600 CONTINUE                                                   21 2237
C
      RETURN                                                     22 1090
      END                                                        23  921

|-----|--|---------|---------|---------|---------|---------|---------|--|------|

      SUBROUTINE SSX2SC(STPSQH,NATOM)                             1 5252
C
C...At time zero, scale initial forces to units of acceleration
C
      IMPLICIT REAL*8 (A-H,O-Z)                                   2 5702
      COMMON /ACCEL / X2(256)    ,Y2(256)    ,Z2(256)            3 7221
      COMMON /FORCE / FX(256)    ,FY(256)    ,FZ(256)            4 6938
        DO 350 I=1,NATOM                                         5 3441
          X2(I)=FX(I)*STPSQH                                     6 3429
          Y2(I)=FY(I)*STPSQH                                     7 4082
          Z2(I)=FZ(I)*STPSQH                                     8 3968
  350   CONTINUE                                                 9 2259
      RETURN                                                    10 1090
      END                                                       11  921

|-----|--|---------|---------|---------|---------|---------|---------|--|------|
```

```
|-----|--|---------|----------|----------|---------|---------|---------|--|------|
       SUBROUTINE SSPRNT                                                  1 2855
C
C          Print initial values of run parameters
C
       IMPLICIT REAL*8 (A-H,O-Z)                                          2 5702
       LOGICAL LG(4), LSHIFT                                             3 4600
       COMMON /INTGRS/ IN(7)                                             4 3711
       COMMON /LGCLS / LG                                               5 2990
       COMMON /REALS / RL(26)                                           6 3496
       EQUIVALENCE (IN(4),NATOM) , (RL(2),CUBE) , (RL(3),CUBEH)          7 2024
       EQUIVALENCE (RL(4),DENSTY), (RL(11),RCUT), (RL(13),RLIST)         8 2327
       EQUIVALENCE (RL(14),STEP) , (RL(20),TEMP), (LG(3),LSHIFT)         9 2754
C
       WRITE(6,900)                                                     10 3373
       WRITE(6,902)                                                     11 3250
       WRITE(6,904)                                                     12 3485
       WRITE(6,906) NATOM                                               13 4093
       WRITE(6,904)                                                     14 3485
         IF (LSHIFT)        WRITE(6,907)                                15 5465
         IF (.NOT.LSHIFT) WRITE(6,909)                                  16 6635
       WRITE(6,904)                                                     17 3485
       WRITE(6,902)                                                     18 3250
       WRITE(6,904)                                                     19 3485
         WRITE(6,908) DENSTY,TEMP                                       20 6309
         WRITE(6,910) STEP                                             21 3924
         WRITE(6,912) CUBE,CUBEH                                       22 5555
         WRITE(6,914) RCUT,RLIST                                       23 5207
       WRITE(6,904)                                                     24 3485
       WRITE(6,902)                                                     25 3250
       WRITE(6,916)                                                     26 2967
       WRITE(6,918)                                                     27 3115
C
C..check that order parameter is unity
       CALL SSORDR(ALAMDA,ALAM1)                                        28 4846
       WRITE(6,921) ALAMDA                                             29 4363
C
   900 FORMAT(1H1///)                                                   30 3700
   902 FORMAT(T20,49('*'))                                              31 6286
   904 FORMAT(T20,'*',T68,'*')                                          32 8019
   906 FORMAT(T20,'*',6X,'MOLECULAR DYNAMICS FOR',I4,' LJ ATOMS',T68,'*') 33 7052
   907 FORMAT(T20,'*',3X,'FIXED DENSITY W/SHIFT-FORCE POTENTIAL',T68,'*') 34 7276
   908 FORMAT(T20,'*',2X,'DENSITY = ',F7.3,T45,'TEMP = ',F7.3,T68,'*')  35 8884
   909 FORMAT(T20,'*',5X,'FIXED DENSITY W/TRUNCATED POTENTIAL',T68,'*')  36 6905
   910 FORMAT(T20,'*',2X,'TIME STEP= ',F6.3,T68,'*')                    37 3137
   912 FORMAT(T20,'*',2X,'CUBE = ',F7.3,T45,'CUBE/2= ',F6.3,T68,'*')    38 7759
   914 FORMAT(T20,'*',2X,'RCUT = ',F7.3,T45,'RLIST = ',F6.3,T68,'*')    39 7401
   916 FORMAT(////)                                                     40 2721
   918 FORMAT(3X,'TSTEP',4X,'TEMP',4X,'<TEMP>',2X,'P ENRGY',2X,         41 5915
      &   '<P ENR>',4X,'PRES',4X,'<PRES>',3X,'ORDER',2X,'H-FUNC',3X,    42 5612
      &   'NABORS',1X,'TTL ENR')                                        43 5746
   921 FORMAT(7X,'0',T65, F6.3)                                         44 6512
C
       RETURN                                                          45 1090
       END                                                             46  921
|-----|--|---------|----------|----------|---------|---------|---------|--|------|
```

```
|-----|--|---------|---------|---------|---------|---------|---------|--|------|

      SUBROUTINE SSRSET                                              1 2721
C
C          At end of equilibration, reinitialize property sums
C
      IMPLICIT REAL*8 (A-H,O-Z)                                      2 5702
      COMMON /RDF   / NGOFR(300), GR(300)                           3 6613
      COMMON /REALS / RL(26)                                        4 3496
      EQUIVALENCE (RL(3),CUBEH)   , (RL(12),RDEL)                   5 8839
      EQUIVALENCE (RL(17),SUMENR), (RL(18),SUMVIR), (RL(19),SUMVSQ) 6 2282
C
      SUMENR=0.D0                                                   7 2811
      SUMVIR=0.D0                                                   8 2507
      SUMVSQ=0.D0                                                   9 2305
C
      NRDELS=CUBEH/RDEL                                            10 4262
      DO 650 I=1,NRDELS                                           11 3788
         NGOFR(I)=0                                               12 2170
  650 CONTINUE                                                    13 2068
C
      WRITE(6,999)                                                14 3610
  999 FORMAT(/10X,'**** EQUILIBRATION COMPLETE ***'//)            15 2260
C
      RETURN                                                      16 1090
      END                                                         17  921

|-----|--|---------|---------|---------|---------|---------|---------|--|------|

      BLOCK DATA                                                   1 2079
C
      IMPLICIT REAL*8 (A-H,O-Z)                                    2 5702
      COMMON /DERIV3/ X3(256)      ,Y3(256)      ,Z3(256)          3 8053
      COMMON /DERIV4/ X4(256)      ,Y4(256)      ,Z4(256)          4 8659
      COMMON /DERIV5/ X5(256)      ,Y5(256)      ,Z5(256)          5 7535
      COMMON /FORCE / FX(256)      ,FY(256)      ,FZ(256)          6 6938
      COMMON /NABLST/ LIST(15000)  ,NPOINT(256)                    7 8097
      COMMON /RDF   / NGOFR(300), GR(300)                          8 6613
      COMMON /REALS / RL(26)                                       9 3496
C
      EQUIVALENCE (RL(17),SUMENR), (RL(18),SUMVIR), (RL(19),SUMVSQ) 10 2282
C
      DATA X3/256*0.D0/,Y3/256*0.D0/,Z3/256*0.D0/                 11 1012
      DATA X4/256*0.D0/,Y4/256*0.D0/,Z4/256*0.D0/                 12 1461
      DATA X5/256*0.D0/,Y5/256*0.D0/,Z5/256*0.D0/                 13  628
      DATA FX/256*0.D0/,FY/256*0.D0/,FZ/256*0.D0/                 14  561
      DATA NGOFR/300*0/                                           15 3711
      DATA LIST/15000*0/   ,NPOINT/256*0/                         16 7377
      DATA SUMENR,SUMVIR,SUMVSQ/3*0.D0/                           17 6747
      END                                                         18  921

|-----|--|---------|---------|---------|---------|---------|---------|--|------|
```

APPENDIX M

A FEW CONSTANTS AND MISCELLANEA

CONVERSION FACTORS

$1 \text{ cm} = 10^8 \text{ Å}$

$1 \text{ N} = 1 \text{ kg m/sec}^2 = 10^5 \text{ g cm/sec}^2 = 10^5 \text{ dyn}$

$1 \text{ J} = 1 \text{ N m} = 10^7 \text{ dyn cm} = 10^7 \text{ ergs}$

$1 \text{ bar} = 0.9869 \text{ atm} = 10^5 \text{ Pa} = 10^5 \text{ N/m}^2 = 10^5 \text{ J/m}^3$

$1 \text{ N sec/m}^2 = 10 \text{ g/(cm sec)} = 10^3 \text{ cP}$

CONSTANTS

Earth's gravitational acceleration $g = 9.8 \text{ m/sec}^2$

Avogadro's number $N_A = 6.022 \times 10^{23}$ molecules/mol

Boltzmann's constant $k = 1.381 \times 10^{-16}$ erg/(molecule K)

Gas constant $R = kN_A = 8.314 \text{ J/(mol K)} = 83.14 \text{ bars cm}^3/\text{(mol K)}$

Planck's constant $h = 6.626 \times 10^{-27}$ erg sec

LENNARD-JONES (12, 6) POTENTIAL PARAMETERS FOR ARGON[†]

$$\sigma = 3.405 \text{ Å} \qquad \varepsilon/k = 120 \text{ K}$$

[†]For a critical discussion of the argon pair potential, see ref 1.

SELECTED PROPERTIES OF SELECTED FLUIDS

	Air	Saturated Liquid Argon	Saturated Liquid Water
Molecular weight	29	40	18
Temperature, K	300	84	293
Pressure[a], bar	1	0.7	0.023
Density[a], g/cm^3	0.0012	1.4	1
C_p, J/mol K	29[a]($\approx 7R/2$)	42[b]	75.3[b]
C_v, J/mol K	21[c]	19[b]	74.8[b]
Adiabatic compressibility, $(10^3 \kappa_s)$, bar^{-1}	720[c]	0.095[b]	0.046[b]
Thermal pressure coefficient, γ_v, bar/K	0.0033[c]	22[b]	4.5[b]
Sonic velocity, m/sec	350[c]	860[b]	1480[b]
Viscosity[a], $(10^3 \eta)$, N sec/m^2	0.018	0.28	1
Self-diffusion coefficient, $(10^5 D)$, cm^2/sec	[21,000][d]	1.8[e]	2.5[f]

[a]From ref. 2.
[b]From ref. 3.
[c]Assuming air is an ideal gas at 300 K, 1 bar.
[d]This value is for pure nitrogen gas at 298 K, from ref. 4.
[e]At 84 K, 13.1 bars, from ref. 5.
[f]At 298 K, 1.01 bar, from ref. 6.

STIRLING'S APPROXIMATION TO $N!$

$$N! = \prod_{k=1}^{N} k$$

$$\ln N! = \sum_{k=1}^{N} \ln k$$

For large N, write the sum as an integral,

$$\ln N! \approx \int_1^N \ln k \, dk = (k \ln k - k)|_1^N = N \ln N - N + 1$$

Then, for $N \gg 1$,

$$\boxed{\ln N! \approx N \ln N - N} \qquad \text{QED}$$

LAW OF LARGE NUMBERS

For a variable x whose values obey some probability distribution, the mean value of x obtained from M observations is given by (2.74),

$$\langle x \rangle = \frac{1}{M} \sum_{k=1}^{M} x_k$$

The expected value of the mean is given by (2.78),

$$E[\langle x \rangle] = \frac{1}{M_s} \sum_{k=1}^{M_s} \langle x \rangle_k$$

Now, provided the observed values x_k are mutually independent and provided there is a bound on the variance [given by (2.79)] of each average $\langle x \rangle_k$ wrt the expected value, the law of large numbers states that as the number of observations M is increased, the average $\langle x \rangle$ approaches the expected value as $1/\sqrt{M}$; that is,

$$\lim_{M \to \text{large}} |\langle x \rangle - E[\langle x \rangle]| \sim \frac{1}{\sqrt{M}}$$

UNITS OF VOLUME IN ENGLISH SYSTEM (at least during the reign of Elizabeth I) [7]

1 tun = 2 butts	= 4 hogsheads	= 8 barrels
= 16 casks	= 32 bushels	= 64 half-bushels
= 128 pecks	= 256 gallons	= 512 pottles
= 1024 quarts	= 4096 pints	= 8192 jills[†]
= 16,384 jacks	= 32,768 jiggers	

[†]The jack and the jill are immortalized in a nursery rhyme.

REFERENCES

[1] G. C. Maitland, M. Rigby, E. B. Smith, and W. A. Wakeham, *Intermolecular Forces*, Oxford University Press, Oxford, 1981, p. 493f.

[2] N. B. Vargaftik, *Tables on the Thermophysical Properties of Liquids and Gases*, 2nd ed, Halsted Press (Wiley), New York, 1975.

[3] J. S. Rowlinson and F. L. Swinton, *Liquids and Liquid Mixtures*, 3rd ed, Butterworths, London, 1982.

[4] E. B. Winn, "The Temperature Dependence of the Self-Diffusion Coefficients of Argon, Neon, Nitrogen, Oxygen, Carbon Dioxide, and Methane," *Phys. Rev.*, **80**, 1024–1027 (1950).

[5] J. Naghizadeh and S. A. Rice, "Kinetic Theory of Dense Fluids. X. Measurement and Interpretation of Self-Diffusion in Liquid Ar, Kr, Xe, and CH_4," *J. Chem. Phys.*, **36**, 2710–2720 (1962).

[6] N. J. Trappeniers, C. J. Gerritsma, and P. H. Oosting, "The Self-Diffusion Coefficient of Water, at 25°C, by Means of Spin-Echo Technique," *Phys. Lett.*, **18**, 256–257 (1965).

[7] H. A. Klein, *The World of Measurements*, Simon and Schuster, New York, 1974, p. 32f.

NOTATION

Roman Uppercase

$A(t)$	Generic dynamic quantity
$\mathbf{A}$	Stability matrix (4.20)
B	Second virial coefficient; Exercise 3.16
$B(t)$	Generic dynamic quantity
$C(t)$	Generic time correlation function (7.1)
C_p	Isobaric heat capacity (6.50)
C_v	Isometric heat capacity (6.27)
C_v^R	Residual isometric heat capacity (6.35)
D	Self-diffusion coefficient (7.45)
E	Total energy (2.10)
E_k	Kinetic energy (2.44)
$\mathbf{F}_i$	Force on atom i (1.4)
F_s	Shifted force (5.8)
$G(\mathbf{r}, t)$	Density–density time correlation function; aka the space–time correlation function (7.89)
G_d	Distinct part of $G(r, t)$ (7.103)
G_s	Self part of $G(r, t)$ (7.91)
H	Kinetic part of Boltzmann's H-function (2.57)
$\mathscr{H}$	Hamiltonian (2.10)
$\mathbf{I}$	Identity matrix
$\mathbf{J}$	Stress tensor (7.36)
L	Length of one edge of the primary cube; Figure 2.13
M	In a hard-sphere (soft-sphere) simulation, the total number of collisions (time-steps) computed
M_∞	Total number of points on a phase-space trajectory; Table 2.1

M_s	Number of samples, each containing M-values, that can be formed from a trajectory of total length M_∞; Table 2.1
N	Total number of atoms
N_A	Avogadro's number
P	Pressure; Appendix B
P_{LR}	Long-range correction to pressure (6.14)
P_{res}	Residual pressure (6.18)
$\mathcal{P}$	Total linear momentum (2.121)
$\mathcal{P}$	Probability (2.61)
Q	NVT partition function (6.54)
$\mathcal{Q}$	Total angular momentum (2.123)
$\mathbf{R}$	In periodic systems, a vector that locates an origin in an image cell; the vector is measured relative to an origin in the primary cell (2.99)
S	Entropy (C.5)
S_{res}	Residual entropy (6.77) or (6.79)
S^2	Trajectory variance (2.76)
T	Absolute temperature (2.43)
U_c	Average configurational internal energy (6.4)
$\mathcal{U}$	Instantaneous total potential energy (1.3)
U_{LR}	Long-range correction to U_c (6.8)
U_t	Test particle potential energy (6.68)
$U_{t,LR}$	Long-range correction to U_t (6.71)
V	System volume
$\langle X \rangle$	Trajectory average of a generic dynamic quantity $x(t)$ (2.75)
Z	Compressibility factor (3.47)
$\mathcal{Z}$	NVT configurational integral (1.6)

Roman Lowercase

a	Length of one edge of the fcc unit cell (3.37)
c	Generic constant
d	A displacement or deviation
e	Errors; especially those in computed phase-space trajectories (4.16)
$f(v)$	Velocity distribution function (2.51)
$g(r)$	Radial distribution function (6.93)
h	Planck's constant (Appendix M)
$\hat{\mathbf{i}}, \hat{\mathbf{j}}, \hat{\mathbf{k}}$	Unit vectors along three, mutually orthogonal, Cartesian axes
k	Boltzmann's constant (Appendix M)
l	Mean free path; Exercise 2.17
m	Atomic mass
m	When sampling, either by coarse graining or by stratified sampling, the M computed points are divided into m segments, each segment composed of n points; Figure 3.12

n	Number of points forming one segment, in either coarse-grain or stratified sampling; Figure 3.12
n_b	Number of bins used in computing chi squared (3.62)
$\mathbf{p}_i$	Momentum of atom i
r	Radial distance
$\mathbf{r}_i$	Position of atom i relative to a space-fixed origin
r_c	Distance at which a pair potential is truncated
r_{tc}	Distance at which a test particle potential is truncated
s	Laplace transform variable; Appendix C
t	Time
t_0	An initial time or time origin
t_c	Collision time
$u(r)$	Pair potential energy function
u_s	Shifted-force potential energy function (5.12)
v	Velocity; for hard spheres, a precollision velocity
$v\!\!\!\!\cdot$	Postcollision velocity
w	Velocity of sound (2.116)
$x(t)$	Generic dynamic quantity (Section 2.8)
x_i, y_i, z_i	Cartesian components of $\mathbf{r}_i$, the position vector for atom i

Greek Uppercase

Γ	Gamma function (C.13)
Δ	*NPT* partition function (6.55)
Λ	Thermal de Broglie wavelength (C.51)
Ψ	Velocity autocorrelation function (7.23)
Ω	*NVE* partition function (C.1)

Greek Lowercase

α	In periodic systems, the cell translation vector; Figure 2.13
α_i	Parameters in Gear predictor–corrector algorithm (4.51–4.56)
β	$1/kT$
γ	Spring constant in the ODHO (1.2)
γ_v	Thermal pressure coefficient (6.44)
δ	Delta symbol; Appendix A
ε	Energy at minimum of Lennard-Jones pair potential (5.4)
ζ	Bulk viscosity
η	Packing fraction (3.34)
η	Shear viscosity (7.69)
κ_s	Adiabatic compressibility (6.39)
κ_T	Isothermal compressibility (6.49)
λ_x	Order parameter (3.42)
μ	Chemical potential (C.36)
μ_{res}	Residual chemical potential (C.54)

ξ	Random number selected from uniform distribution on $[-1, 1]$
ρ	Number density N/V
σ	For hard spheres, the diameter of one sphere; for soft spheres, the distance to the zero in the pair potential (5.4)
σ^2	Variance, aka the second moment, of a Gaussian distribution (2.77)
$\sigma_{\langle x \rangle}$	Standard deviation of a sample average $\langle x \rangle$ with respect to the trajectory average $\langle X \rangle$ (2.96)
τ	Dummy variable in integrations over time
τ	Relaxation time, such as in (7.34)
τ_{pbc}	In periodic systems, the time needed for spatial translations to transverse the primary cell (2.115)
χ^2	Chi squared; a measure of how a computed distribution deviates from a known distribution (3.62)
ω	Frequency, such as in periodic motion of the ODHO
ω	NVE phase-space density of states (C.3)

Abbreviations

aka	Also known as
bcc	Body-centered cubic
fcc	Face-centered cubic
ge	Global error
gre	Global round-off error
gte	Global truncation error
hcp	Hexagonal close packed
lhs	Left-hand side
MC	Monte Carlo
MD	Molecular dynamics
msd	Mean-square displacement
mse	Mean-square error
ODHO	One-dimensional harmonic oscillator
pbc	Periodic boundary conditions
QED	*Quod erat demonstrandum*
rhs	Right-hand side
rms	Root-mean-square
te	Truncation error
vacf	Velocity autocorrelation function
wrt	With respect to

Operators

(To distinguish the operators E, L, and V from the properties E, L, and V, the arguments of the operators E, L, and V are always placed in square brackets.)

E Expected value of a sampling distribution (2.78)

L Laplace transform:

$$L[F] = f(s) = \int_0^\infty e^{-st} F(t)\, dt$$

L^{-1} Inverse Laplace transform: $L^{-1}[f(s)] = F(t)$

V Variance (2.76)

Δ Change or difference: $\Delta x = x_{final} - x_{initial}$

δ Fluctuation or displacement of an instantaneous value about the average: $\delta x = x - \langle x \rangle$. Otherwise, the delta symbol (Appendix A).

θ Unit step function:

$$\theta(x) = \begin{cases} 0 & x < 0 \\ 1 & x > 0 \end{cases}$$

∇ Gradient; it operates on a scalar to produce a vector. In Cartesian coordinates

$$\nabla[\ ] = \frac{\partial[\ ]}{\partial \mathbf{r}} = \hat{\mathbf{i}}\frac{\partial[\ ]}{\partial x} + \hat{\mathbf{j}}\frac{\partial[\ ]}{\partial y} + \hat{\mathbf{k}}\frac{\partial[\ ]}{\partial z}$$

∇^2 Laplacian; as used in this book, the Laplacian is the divergence of the gradient of a scalar. It produces a scalar

$$\nabla^2[\ ] = \nabla \cdot \nabla[\ ]$$

$$= \hat{\mathbf{i}}\frac{\partial}{\partial x} \cdot \nabla[\ ] + \hat{\mathbf{j}}\frac{\partial}{\partial y} \cdot \nabla[\ ] + \hat{\mathbf{k}}\frac{\partial}{\partial z} \cdot \nabla[\ ]$$

$$\nabla^2[\ ] = \frac{\partial^2[\ ]}{\partial x^2} + \frac{\partial^2[\ ]}{\partial y^2} + \frac{\partial^2[\ ]}{\partial z^2}$$

$\dot{x}$ Total time derivative, dx/dt

$\ddot{x}$ Second time derivative, d^2x/dt^2

$x^{(iii)}$ (iii)th time derivative, $d^{(iii)}x/dt^{(iii)}$

$\langle x \rangle$ Time average of dynamic quantity $x(t)$ (1.8)

BIBLIOGRAPHY[†]

The following citations should give you a feel for the diverse problems to which molecular simulation can be applied and, perhaps, provide you with an entrée to the literature that concerns your particular application. Part I cites books devoted to molecular simulation and closely related topics. Part II cites articles, primarily reviews. No attempt has been made for completeness; rather, papers were selected to demonstrate the breadth of applications. Part III provides a topical index to Parts I and II.

I. BOOKS

1. Allen, M. P. and D. J. Tildesley, *Computer Simulation of Liquids*, Oxford University Press, Oxford, 1987.

2. Angell, C. A. and M. Goldstein, Eds., *Dynamic Aspects of Structural Change in Liquids and Glasses*, Vol. 484 of *Ann. NY Acad. Sci.*, New York Academy of Sciences, New York, 1986.

3. Beeler, J. R., Jr., *Radiation Effects: Computer Experiments*, North-Holland, Amsterdam, 1983.

4. Beveridge, D. L. and W. L. Jorgensen, Eds., *Computer Simulation of Chemical and Biomolecular Systems*, Vol. 482 of *Ann. NY Acad. Sci.*, New York Academy of Sciences, New York, 1986.

5. Binder, K. and D. W. Heermann, *Monte Carlo Simulation in Statistical Physics: An Introduction*, Springer-Verlag, Berlin, 1988.

[†]W. Thomson, Lecture IX, p. 87: "I got another quarter hundred weight of books on the subject last night. I have not read them all through."

6. Binder, K., Ed., *Monte Carlo Methods in Statistical Physics*, 2nd ed., Springer-Verlag, Berlin, 1986.

7. Binder, K., Ed., *Applications of the Monte Carlo Method in Statistical Physics*, Springer-Verlag, Berlin, 1984.

8. Brooks, C. L. III, M. Karplus, and B. M. Pettitt, *Proteins: A Theoretical Perspective of Dynamics, Structure, and Thermodynamics*, Vol. 71 of *Adv. Chem. Phys.*, Wiley, New York, 1988.

9. Broughton, J., W. Krakow, and S. T. Pantelides, Eds., *Computer-based Microscopic Description of the Structure and Properties of Materials*, Materials Research Society, Pittsburgh, PA, 1986.

10. *Brownian Motion, Faraday Discussions*, **83**, Chemical Society, London, 1987.

11. Catlow, C. R. A. and W. C. Mackrodt, *Computer Simulation of Solids*, Springer-Verlag, Berlin, 1982.

12. Catlow, C. R. A., S. C. Parker, and M. P. Allen, Eds., *Computer Modelling of Fluids, Polymers, and Solids*, Kluwer Academic, Norwell, MA, 1989.

13. Ciccotti, G. and W. G. Hoover, Eds., *Molecular-Dynamics Simulation of Statistical–Mechanical Systems*, North-Holland, Amsterdam, 1986.

14. Ciccotti, G., D. Frenkel, and I. R. McDonald, Eds., *Simulation of Liquids and Solids*, North-Holland, Amsterdam, 1987.

15. Clementi, E. and R. H. Sarma, Eds., *Structure and Dynamics: Nucleic Acids and Proteins*, Adenine, Guilderland, New York, 1983.

16. Evans, D. J. and G. Morriss, *Statistical Mechanics of Nonequilibrium Liquids*, Academic, New York, 1990.

17. Hockney, R. W. and J. W. Eastwood, *Computer Simulation Using Particles*, McGraw-Hill, New York, 1981.

18. Hoover, W. G., *Molecular Dynamics*, Springer-Verlag, Berlin, 1986.

19. Hoover, W. G., *Computational Statistical Mechanics*, Elsevier, Amsterdam, 1991.

20. Kalos, M. H., Ed., *Monte Carlo Methods in Quantum Problems*, NATO ASI Series C, Vol. 125, Reidel, Dordrecht, Holland, 1984.

21. Kalos, M. H. and P. A. Whitlock, *Monte Carlo Methods*, Vol. 1, *Basics*, Wiley-Interscience, New York, 1986.

22. Kocak, H., *Differential and Difference Equations through Computer Experiments*, 2nd ed., Springer-Verlag, New York, 1990.

23. Lykos, P., Ed., *Computer Modeling of Matter*, ACS Symposium Series, Vol. 86, American Chemical Society, Washington, DC, 1978.

24. McCammon, J. A. and S. C. Harvey, *Dynamics of Proteins and Nucleic Acids*, Cambridge University Press, Cambridge, 1987.

25. Mouritsen, O. G., *Computer Studies of Phase Transitions and Critical Phenomena*, Springer-Verlag, Berlin, 1984.

26. Vesely, F., *Computerexperimente an Flüssigkeitsmodellen*, Physik-Verlag, Weinheim, 1978.

27. Whitney, C. A., *Random Processes in Physical Systems: An Introduction to Probability-based Computer Simulations*, Wiley, New York, 1990.

II. ARTICLES

28. Adams, D. and G. Hills, "The Computer Simulation of Ionic Liquids" in *Ionic Liquids*, D. Inman and D. G. Lovering, Eds., Plenum, New York, 1981, pp. 27–55.

29. Alder, B. J. and W. G. Hoover, "Numerical Statistical Mechanics" in *Physics of Simple Liquids*, H. N. V. Temperley, J. S. Rowlinson, and G. S. Rushbrooke, Eds., North-Holland, Amsterdam, 1968, Chapter 4.

30. Allen, M. P., D. Frenkel, and J. Talbot, "Molecular Dynamics Simulation Using Hard Particles," *Comp. Phys. Repts.*, **9**, 301–353 (1989).

31. Angell, C. A., J. H. R. Clarke, and L. V. Woodcock, "Interaction Potentials and Glass Formation: A Survey of Computer Experiments," *Adv. Chem. Phys.*, **48**, 397–453 (1981).

32. Barker, J. A. and D. Henderson, "What is 'Liquid'? Understanding the States of Matter," *Rev. Mod. Phys.*, **48**, 587–671 (1976).

33. Barnes, P., "Machine Simulation of Water" in *Progress in Liquid Physics*, C. A. Croxton, Ed., Wiley, New York, 1978, Chapter 9.

34. Baumgärtner, A., "Simulation of Polymer Motion," *Ann. Rev. Phys. Chem.* **35**, 419–435 (1984).

35. Berne, B. J. and G. D. Harp, "On the Calculation of Time Correlation Functions," *Adv. Chem. Phys.*, **17**, 63–227 (1970).

36. Berendsen, H. J. C. and W. F. van Gunsteren, "Molecular Dynamics Simulations: Techniques and Approaches" in *Molecular Liquids: Dynamics and Interactions*, A. J. Barnes, W. J. Orville-Thomas, and J. Yarwood, Eds., Reidel, Boston, 1984, pp. 475–500.

37. Berendsen, H. J. C., "Dynamic Simulation as an Essential Tool in Molecular Modeling," *J. Comput.-aided Mol. Des.*, **2**, 217–221 (1988).

38. Berne, B. J. and D. Thirumalai, "On the Simulation of Quantum Systems: Path Integral Methods," *Ann. Rev. Phys. Chem.*, **37**, 401–424 (1986).

39. Beveridge, D. L., M. Mezei, P. K. Mehrotra, F. T. Marchese, G. Ravi-Shanker, T. Vasu, and S. Swaminathan, "Monte Carlo Computer Simulation Studies of the Equilibrium Properties and Structure of Liquid Water," *ACS Adv. Chem. Ser.*, **204**, 297–352 (1983).

40. Beveridge, D. L. and F. M. DiCapua, "Free-Energy via Molecular Simulation: Applications to Chemical and Biomolecular Systems," *Ann. Rev. Biophys. Biophys. Chem.*, **18**, 431–492 (1989).

41. Binder, K., "Monte Carlo Computer Experiments on Critical Phenomena and Metastable States," *Adv. Phys.*, **23**, 917–939 (1974).

42. Binder, K. and D. P. Landau, "Monte Carlo Calculations on Phase Transitions in Adsorbed Layers," *Adv. Chem. Phys.*, **76**, 91–152 (1989).

43. Binder, K. "Computer Simulation of Macromolecular Materials," *Colloid Polym. Sci.*, **266**, 871–885 (1988).

44. Bopp, P., "Molecular Dynamics Computer Simulations of Solvation in Hydrogen Bonded Systems," *Pure Appl. Chem.*, **59**, 1071–1082 (1987).

45. Bopp, P., "Molecular Dynamics Simulations of Aqueous Ionic Solutions," *NATO ASI Ser. C*, **205**, 217–243 (1987).

46. Brady, J. F. and G. Bossis, "Stokesian Dynamics," *Ann. Rev. Fluid Mech.*, **20**, 111–157 (1988).

47. Brenner, D. W. and B. J. Garrison, "Gas-Surface Reactions: Molecular Dynamics Simulations of Real Systems," *Adv. Chem. Phys.*, **76**, 281–334 (1989).

48. Brooks, B. R., "Molecular Dynamics for Problems in Structural Biology," *Chem. Scr.*, **29A**, 165–169 (1989).

49. Ceperley, D., "Quantum Monte Carlo Simulations of Systems at High Pressure," *NATO ASI Ser. B*, **186**, 477–489 (1989).

50. Clarke, J. H. R., "Computer Simulation of Liquids and Related Systems," *Ann. Rep. Prog. Chem., Sect. C.*, **84**, 273–301 (1987).

51. Clementi, E., S. Chin, G. Corongiu, J. H. Detrich, M. Dupuis, D. Folsom, G. C. Lie, D. Logan, and V. Sonnad, "Supercomputing and Supercomputers for Science and Engineering in General and for Chemistry and Biosciences in Particular," *Int. J. Quant. Chem.*, **35**, 3–89 (1989).

52. Copley, J. R. D. and S. W. Lovesey, "The Dynamic Properties of Monatomic Liquids," *Rep. Prog. Phys.*, **38**, 461–563 (1975).

53. Croxton, C. A., *Statistical Mechanics of the Liquid Surface*, Wiley, New York, 1980, Chapter 10.

54. de Leeuw, S. W., J. W. Perram, and E. R. Smith, "Computer Simulation of the Static Dielectric Constant of Systems with Permanent Electric Dipoles," *Ann. Rev. Phys. Chem.*, **37**, 245–270 (1986).

55. Doan, N. V., "Molecular Dynamics and Defects in Metals in Relation to Interatomic Force Laws," *Philos. Mag. A*, **58**, 179–192 (1988).

56. Dodson, B. W., "Molecular Dynamics Modeling of Vapor-Phase and Very-Low-Energy Ion-Beam Crystal Growth Processes," *Crit. Rev. Solid St. Mat. Sci.*, **16**, 115–130 (1990).

57. Doll, J. D. and A. F. Voter, "Recent Developments in the Theory of Surface Diffusion," *Ann. Rev. Phys. Chem.*, **38**, 413–431 (1987).

58. Dove, M., "Molecular Dynamics Simulations in the Solid State Sciences," *NATO ASI Ser. C*, **225**, 501–590 (1988).

59. Erpenbeck, J. J. and W. W. Wood, "Molecular Dynamics Techniques for Hard-Core Systems" in *Statistical Mechanics*, Part B, B. J. Berne, Ed., Plenum, New York, 1977, Chapter 1.

60. Erpenbeck, J. J., "Non-equilibrium Molecular Dynamics Calculations of the Shear Viscosity of Hard Spheres" in *Nonlinear Fluid Behavior*, H. J. M. Hanley, Ed., North-Holland, Amsterdam, 1983 (reprint of *Physica*, **118A**), pp. 144–156.

61. Evans, D. J., "Molecular Dynamics Simulations of the Rheological Properties of Simple Fluids" in *Nonlinear Fluid Behavior*, H. J. M. Hanley, Ed., North-Holland, Amsterdam, 1983 (reprint of *Physica*, **118A**), pp. 51–68.

62. Evans, D. J., H. J. M. Hanley, and S. Hess, "Non-Newtonian Phenomena in Simple Fluids," *Phys. Today*, **37**(1), 26–33 (1984).

63. Evans, D. J. and G. P. Morriss, "NonNewtonian Molecular Dynamics," *Comp. Phys. Repts.*, **1**, 297–343 (1984).

64. Evans, D. J. and W. G. Hoover, "Flows Far from Equilibrium via Molecular Dynamics," *Ann. Rev. Fluid Mech.*, **18**, 243–264 (1986).

65. Evans, G. T., "Liquid State Dynamics of Alkane Chains," *ACS Adv. Chem. Ser.*, **204**, 423–444 (1983).

66. Fincham, D., "Parallel Computers and Molecular Simulation," *Molec. Simul.*, **1**, 1–46 (1987).

67. Fisher, I. Z., "Applications of the Monte Carlo Method in Statistical Physics," *Soviet Phys. Uspekhi*, **2**, 783–796 (1960). [Translation of *Usp. Fiz. Nauk*, **69**, 349–369 (1959).]

68. Frenkel, D., "Intermolecular Spectroscopy and Computer Simulations" in *Intermolecular Spectroscopy and Dynamical Properties of Dense Systems*, J. Van Kranendonk, Ed., North-Holland, Amsterdam, 1980, pp. 156–201.

69. Frenkel, D. and J. P. McTague, "Computer Simulations of Freezing and Supercooled Liquids," *Ann. Rev. Phys. Chem.*, **31**, 491–521 (1980).

70. Gubbins, K. E., K. S. Shing, and W. B. Streett, "Fluid Phase Equilibria: Experiment, Computer Simulation, and Theory," *J. Phys. Chem.*, **87**, 4573–4585 (1983).

71. Gubbins, K. E., "The Role of Computer Simulation in Studying Fluid Phase Equilibria," *Molec. Simul.*, **2**, 223–252 (1989).

72. Halicioglu, T. and C. W. Bauschlicher, Jr., "Physics of Microclusters," *Rep. Prog. Phys.*, **51**, 883–921 (1988).

73. Hanley, H. J. M. and D. J. Evans, "Non-Newtonian Molecular Dynamics and Thermophysical Properties," *Int. J. Thermophys.*, **11**, 381–398 (1990).

74. Hansen, J. P., "Recent Developments in the Theory of Ionic Melts," *Nuovo Cimen. D*, **12**, 703–717 (1990).

75. Harris, A. L., J. K. Brown, and C. B. Harris, "The Nature of Simple Photodissociation Reactions in Liquids on Ultrafast Time Scales," *Ann. Rev. Phys. Chem.*, **39**, 341–366 (1988).

76. Harris, C. B., D. E. Smith, and D. J. Russell, "Vibrational Relaxation of Diatomic Molecules in Liquids," *Chem. Rev.*, **90**, 481–488 (1990).

77. Heyes, D. M., "Molecular, Brownian and Diffusive Dynamics: Applications to Viscous Flow," *Comp. Phys. Repts.*, **8**, 71–108 (1988).

78. Hoheisel, C. and R. Vogelsang, "Thermal Transport Coefficients for One- and Two-Component Liquids from Time Correlation Functions Computed by Molecular Dynamics," *Comp. Phys. Repts.*, **8**, 1–69 (1988).

79. Hoover, W. G. and W. T. Ashurst, "Nonequilibrium Molecular Dynamics" in *Theoretical Chemistry*, Vol. 1, H. Eyring and D. Henderson, Eds., Academic, New York, 1975, pp. 1–51.

80. Hoover, W. G., "Atomistic Nonequilibrium Computer Simulations" in *Nonlinear Fluid Behavior*, H. J. M. Hanley, Ed., North-Holland, Amsterdam, 1983 (reprint of *Physica*, **118A**), pp. 111–122.

81. Hoover, W. G., A. J. C. Ladd, and V. N. Hoover, "Historical Development and Recent Applications of Molecular Dynamics Simulation," *ACS Adv. Chem. Ser.*, **204**, 29–46 (1983).

82. Hoover, W. G., "Nonequilibrium Molecular Dynamics," *Ann. Rev. Phys. Chem.*, **34**, 103–127 (1983).

83. Hoover, W. G., "Computer Simulation of Many-Body Dynamics," *Phys. Today*, **37**(1), 44–50 (1984).

84. Jorgensen, W. L., "Free Energy Calculations: A Breakthrough for Modeling Organic Chemistry in Solution," *Acc. Chem. Res.*, **22**, 184–189 (1989).

85. Karplus, M., "Molecular Dynamics Simulations of Proteins," *Phys. Today*, **40**(10), 68–72 (1987).

86. Klein, M. L., "Computer Simulation Studies of Solids," *Ann. Rev. Phys. Chem.*, **36**, 525–548 (1985).

87. Klein, M. L., "Intermolecular Potentials and Computer Simulation Studies of Molecular Crystals," *Stud. Phys. Theor. Chem.*, **46**, 659–675 (1987).

88. Klein, M. L. and L. J. Lewis, "Simulation of Dynamical Processes in Molecular Solids," *Chem. Rev.*, **90**, 459–479 (1990).

89. Kollman, P., "Molecular Modeling," *Ann. Rev. Phys. Chem.*, **38**, 303–316 (1987).

90. Kremer, K. and K. Binder, "Monte Carlo Simulations of Lattice Models for Macromolecules," *Comp. Phys. Repts.*, **7**, 259–310 (1988).

91. Kushick, J. and B. J. Berne, "Molecular Dynamics Methods: Continuous Potentials" in *Statistical Mechanics*, Part B, B. J. Berne, Ed., Plenum, New York, 1977, Chapter 2.

92. Ladd, A. J. C., "Molecular Dynamics," *NATO ASI Ser. C*, **293**, 55–82 (1990).

93. Lester, W. A., Jr. and B. L. Hammond, "Quantum Monte Carlo for the Electronic Structure of Atoms and Molecules," *Ann. Rev. Phys. Chem.*, **41**, 283–311 (1990).

94. Liska, M., B. Hatalova, and G. G. Bojko, "Molecular Dynamics and Its Applications in Studies of the Structure and Properties of Glasses and Glass-forming Melts," *Silikaty* (Prague), **32**, 359–380 (1988).

95. MacElroy, J. M. D. and S. H. Suh, "Simulation Studies of a Lennard-Jones Liquid in Micropores," *Molec. Simul.*, **2**, 313–351 (1989).

96. McCammon, J. A. and M. Karplus, "Simulation of Protein Dynamics," *Ann. Rev. Phys. Chem.*, **31**, 29–45 (1980).

97. McCammon, J. A., "Computer-aided Molecular Design," *Science*, **238**, 486–491 (1987).

98. Mezei, M. and D. L. Beveridge, "Free Energy Simulations" in *Computer Simulation of Chemical and Biomolecular Systems*, D. L. Beveridge and W. L. Jorgensen, Eds., New York Academy of Sciences, New York, 1986, pp. 1–23.

99. Mulla, D. J., "Simulating Liquid Water Near Mineral Surfaces: Current Methods and Limitations," *ACS Symp. Ser.*, **323**, 20–36 (1986).

100. Nezbeda, I., S. Labik, and A. Malijevsky, "Structure of Hard-Body Fluids. A Critical Compilation of Selected Computer Simulation Data," *Coll. Czech Chem. Comm.*, **54**, 1137–1202 (1989).

101. Nicholson, D. and N. G. Parsonage, *Computer Simulation and the Statistical Mechanics of Adsorption*, Academic, London, 1982, Chapter 4.

102. Panagiotopoulos, A. Z., "Gibbs Ensemble Monte Carlo Simulations of Phase Equilibria in Supercritical Fluid Mixtures," *ACS Symp. Ser.*, **406**, 39–51 (1989).

103. Parlinski, K., "Molecular Dynamics Simulation of Incommensurate Phases," *Comp. Phys. Repts.*, **8**, 153–219 (1988).

104. Pontikis, V., "Grain Boundary Structure and Phase Transformations: A Critical Review of Computer Simulation Studies and Comparison with Experiments," *J. Phys., Colloq.*, **C5**, C5-327–C5-336 (1988).

105. Rao, C. N. R. and S. Yashonath, "Computer Simulation of Transformations in Solids," *J. Solid State Chem.*, **68**, 193–213 (1987).

106. Rapaport, D. C., "Large-Scale Molecular Dynamics Simulation Using Vector and Parallel Computers," *Comp. Phys. Repts.*, **9**, 1–53 (1988).

107. Ray, J. R., "Elastic Constants and Statistical Ensembles in Molecular Dynamics," *Comp. Phys. Repts.*, **8**, 111–151 (1988).

108. Ree, F. H., "Computer Calculations for Model Systems" in *Physical Chemistry — An Advanced Treatise*, Vol. 8A, H. Eyring, D. Henderson, and W. Jost, Eds., Academic, New York, 1971, Chapter 3.

109. Rossky, P. J., "The Structure of Polar Molecular Liquids," *Ann. Rev. Phys. Chem.*, **36**, 321–346 (1985).

110. Saboungi, M.-L., W. Geertsma, and D. L. Price, "Ordering in Liquid Alloys," *Ann. Rev. Phys. Chem.*, **41**, 207–244 (1990).

111. Scheek, R. M., W. F. van Gunsteren, and R. Kaptein, "Molecular Dynamics Simulation Techniques for Determination of Molecular Structures from Nuclear Magnetic Resonance Data," *Methods Enzym.*, **177**, 204–218 (1989).

112. Sharp, K. A. and B. Honig, "Electrostatic Interactions in Macromolecules: Theory and Applications," *Ann. Rev. Biophys. Biophys. Chem.*, **19**, 301–332 (1990).

113. Sheykhet, I. I. and B. Y. Simkin, "Monte Carlo Method in the Theory of Solutions," *Comp. Phys. Repts.*, **12**, 67–133 (1990).

114. Shing, K. S. and K. E. Gubbins, "A Review of Methods for Predicting Fluid Phase Equilibria: Theory and Computer Simulation," *ACS Adv. Chem. Ser.*, **204**, 73–106 (1983).

115. Skolnick, J. and A. Kolinski, "Dynamics of Dense Polymer Systems: Computer Simulations and Analytic Theories," *Adv. Chem. Phys.*, **78**, 223–278 (1990).

116. Stillinger, F. H., "Theory and Molecular Models for Water," *Adv. Chem. Phys.*, **31**, 1–101 (1975).

117. Strandburg, K. J., "Two-Dimensional Melting," *Rev. Mod. Phys.*, **60**, 161–207 (1988).

118. Stoltze, P., "Simulations of the Premelting of Al(110)," *J. Chem. Phys.*, **92** 6306–6321 (1990).

119. Streett, W. B. and K. E. Gubbins, "Liquids of Linear Molecules: Computer Simulation and Theory," *Ann. Rev. Phys. Chem.*, **28**, 373–410 (1977).

120. Tildesley, D. J., "Towards a More Complete Simulation of Small Polyatomic Molecules" in *Molecular Liquids: Dynamics and Interactions*, A. J. Barnes, W. J. Orville-Thomas, and J. Yarwood, Eds., Reidel, Boston, 1984, pp. 519–560.

121. Turq, P., "Computer Simulation of Electrolyte Solutions," *NATO ASI Ser. C*, **205**, 409–415 (1987).

122. Valleau, J. P. and S. G. Whittington, "A Guide to Monte Carlo for Statistical Mechanics: 1. Highways" in *Statistical Mechanics*, Part A, B. J. Berne, Ed., Plenum, New York, 1977, Chapter 4.

123. Valleau, J. P. and G. M. Torrie, "A Guide to Monte Carlo for Statistical Mechanics: 2. Byways" in *Statistical Mechanics*, Part A, B. J. Berne, Ed., Plenum, New York, 1977, Chapter 5.

124. Visscher, W. M. and J. E. Gubernatis, "Computer Experiments and Disordered Solids," *Dyn. Prop. Solids*, **4**, 63–155 (1980).

125. Watanabe, K. and M. L. Klein, "Shape Fluctuations in Ionic Micelles," *J. Phys. Chem.*, **93**, 6897–6901 (1989).

126. Wood, D. W., "Computer Simulation of Water and Aqueous Solutions" in *Water — A Comprehensive Treatise*, Vol. 6, F. Franks, Ed., Plenum, New York, 1979, Chapter 6.

127. Wood, W. W., "Monte Carlo Studies of Simple Liquid Models" in *Physics of Simple Liquids*, H. N. V. Temperley, J. S. Rowlinson, and G. S. Rushbrooke, Eds., North-Holland, Amsterdam, 1968, Chapter 5.

128. Wood, W. W., "A Review of Computer Studies in the Kinetic Theory of Fluids," *Acta. Phys. Austr.*, Suppl. X, 451–490 (1973).

129. Wood, W. W. and J. J. Erpenbeck, "Molecular Dynamics and Monte Carlo Calculations in Statistical Mechanics," *Ann. Rev. Phys. Chem.*, **27**, 319–348 (1976).

130. Woodcock, L. V., "Predicting the Rheology of Complex Fluids," *Molec. Simul.*, **2**, 253–279 (1989).

III.   TOPICAL INDEX TO PARTS I AND II OF BIBLIOGRAPHY[†]

[†]Numbers refer to the reference numbers in Parts I and II.

INDEX